D0965852

Techniques of Combined Gas Chromatography/Mass Spectrometry:

Applications in Organic Analysis

WILLIAM MCFADDEN

Space Sciences Laboratory
University of California, Berkeley

A Wiley-Interscience Publication
JOHN WILEY AND SONS
New York   London   Sydney   Toronto

Library of Congress Cataloging in Publication Data

McFadden, William H.   1927 –
    Techniques of combined gas chromatography/mass spectrometry.

    "A Wiley-Interscience publication."
    1. Gas chromatography. 2. Mass spectrometry
3. Chemistry, Organic. I. Title.
QD272.C44M3   547$'$.34$'$926   73–6916
ISBN 0–471–58388–X

Printed in the United States of America

10   9   8   7   6   5   4   3

Preface

It is now sixteen years since the first attempt to couple gas chromatography with mass spectrometry.[1] Both of these techniques were proven, powerful analytical tools and their merger was inevitable. Many technical and financial obstacles had to be overcome, but developments of the past decade have successfully reduced these problems. Today, GCMS analysis is applied in every branch of organic chemistry.[2, 3]

To understand GCMS, the chemist must be knowledgeable in gas chromatography, mass spectrometry, vacuum technology, and computer science. In addition, he must have a feeling for interface problems and the necessary operational modifications. Many excellent textbooks are available for each of the individual disciplines. Unfortunately, these texts give scant consideration to the interface between the combined instruments, and necessary operational modifications are seldom discussed. Vacuum technology, for example, is very important in GCMS, but it is glossed over in almost every standard mass spectrometry text.

The purpose of this book is to provide the organic chemist with a single volume that gives the necessary background for understanding GCMS. Each of the important technologies is considered. Special emphasis is given to the GCMS interface and to the problems and restraints demanded in the combined system.

The text material has been carefully selected from the many questions that have been posed during the ten years in which I have lectured on GCMS. It has always been obvious that a beginner in GCMS is familiar with the problems of gas chromatography but not with those of mass spectrometry. Consequently, the discussion of mass spectrometric instrumentation (Chapter 2) is considerably more extensive than that of gas chromatographic instrumentation. At times, the reader may even feel that Chapter 2 is not directed toward GCMS per se. Nevertheless, every section illustrates some important facet of GCMS. For example, comparison of the two popular double-focusing mass spectrometers (Mattauch-Herzog and Nier-Johnson) is given because of the significant difference that the double-focus mode plays in recording high-resolution mass spectra. The other sections are similarly chosen because of a direct relationship to the overall theme of GCMS.

PREFACE

My experience in computerization of the GCMS system has been as a user rather than as a developer. However, Chapter 7 is not written with the purpose of teaching computer technology, but rather with illustrating the importance of automatic data treatment in GCMS, and evaluating various output options in current use. Throughout the chapter, the reader is continually reminded of the cost of this tool relative to the anticipated use and is urged to evaluate fully both the advantages and disadvantages of automation.

Interpretation of mass spectral data is outside the scope of the present volume and except for a discussion on computer interpretation of GCMS data, the subject has been avoided. There are many excellent textbooks on mass spectral interpretation[4-11] which the reader will find more entertaining than a hurried chapter in a book primarily devoted to GCMS techniques.

The final chapter presents a selection of GCMS applications from several fields of organic chemistry and biochemistry. Originally, I had intended to write this section by the conventional method of reviewing significant papers from the literature. However, such an approach can lead to a sterile review that omits the more exciting, personal facets of a research program. To avoid such flatness, I invited several GCMS users to write a two- or three-page review of their favorite application. The response was extremely gratifying. Most of the applications have thus been described by the original workers and reflect their interests and enthusiasms. For my part, editorial changes have been minimal, although comments were added to some examples to emphasize specific points. The selfless cooperation of these coauthors has provided an unexpected personal reward, and I will always be indebted for this display of true scientific devotion.

The author also acknowledges the help and encouragement given by his many professional friends and coworkers. Special thanks are due to Dr. A. L. Burlingame for urging me to write this volume and for his support during my stay at the Space Sciences Laboratory. I wish to thank members of the Space Sciences Mass Spectrometry Group who have provided me with data, information, and stimulating discussion.

It is impossible to mention everyone who has worked with me over the past decade and contributed to my appreciation of GCMS, but it is necessary to thank my early coworkers, Drs. Roy Teranishi and Ron Buttery, for contributing indispensable gas chromatographic expertise. Mr. Dale Black, who worked with me from 1961 to 1967 in the mass spectrometry laboratory at Western Regional Laboratories (USDA, Albany, California), deserves special mention for his tireless efforts that assured the success of each new project.

PREFACE

There is an endless list of friends who have given encouragement, have read and criticized parts of the book, have made innumerable suggestions, and have supplied data and information. I thank each of them most sincerely.

Berkeley, California   W. H. McFadden

List of Contributors

Klaus Biemann
Department of Chemistry
Massachusetts Institute of Technology
Cambridge, Mass. 02139

E. J. Bonelli
Finnigan Corporation
Sunnyvale, California 94086

R. G. Buttery
Western Regional Research Laboratory
U. S. Department of Agriculture
Berkeley, California 94710

J. R. Chapman
AEI Scientific Apparatus Ltd.
Manchester, England

C. E. Costello
Department of Chemistry
Massachusetts Institute of Technology
Cambridge, Mass. 02139

J. N. Damico
Division of Chemistry and Physics
Office of Science, Bureau of Foods
Food and Drug Administration
U. S. Department of Health, Education, and Welfare
Washington, D. C. 20204

I. Dzidic
Institute for Lipid Research
Baylor College of Medicine
Houston, Texas 77025

LIST OF CONTRIBUTORS

B. S. Finkle
Santa Clara County
Laboratory of Criminalistics
875 North San Pedro Street
San Jose, California 95110

R. A. Flath
Western Marketing and Nutrition Division
U. S. Department of Agriculture
Albany, California 94710

H. J. Förster
Department of Chemistry
Massachusetts Institute of Technology
Cambridge, Mass. 02139

E. J. Gallegos
Chevron Research Company
Richmond, California 94802

K. Habfast
Varian MAT
GMBH
Bremen, Germany

J. Hoffman
Varian MAT
GMBH
Bremen, Germany

E. C. Horning
Institute for Lipid Research
Baylor College of Medicine
Houston, Texas 77025

M. G. Horning
Institute for Lipid Research
Baylor College of Medicine
Houston, Texas 77025

LIST OF CONTRIBUTORS

J. A. Kelley
Department of Chemistry
Massachusetts Institute of Technology
Cambridge, Mass. 02139

K. E. Matsumoto
Stanford University
Stanford, California 94305

K. H. Maurer
Variant MAT
GMBH
Bremen, Germany

T. R. Mon
Western Marketing and Nutrition Division
U. S. Department of Agriculture
Albany, California 94710

H. Nau
Department of Chemistry
Massachusetts Institute of Technology
Cambridge, Mass. 02139

N. Nicolaides
Department of Biochemistry
University of Southern California
School of Medicine
Los Angeles, California 90033

Linus Pauling
Stanford University
Stanford, California 94305

A. B. Robinson
Stanford University
Stanford, California 94305

LIST OF CONTRIBUTORS

T. Sakai
Department of Chemistry
Massachusetts Institute of Technology
Cambridge, Mass. 02139

Roy Teranishi
Western Marketing and Nutrition Division
U. S. Department of Agriculture
Albany, California 94710

Paul Vallon
Givaudan Corporation
100 Delawanna Avenue
Clifton, New Jersey 07014

J. P. Walradt
International Flavors and Fragrances
Union Beach, New Jersey 07735

J. Throck Watson
Department of Pharmacology
Vanderbilt University
Nashville, Tenn. 37232

Contents

CONTENTS

CONTENTS

CONTENTS

CONTENTS

1
Introduction

Before the advent of combined gas chromatography/mass spectrometry (GCMS),[1, 12-14] it was impractical to consider a complete qualitative analysis on a complex organic mixture of 20 components or more. Even the separation and identification of two or three major components could require more than a year's work using classical analytical methods. The minimum quantity of sample needed for each component was usually several milligrams. In contrast, today an organic chemist may submit a mixture containing 100 components (to the 0.1% level) to a GCMS laboratory and expect, within a day or two, some degree of qualitative identification for nearly every peak. Furthermore, this analysis will be performed on less than a milligram of the total mixture. The chromatogram shown in Figure 1.1 provides a typical example. Each numbered peak, including even such miniscule components (less than 0.1%) as peaks 45, 50, 70, 75, etc., was identified.[15]

Gas-liquid partition chromatography was first introduced in 1952.[16] The analytical applications of this powerful separation method were recognized immediately, and within three or four years, gas chromatography was used in many laboratories for both qualitative and quantitative organic analysis. Excellent results were easily obtained. The initial success was so exciting that many chemists jested that the other methods of instrumental analysis were obsolete. Quite the contrary. Like any other good research tool, gas chromatography opened up many new analytical problems as it solved old ones.

In its simplest form, qualitative analysis by gas chromatography is performed using the retention time of the unknown as a means of identification. This direct approach may be the best for easy straight-forward analyses, but

1

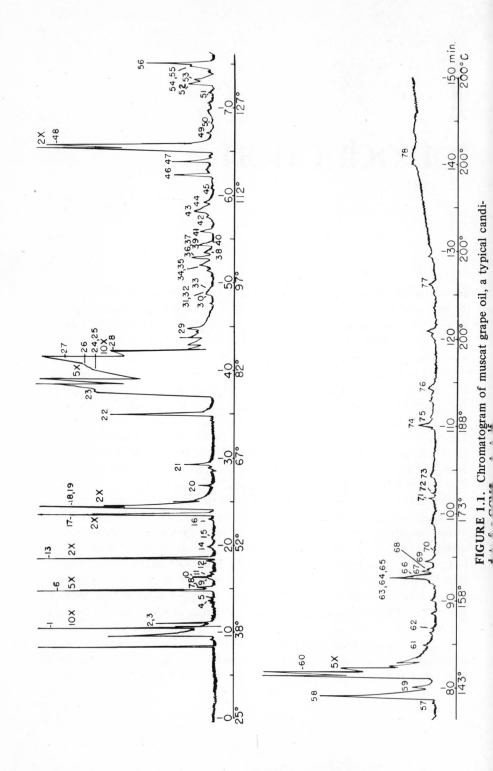

FIGURE 1.1. Chromatogram of muscat grape oil, a typical candi-

it is limited by two very significant drawbacks. First, the time resolution of a retention index system seldom gives more than about 600 windows which must contain all possible volatile compounds. The second limitation is that with so many compounds assignable to any given time point, the chemist must know in advance what type of compound he is looking for. This is equivalent to the contradictory suggestion that doing qualitative analysis by matching retention indexes necessitates knowing what the compound is before starting.

For sophisticated qualitative analysis, gas chromatography must be coupled to other analytical systems.[17, 18] Mass spectrometry has, for many reasons, proven to be the most effective of these ancillary devices. However, no single technique provides all necessary analytical data, and the chemist must consider the pros and cons of each available method. Frequently, an inexpensive specific detector gives information that is not attainable from any number of expensive instruments, and a versatile laboratory should have at least two or three of these qualitative detectors.[19]

Of the many available qualitative detectors and analytical instruments, mass spectrometry offers the best compromised combination of compatibility, sensitivity, and specific data. Unfortunately, it is the most expensive and demands considerable operational skill.

A comparison of several of these qualitative gas chromatographic detectors is given in Table 1.1. Comparison of the specificity of different methods is not always meaningful since the results are strongly dependent on the compound being analyzed. It is obvious from Table 1.1, however, that mass spectrometry, used in a GCMS mode, is the only common technique with a sensitivity comparable to the pure chromatographic detectors (i.e., hydrogen flame, electron capture, flame photometer). This advantage, plus other compatible features, outweighs the additional cost for many GCMS applications.

The most significant problem of the combined GCMS system is illustrated schematically in Figure 1.2. For most work, the chromatograph is operated with an outlet pressure of one atmosphere (760 Torr). The mass spectrometer should be operated at a pressure at least as low as 10^{-5} Torr. Thus, the interface system must reduce the pressure of the carrier gas by eight orders of magnitude and still convey a usable fraction of the organic sample to the mass spectrometer.

Because of this considerable pressure differential, the gas chromatograph and mass spectrometer have often been termed incompatible for combined operation. In actuality, both of these systems are gas phase, flow methods of analysis that are well matched in almost every operational mode. This important fact is best appreciated by comparison of gas chromatographic

TABLE 1.1. Comparison of several specific gas chromatographic detectors.

Method	Approximate Cost, $	Approximate Detection Limit, gm	Specificity or Common Uses
Gas Chromatography			
Retention indices	Compound file, extra columns	10^{-10} (H_2 flame)	Confirm any compound
Electron capture	1200	10^{-12}	Halides, conj. carbonyls, nitriles, di and trisulfides
Flame photometer	4500	10^{-9} (S), 10^{-11} (P)	Phosphorous, sulfur
Chemical Methods			
Pyrolysis	5000	10^{-9}	Skeletal structure, functional determination
Chemical reagents	~100	10^{-6}	Classical functionality determination
Electrolytic systems	3000	10^{-8}	Sulfur, nitrogen, halogens
Instrumentation			
Infrared–grating –interferometer	5–15000	10^{-6}	Functional groups
Ultraviolet	60000	10^{-7}	Functional groups
Proton magnetic resonance	3–10000	Variable to 10^{-10}	Aromatics, conj. carbonyls
Mass spectrometer	20–80000	10^{-5}	Excellent for function, some MW data
Batch inlet	25–100000	10^{-7}	Best for complete identification, MW, structure, and function.
GCMS mode		10^{-11}	
Multiple ion detection		10^{-12}	Confirm any compound.

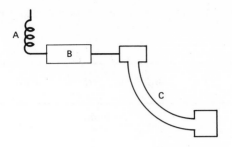

FIGURE 1.2. Schematic diagram of the GCMS system illustrating pressure differential. A = Gas chromatograph at 1 atmosphere exit pressure (760 Torr), B = Pressure reducing interface, C = Mass spectrometer at operating pressure of 10^{-5} to 10^{-4} Torr.

operating conditions with those of other common instrumental methods. In Table 1.2, it is seen that every characteristic of the gas chromatograph is matched by the mass spectrometer except for ambient pressure operation. In contrast, the other instruments have at best only one or two common features. Combined operation of these with gas chromatography thus involves more problems than are encountered in GCMS.

The pressure drop of six to seven orders of magnitude demanded across the GCMS interface is a frequent source of trouble. The difficulty can be reduced, however, by proper understanding of gas flow behavior and vacuum technology. In actuality, the magnitude of the pressure drop is not important. The important point is to compare the quantity of gas flow needed for proper chromatography to the quantity that can be pumped by the vacuum system at the mass spectrometer operating pressure. Depending upon the operating conditions in both the mass spectrometer and gas chromatograph, the ratio of the gas flow through each machine may be higher than 100/1 or close to 1/1. When this ratio has been established for the specified operating conditions, the problems of the interface can be clearly defined. Proper application of vacuum skills, use of sample enrichment devices, and optimization of the gas chromatographic flow system leads to high sample utilization even when the ratio of carrier gas flow is unfavorable.

High sample utilization is only one of several reasons for using mass spectrometry with gas chromatography in the combined GCMS mode. Another important factor favoring direct GCMS is speed and convenience. Consider the problem of trapping the components of a complex mixture

TABLE 1.2. Compatibility of gas chromatography with other instrumental analytical methods.

Operational Parameter	GC	MS	IR	PMR	UV
Gas Phase	yes	yes	undesirable	no	no
Nanogram Sensitivity	yes	yes	no	no	depends on sample
Compatible Scan Time	—	yes	yes[1]	yes[2]	no
Continuous Flow	yes	yes	no	no	no
Compatible Temperature	—	yes	no	no	no
Ambient Pressure	yes	no	yes	yes	yes

[1] Interferometric IR.
[2] Fourier Transform Pulse Method

showing 50 peaks down to a level of 0.5%. To collect and submit each of these components for individual mass spectral analysis will take one to two weeks to collect the fractions and well over a week for the mass spectrometry. The same data can be obtained in one or two hours by GCMS. Furthermore, where peaks overlap, collected fractions are of doubtful purity. Conventional GCMS scanning methods usually provide spectra for each component with minimal cross contamination.

In addition to being slow and inconvenient, trapping methods are inefficient for small samples, and often impossible due to aerosol effects, total volume considerations, and general handling and transfer problems. Few chemists care to attempt sample trapping on submicrogram quantities although some special techniques are effective when properly used.[21-23] Trapping is particularly difficult on the effluent from low-load capillary columns in which the peaks follow one another with only a few seconds separation.

Another problem encountered with collected samples is oxidation or polymerization of the pure component. In a dilute solution, active compounds are protected from autoxidation or self-polymerization, but in the pure state, they react and thus destroy their own identity. Direct GCMS obviates all of the difficulties associated with sample collection and subsequent qualitative analysis.

2

The Relationship of Components of a Mass Spectrometer to the Requirements of GCMS Analysis

Mass spectrometry is recognized today as an important analytical tool in all fields of organic and biochemistry. Many textbooks are available on mass spectral instrumentation in organic chemistry,[24-28] on the interpretation of organic mass spectra,[4-11] and on new developments in applications and instrumentation.[29-32] Two mass spectrometry periodicals are now published. One of these is devoted exclusively to organic mass spectrometry[33] and the other to general mass spectrometry and related fields.[34]

The GCMS technique has contributed significantly to the important role of modern mass spectrometry. The applications of this specialized technique are discussed in numerous review articles,[35-47] and a special abstract periodical is available dealing solely with GCMS papers.[48]

The important features to consider in selecting a mass spectrometer for GCMS are:

(1) basic mass spectrometer sensitivity,
(2) sample utilization or pumping efficiency,
(3) mass resolution,
(4) reproducible spectral pattern without mass discrimination, and
(5) computer compatibility.

Each of the four common types of mass spectrometer used in GCMS (i.e., single-focusing magnetic deflection, double-focusing electrostatic sector/magnetic deflection, time-of-flight, and quadrupole) possesses operational characteristics that influence the performance. To understand and appreciate the difference between the various mass spectrometers, one must examine each part of the system and determine how it contributes to the final mass spectral pattern.

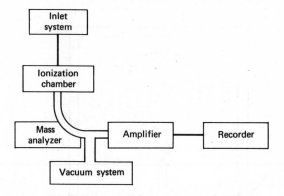

FIGURE 2.1. Basic components of a mass spectrometer.

The basic components of a mass spectrometer are shown schematically in Figure 2.1. Except for the vacuum system, the role of each component will be studied with particular concern for adaptation of the total system to the rigorous demands of GCMS analysis. The vacuum system, which plays a special role in the high gas flow combination of GCMS, is considered separately in Chapter 4.

A. INLET SYSTEMS (OTHER THAN GCMS)

1. The Standard Batch Inlet

Every mass spectrometer used for organic analysis should be equipped with a standard batch inlet system for the introduction of liquid samples. Such an inlet is convenient for mass analysis of collected gas chromatographic fractions. It is useful for a quick inspection of a reaction product, a reagent, or to establish the complexity of an unknown. The batch inlet is needed for running authentic samples for spectral files. It also provides the most convenient and accurate method of metering a sample to the mass spectrometer for determination of sensitivity, which is an essential part of the evaluation of the GCMS interface enrichment devices.

An example of a simple batch inlet is shown in Figure 2.2. A silicone rubber septum is used for liquid sample injection, although other methods are available which utilize probes, mercury-sealed glass frits, Teflon plugs, etc. Vacuum valves isolate and protect the various compartments. An expansion volume of 1 to 2 liters assures a constant sample pressure over a long sampling period. The molecular leak meters the sample to the ion chamber at a constant volume rate. Most inlet systems are constructed of stainless steel

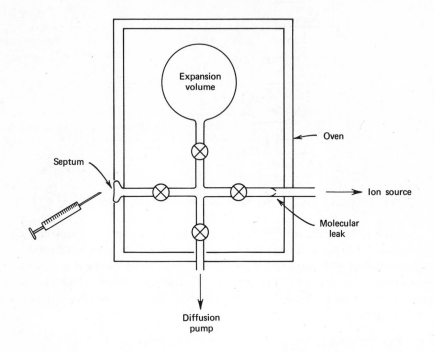

FIGURE 2.2. Standard batch inlet for mass spectrometry sample introduction.

for operation in the temperature range of 150–200°C. For operation at temperatures up to 350°C, glass is used to minimize thermal decomposition.

Volume sample flow rate to the ionization chamber is constant if the leak is of the type that operates with molecular flow conditions rather than viscous flow conditions. (See Chapter 4, page 124.) Consequently, except for isotope analysis, organic mass spectrometer inlets are equipped with a molecular leak. The leak conductance for normal operation should be in the range 0.1–0.4 cc/sec. If the conductance value is obtained from a manufacturer's specification, it will refer to the molecular conductance for nitrogen at 25°C, unless otherwise noted. Since the molecular flow rate is inversely proportional to the square root of the molecular weight (see Chapter 4, equations 4.10, 4.11), the conductance will be lower for higher molecular weight compounds.

Suppose the molecular conductance of a leak is specified as 0.4 cc/min for N_2 at 25°C. For molecular weight 200, the conductance will be

0.4 cc/min x $\sqrt{28/200}$ = 0.15 cc/min at 25°C. Since the effusion rate is proportional to the square root of the absolute temperature (equation 4.10), at 150°C (423°K) the conductance for molecular weight 200 will be 0.15 cc/min x $\sqrt{423/298}$ = 0.21 cc/min. (The area of the effusion holes will increase less than 1% over a temperature span of 150°C, so this effect can be ignored.) This value can be used to determine the rate of sample flow to the mass spectrometer.

If 2 x 10^{-4} gm of ethyl decanoate, molecular weight 200, is injected into a typical 1-liter inlet at 150°C, the inlet pressure, as calculated from PV = nRT, will be 2.6 x 10^{-2} Torr. (1 Torrecelli = 1 mm of Hg pressure.) The rate of sample flow or effusion to the mass spectrometer can be calculated from a variation of the gas formula in which the static volume V is replaced by the volume flow rate C. Thus:

$$n \text{ (moles/sec)} = \frac{PC}{RT} \tag{2.1}$$

where

P = pressure, Torr
C = conductance, liters/sec
R = gas constant, 62.3 liter Torr/mole °K
T = absolute temperature.

For P = 2.6 x 10^{-2} Torr, C = 0.21 cc/min, and T = 423°K, the number of moles of ethyl decanoate effusing to the mass spectrometer per second will be 2.4 x 10^{-10} mole/sec or 4.8 x 10^{-8} gm/sec. This amount of sample will give an intense mass spectral pattern in most mass spectrometers operating at a resolution of 500–1000. More sample per unit time is seldom necessary.

As calculated above, the quantity of gas Q is given in the convenient chemical unit of moles/sec. For vacuum calculations, the unit of liter Torr/sec is often preferred. Equation 2.1 is then simplified to the form

$$Q = PC \tag{2.2}$$

at constant temperature. For the above example, Q would be equal to 5.5 x 10^{-6} liter Torr/sec.

If the conductance of the leak is not known, it can be determined by measuring the decrease in the intensity of the spectral pattern as a function of time. As sample effuses to the mass spectrometer, the sample pressure in the inlet system decreases exponentially according to the equation

$$P = P_oe^{-\frac{C}{V}t} \tag{2.3}$$

where

P_o = initial pressure at t = 0
C = conductance, cc/sec
V = volume, cc
t = time.

By algebraic rearrangement

$$C = V/t \ln(P_o/P). \tag{2.4}$$

The volume V must include both the 1-liter expansion volume and the manifold volume of about 100 cc. If in the above example, the mass 88 peak was observed to change from an initial peak height of 8200 mm deflection to 5810 mm after 30 minutes, the conductance for a molecular weight of 200 at 150°C would be

$$C = \frac{1100}{30 \times 60} \frac{cc}{sec} \ln\frac{8200}{5810}$$

$$= 0.21 \text{ cc/sec.}$$

When the manifold volume is known with sufficient accuracy (5–10%), the expansion volume can be closed off to decrease the time required to attain the same reduction in mass peak intensity or pressure.

This method of measuring and calculating sample consumption is the most convenient method of obtaining the mass spectrometer sensitivity. In Chapter 5, the method is discussed further in the section on measurement of separator efficiencies (page 210).

2. The Direct Probe Inlet

Many samples submitted for mass analysis are too involatile to be introduced through a standard batch inlet. In addition, samples available in submicrogram quantities will not give a sufficient pressure in the batch inlet to get a suitable sample pressure in the ion chamber. These obstacles are avoided by use of a direct introduction probe that enters the source housing through a vacuum lock and is inserted directly to a hole in the ionization chamber. An example of this type of introduction system is shown in Figure 2.3 Sample is placed in a small glass capillary in the probe tip and the probe is inserted into the vacuum lock. After initial evacuation, the isolation valve is opened and the probe is pushed up to a small hole in the ion chamber. Rate of evaporation is controlled by varying the temperature. Excellent sample utilization is possible provided the sample capillary makes a closed seal at the entrance to the ion chamber.

Several probe/ion chamber configurations are shown in Figure 2.4. In (a), the ceramic probe tip is butted solidly against the ion chamber and, provided there are no openings between the probe and orifice, 100% of the sample will pass through the ion chamber. In case (b), the glass capillary is pushed tight against the entrance hole and if the glass lip is completely flush, 100% of the evaporated sample will again go into the ion chamber. However, often the capillary is not tight against the ion source and vaporized sample effuses back away from the ion chamber. The amount of lost sample cannot be easily determined since it depends upon the ratio of the area of the entrance hole and the area of the "misfit" hole opening. In unfavorable circumstances (such as a jagged irregular lip on the glass capillary), sample loss may be as high as 30–50%. For compounds of higher vapor pressure, the evaporation rate is often too high for all the sample to pass through the ion chamber, and the probe tip may be partially withdrawn as illustrated in Figure 2.4c. This can lead to sample utilization as low as 1–5%. In some cases, high evaporation rates are reduced by use of a watercooled or refrigerated probe.

Quantitative evaporation of sample from a probe inlet is often used to calibrate the mass spectrometer sensitivity for evaluation of the GCMS separator efficiency. From the foregoing discussion, it is seen that if the probe is not tightly butted against the ion chamber, a low value will be obtained for the mass spectrometer sensitivity. When this low value is compared with the integrated signal from the GCMS separator system, the calculated separator efficiency will be too high. Unfortunately, this kind of error is very difficult to detect, particularly after the event; consequently, some separator yield determinations may be overevaluated.

The direct probe can be modified to serve as an inlet for the effluent from the GCMS interface. This technique was first introduced by McCloskey[49] and offers a convenient method for modifying a mass spectrometer that is not specifically designed for GCMS (Figure 2.5). It also provides a versatile meth-

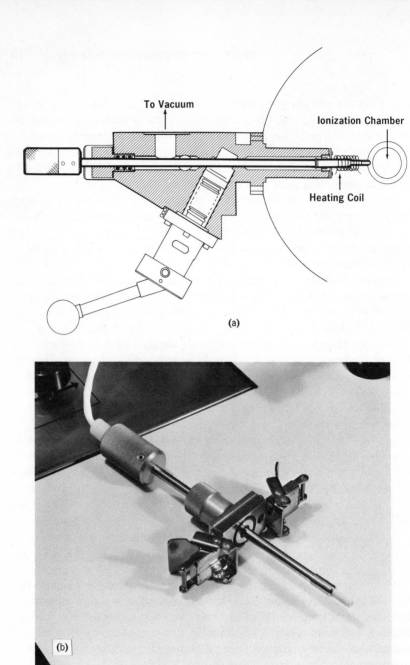

FIGURE 2.3. Direct insertion probe for mass spectrometry sample introduction (Du Pont Instrument Products Division). (a) Schematic diagram showing vacuum lock mechanism. (b) Photograph of Du Pont 21–490 series probe.

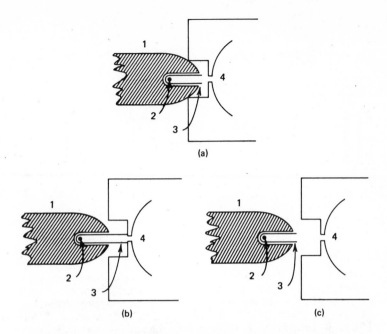

FIGURE 2.4. Possible configurations of probe sample relative to ion chamber entrance port. (1) Ceramic probe tip. (2) Sample. (3) Glass capillary. (4) Ion chamber entrance port.

od for modification of interface hardware. Care must be taken to heat all parts of the probe line and, of course, a tight butt seal at the ion chamber is essential to prevent loss of sample. A commercial adaptation of this technique is now available from the Varian-MAT Instrument Corporation.

B. IONIZATION METHODS

1. Electron Impact

Many different methods have been devised for the production of ions in mass spectrometry, but only the electron impact process is widely applied in organic analysis. Two other methods, field ionization and chemiionization, have specific applications that will be discussed later. Other processes, such as spark discharge, high voltage glow discharge, and thermionic emission, have special applications and are not practical in organic mass spectrometry.

Several factors contribute to the popularity of the electron impact process: stability, ease of operation, precise beam intensity control, relatively

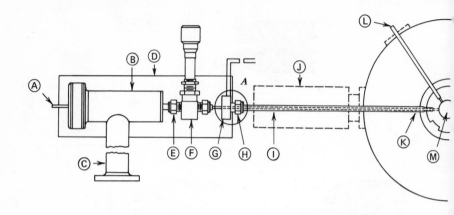

FIGURE 2.5. Modification of probe inlet for GCMS sample introduction. A. Capillary from GC. B. Effusion separator (page 169). C. Pump flange. D. Aluminum block. E, H. Swagelok fittings. F. Metering valve. G. Probe connector. I. Hollow probe. J. Vacuum lock. K. Teflon tip. L. Conventional inlet. M. Ion chamber.

high efficiency of ionization, simplicity of construction, convenient control of temperature, and narrow kinetic energy spread in the ions formed. The mass spectral pattern obtained is specific and characteristic of the chemical structure of the sample. Some groups of compounds give very similar mass spectra, but as a rule, the process provides a unique fingerprint. In addition, the only significant compilations of organic mass spectral data are from electron impact studies so that to utilize these spectral catalogues, it is necessary to have data from an electron impact source.

Figure 2.6 shows two forms of an electron impact source, a tight source and an open source. Electrons are emitted from a hot filament and accelerated through the ion chamber (perpendicular to plane of paper) toward an anode which measures the intensity of the electron beam. Currents in the range of 50 to 250 μamps are commonly used. The acceleration voltage or ionization energy of the electrons is varied by changing the potential between the filament and ionization chamber (often called the block).

Interaction between the electron beam and the organic molecules results in an energy exchange of around 10 to 20 eV which is sufficient to cause ionization of the molecule and, in many cases, decomposition to smaller fragment ions. The assembly of molecular and fragment ions in the chamber

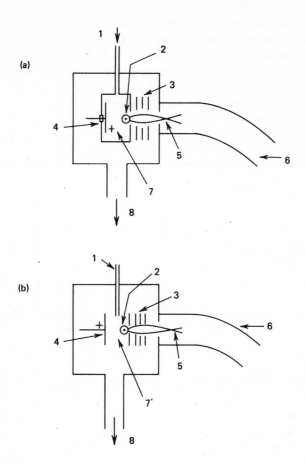

FIGURE 2.6. Variations of ion source. (a) Tight. (b) Open.
(1) Sample inlet. (2) Electron beam directed to plane of diagram.
(3) Accelerating and focusing electrodes. (4) Repeller electrode.
(5) Ion beam. (6) Analyzer. (7) Ion Source; in (a) source conductance is in range 0.3–2 liters/sec; in (b) source conductance is same
as vacuum conductance. (8) Pumping system; vacuum conductance
in range 30–100 liters/sec.

is accelerated through the ion slits into the mass spectrometer analyzer by a
positive potential on the ion repellers or by the field penetration of the high
voltage on the focusing electrodes.

The total amount of positive ion current and the nature of the fragmentation pattern depend upon the energy of the bombarding electron beam. Figure 2.7 shows the variation in mass spectral pattern of ethyl acetate observed when using nominal electron energies of 14, 20, and 60 eV. Because existing spectral catalogs contain mass spectra obtained with 60–80 eV electrons, this energy range is used for most organic analysis.

In GCMS analysis, it is often convenient to use a fraction of the ion current produced in the ion chamber as a means of monitoring the chromatographic run (see page 257). The process is made difficult by the large ion

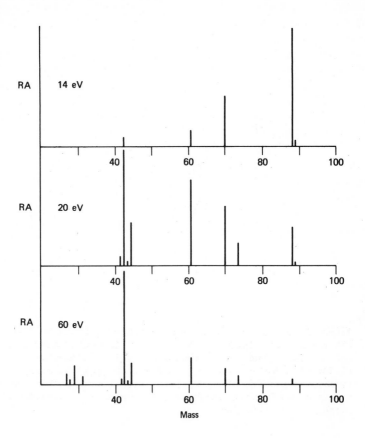

FIGURE 2.7. Mass spectrum of ethyl acetate as a function of electron bombarding energy. (a) 14 eV. (b) 20 eV. (c) 60 eV.

current due to He⁺ from helium carrier gas. To avoid this annoyance, the
mass spectrometer is often operated at 20 eV, which is about 4 eV below the
ionization potential of helium. Unfortunately, as is evidenced by Figure 2.7b,
the mass spectral pattern is greatly modified so that comparison with the catalogued spectra is difficult. A useful compromise is obtained by using 20 eV
electrons during the monitor phase of a run and switching to 70 eV electrons
for the few seconds that the spectrum is scanned. This, and other monitor
processes are discussed more fully in Chapter 6.

The mass spectral pattern in Figure 2.7a for a nominal electron energy
of 14 eV shows a high percentage of the total ion current in the parent ion.
In some analytical applications, this phenomenon has been used to simplify
the pattern and enhance the parent ion signal with respect to the background.
However, the absolute intensity is considerably decreased as shown in Figure
2.8, and the process is not applicable to increase a low or absent parent ion
current. Chemical ionization or field ionization methods are more practical for
this purpose. GCMS runs are seldom performed at a low electron energy
because of loss of sensitivity and because the simplified pattern also means
loss of structural information.

Ion sources encountered in commercial mass spectrometers are often
classified as open sources or tight sources. A comparison of these two different constructions is given in Figure 2.6. The detail of two tight commercial ion sources is shown in Figure 2.9.

In an open ion source (Figure 2.6b) the sample is directed from the
inlet system (probe, batch inlet, GCMS interface) into the region of the elec-

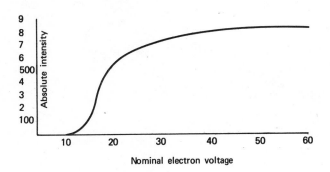

FIGURE 2.8. Ionization efficiency of ethyl acetate as a function
of electron bombarding energy.

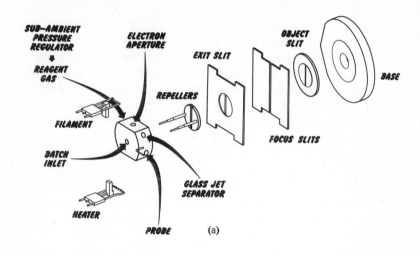

SUB-AMBIENT
PRESSURE
REGULATOR
↓
REAGENT
GAS

ELECTRON
APERTURE

EXIT SLIT

OBJECT
SLIT

BASE

FILAMENT

REPELLERS

BATCH
INLET

FOCUS SLITS

GLASS JET
SEPARATOR

HEATER

PROBE   (a)

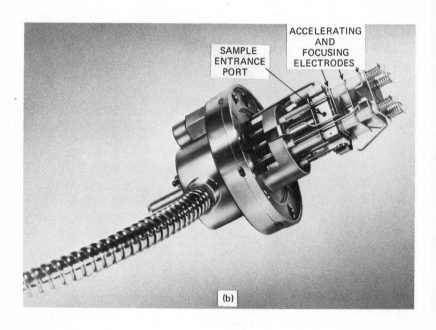

FIGURE 2.9. Commercial ion sources. (a) DuPont Instrument
Company source for either CI or EI mass spectrometry. (b) Varian MAT III source.

tron beam and ion accelerating electrodes. Because there is no resistance to flow in the active region of the electron beam, the sample undergoes considerable expansion into the volume of the mass spectrometer tube with a subsequent decrease of pressure or effective concentration. The rate at which sample is conducted away from the ionizing region is essentially the pumping speed of the vacuum system and typically, might be 30–100 liters/sec. The partial pressure of sample in the electron beam can be calculated from equation 2.2 ($Q = PC$) if the quantity of sample is known. Slower pumping speed will result in increased sample pressure and hence higher sensitivity, but in standard practice, the slower pumping speed would also introduce disadvantages such as slow initial pump-out and high background. In GCMS work, the slower pumping speed would also mean a corresponding decrease in the total amount of effluent allowed into the mass spectrometer so that no easy gain of sensitivity or (more correctly) sample utilization is attained.

With a tight ion source, the sample pressure in the region of the electron beam is increased while fast pumping is maintained in the rest of the system. This is illustrated in Figure 2.6a. The gas entering the ionization chamber is restricted to the ion chamber and can escape only through the slits provided for the electron beam and the ion beam. Assuming 1 liter/sec as a typical ion source conductance (see also page 126), the pressure in the ionizing region can be calculated from equation 2.2. If Q is the same for a closed source as for an open source, an increase of one or two orders of magnitude in sample pressure is obtained in the closed source with a corresponding increase in sensitivity.

From this comparison, it appears that a tight ion source should always be used but there are cases, particularly for fast kinetic studies, where the open fast-pumped source is a necessity. Furthermore, for mass spectrometers that do not require a slit system (the time-of-flight, quadrupole, or RF mass analyzer), the ion source can be designed and constructed more easily as an open system than an enclosed one. The resulting loss in sensitivity has generally been glossed over and only a few attempts have been made to improve the sample utilization. The sensitivity of the time-of-flight mass spectrometer has been increased by addition of a baffle with a small hole to allow passage of the ion beam. This baffle is placed just past the final accelerating grid and gives close to one order of magnitude improvement in sample utilization.[50] Recent advances in the design of the ion sources for the quadrupole and the time-of-flight mass spectrometers have resulted in application of a tight source. These sources were intended primarily for chemiionization studies, but improved GCMS analyses have also been obtained as an important consequence.[51, 52]

Another important consideration in GCMS studies is the time constant that governs the rate of pump-out of the effluent peak from the ion source. Equations 2.3 and 2.4 can be applied for a first approximation so that the time for sample partial pressure to be reduced to a specified level is given by $t = V/C \ln(P_0/P)$. The conduction of a typical source for a molecular weight 200 compound is around 0.4 liter/sec (see page 128) and the volume is about 7.5 cc, including the volume of sample lines. The half-life for sample removal will thus be given by

$$t_{1/2} = V/C \ln 2$$
$$= (7.5 \text{ cc}/400 \text{ cc per sec}) \times 0.69 \text{ sec}$$
$$= 0.013 \text{ sec.}$$

Consequently, the time required to reduce the sample level to 0.1% of its initial value is about 0.1 sec.

Table 2.1 summarizes calculations of this type for different ion sources of different conductance values. The volume of the three closed sources will be the same for each. For an open source, the conductance is the conductance of the pumping system and might be as high as 50 liters/sec. However, the volume of an open source system would be the volume of the ion source housing and the analyzer tube, assuming the system is not differentially pumped. For a time-of-flight mass spectrometer this volume is close to 5000 cc. As a consequence, the open ion source does not have a faster pump-out as is commonly assumed. The half-life would be more favorable for a system with a smaller volume such as exists for a quadrupole mass spectrometer, but the value would still not be appreciably faster than can be attained with an average closed source. On the other hand, the open source suffers a significant loss in ionization efficiency.

This discussion of pump-out rate is strictly correct for material in the vapor state. Caution must be exercised in applying these concepts to a situation in which a large quantity of sample is passed quickly into the ionization chamber, especially at high chromatographic temperatures. In such cases, noticeable adsorption or condensation can occur in the ionization chamber, particularly with polar materials, and the actual rate at which a source is cleaned out may be much slower than is predicted in Table 2.1. It is not uncommon to observe tailing of overloaded chromatographic peaks for several minutes even though other peaks are observed to come in and out of the ion chamber as sharply as they are observed on the chromatographic trace. Even relatively volatile compounds in large, but not excessive, amounts may show unreasonable tailing if the material is fairly polar.

TABLE 2.1. Comparison of conductance characteristics for various ion source configurations.

	Source Conductance, MW 200, liter/sec	Relative Ionization Efficiency for Same Quantity of Sample	Source Volume, liters	Half-Life $t_{1/2}$, sec	Time to Attain 0.1% Level, sec
Open Source	50	1	5	0.07	0.7
Average Source	0.4	120	7.5×10^{-3}	0.013	0.13
Tight Source	0.1	480	7.5×10^{-3}	0.05	0.5
Very Tight Chemiionization Source	0.003	(not applicable)	7.5×10^{-3}	0.17	1.7

2. Chemical Ionization

Chemical ionization mass spectrometry has recently emerged as an important new technique to obtain additional information not provided by electron impact methods.[52-60] One of the most annoying shortcomings of conventional electron impact data is that many types of compounds give a very weak signal for the molecular ion. Even when the molecular ion is as high as 1 or 2% relative abundance, this often means two orders of magnitude higher sample requirement if the molecular weight is an important piece of data. (And it always is.)

The extensive fragmentation observed in the electron impact spectra of many compounds results from the fact that during the initial electron/molecule interaction, many molecules receive considerable energy above the ionization voltage. Potentially, the molecule-ion can undergo one or more bond breaks thus reducing the concentration of the parent ion. No single feature in the interpretation of mass spectral data causes the mass spectrometrist more embarrassment than the inability to state unequivocally the molecular weight.

On the other hand, the chemical ionization process occurs with a much lower transfer of energy, and as a natural consequence, the fragmentation process is modified and greatly reduced. The parent ion is not usually abundant, but a quasi-molecular ion formed by loss or gain of one hydrogen is often the most prominent ion in the spectrum. In this manner, chemical ionization spectra provide information about the molecular weight and in addition, the fragmentation pattern may differ sufficiently from the electron impact pattern to reveal other structural features not indicated by the conventional mass spectrum. The difference between typical electron impact (EI) and chemical ionization (CI) spectra is shown in Figure 2.10[52,55] For ortal, the quasi-molecular ion at mass 241 gives considerable additional information in the parent region. The other example, alanylvaline, is chosen to illustrate that chemical ionization does not always improve the data in the parent region. In this case, both ionization modes give a quasi-molecular ion at mass 189.

Chemical ionization mass spectra result from the ion-molecule reaction that occurs between a low-pressure sample gas and the primary ions of a high-pressure reactant gas. Typical ion chamber pressures will be 0.3-3 Torr for the reactant gas and 10^{-6} Torr or lower for the sample. Both gases are introduced into the ion chamber where they are bombarded by an electron beam, but because of the very low abundance of the sample, virtually all primary ionization due to electron bombardment occurs to the reactant gas. The ionized reactant gas undergoes ion-molecule reactions with itself to form a

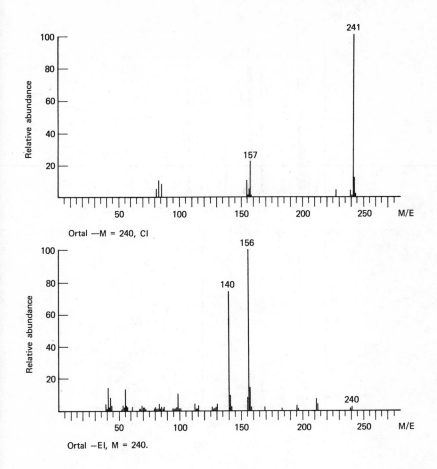

FIGURE 2.10. Comparison of chemical ionization and electron impact spectra. (a) Ortal, MW 240.[55]

steady-state plasma which in turn reacts chemically with the dilute sample gases. The process results in fragment and product ions characteristic of the unknown sample.

A variety of reactant gases have been proposed for chemical ionization, but the most common to date are the simple hydrocarbons methane and isobutane. If methane is the reactant gas, the most important ions in the reaction plasma are CH_5^+ and $C_2H_5^+$ which make up 90% of the ionic content. These are formed by reaction of the normal electron impact products with the excess

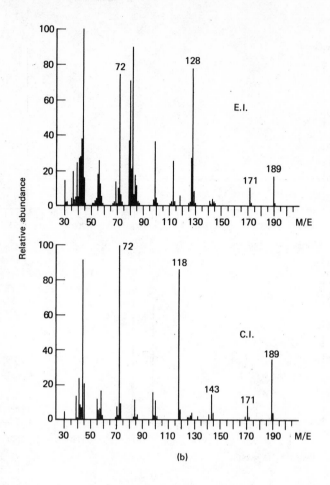

FIGURE 2.10. Comparison of chemical ionization and electron impact spectra. (b) Alanylvaline, MW 188.[56]

of CH_4 in the ion chamber. Thus:

$$CH_4 + e \longrightarrow CH_4^+ + 2e$$

$$CH_4^+ \longrightarrow CH_3^+ + H$$

$$CH_4^+ + CH_4 \longrightarrow CH_5^+ + CH_3$$

$$CH_3^+ + CH_4 \longrightarrow C_2H_5^+ + H_2$$

In the presence of a good proton acceptor, the ions CH_5^+ and $C_2H_5^+$ act as Brönsted acids and protonate the sample molecule:

$$CH_5^+ + BH \longrightarrow BH_2^+ + CH_4$$
$$C_2H_5^+ + BH \longrightarrow BH_2^+ + C_2H_4$$

These reactions are typical of those observed for alcohols, aldehydes, esters, etc., and also for many biochemical compounds typically encountered in recent chemical ionization applications.

If the sample material is not a good proton acceptor, the chemical ionization process will occur as a hydride ion abstraction or as a dissociative proton transfer. For example, with decane (MW 140) the main ion peak is observed at mass 139[53] due to the processes

$$C_2H_5^+ + C_{10}H_{22} \longrightarrow C_{10}H_{21}^+ + C_2H_6$$
$$CH_5^+ + C_{10}H_{22} \longrightarrow C_{10}H_{21}^+ + CH_4 + H_2$$

Additional fragmentation occurs to give a mass spectral pattern that is similar in appearance to the electron impact spectrum of a hydrocarbon, but the abundance of the quasi-parent ion is greatly increased relative to the fragment ions.

The increased relative abundance of the quasi-molecular ion has proven to be of great value in many studies, particularly with relatively complex bioorganic molecules.[55, 59] As studies of the fragmentation patterns are extended and the various effects of different reaction gases are understood and applied, chemical ionization spectra will prove to have increased value. One of the important reasons that this increased quasi-parent ion abundance is so useful is that the overall mass spectral sensitivity is of the same order of magnitude as the sensitivity obtained for electron impact spectra. High sensitivity occurs due to the fact that the electron beam is fully utilized because of the considerable increase in the partial pressure of reactant gas in the ion chamber. The reactant gas or ion plasma, in turn, has a high probability of reactive collision with the unknown sample; thus the overall sensitivity is quite comparable to that obtained by conventional electron impact.

In a chemical ionization mass spectrometer, it is essential that the vacuum system is optimized to pump a large volume of gas. Typically, 2–5 cm^3 atm/min of reactant gas passes into the ion chamber corresponding to a quantity Q equal to 2.5–6.3 x 10^{-2} liter Torr/sec. If a pressure of 0.5 Torr is desired in the ion chamber, the ion chamber conductance would have to be about 0.05 liter/sec (calculated using equation 2.2).

Chemical ionization systems must use large fast-pumping units with different pumping between the ion source housing and analyzer section. The same technology should also be applied to standard GCMS systems. (The details of correct vacuum practice are discussed in Chapter 4.) To operate a chemical ionization mass spectrometer in a GCMS mode does not require any hardware modifications or any change in the method of operation. The only variation is to replace the large quantity of reactant gas with a selected fraction of the GC effluent stream. It should be emphasized here, that the common conception that "application of the chemical ionization technique to GCMS avoids problems inherent in normal GCMS work" is not strictly correct.[54] Both GCMS and chemical ionization mass spectrometry are systems that pump relatively large quantities of gas (in the range 10^{-3} to 10^{-1} liter Torr/sec) through the mass spectrometer. There is nothing different in the chemical ionization vacuum methods that is not completely applicable to GCMS.

Some consideration must be given to choice of reactant gas. Studies have shown that the volatile hydrocarbon gases are suitable for most unknown samples, and their use appears to assure maximum probability of obtaining useful data. For GCMS studies, the conventional helium carrier gas may be replaced with methane, isobutane, or some other acceptable hydrocarbon gas. The reactant does not need to be a gas at ambient temperature, but it must have a low molecular weight in order that the mass spectrum will not be obscured by the fragmentation from the reactant. The less common carrier gases such as nitrogen, hydrogen, or water can also be used in chemical ionization GCMS studies. One convenient way to select the carrier/reactant gas is to set up a manifold (as shown in Figure 2.11) which makes a choice of several gases easily available. This technique has been used by Flotz[58] and by Arsenault[60] to develop chemical ionization methods in GCMS analysis.

Chemical ionization methods are most effective when used in conjunction with conventional electron impact analysis. Recognizing this, Arsenault[59] designed a dual ionization source that provides both chemical ionization and electron impact (Figure 2.12). The system can be used with each source operating by itself or with both sources operating simultaneously. An example of a spectrum obtained with a dual source is shown in Figure 2.13.[60] The main advantage of this dual configuration is that both types of spectra are obtained from a chromatographic peak within a few seconds.

One of the questions that occurs when considering the combination of chemical ionization and GCMS is whether or not separation of the compo-

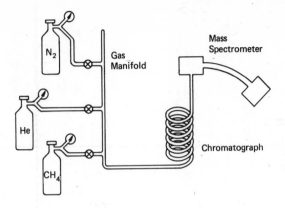

FIGURE 2.11. Gas selection manifold for chemical ionization GCMS.[58]

nents of a mixture is necessary when the chemical ionization produces primarily quasi-molecular ions. The answer can be determined only by consideration of each individual analysis. Frequently, a chemical ionization analysis

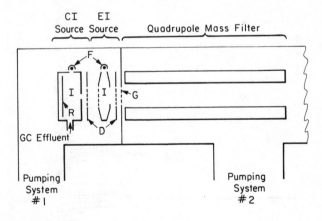

FIGURE 2.12. Schematic diagram of dual CI/EI source system for use with a quadrupole mass spectrometer.[59]

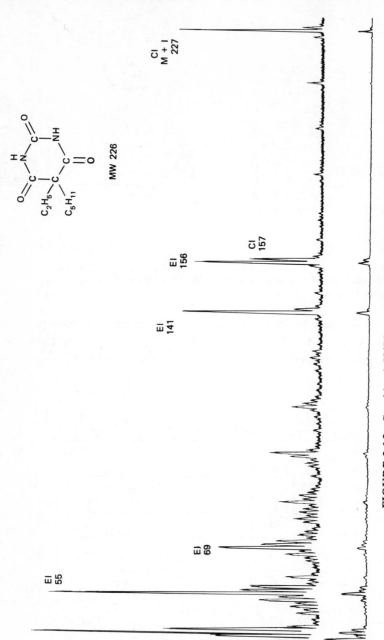

FIGURE 2.13. Combined CI/EI spectrum of amobarbital using methane as a reactant/carrier gas.[60] Major peaks due to each mode of ionization are indicated by CI or by EI.

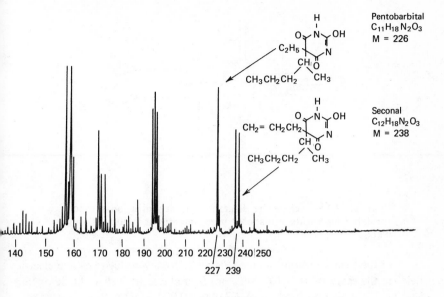

FIGURE 2.14. Identification of extract of gastric contents by chemical ionization of total mixture.[55]

on a complex mixture will give the information in a rapid and convenient fashion without separation. Such an example is shown in Figure 2.14 which illustrates the rapid and more or less certain identification of barbiturates in the gastric contents of a drug victim.[55] However, many other peaks are observed in this spectrum, and the "identification" is really only a strong confirmation of a fact already suspected. In general, any spectroscopic, chemical, or physical method of analysis is best performed on a pure substance, and it follows that combined operation in a GCMS mode will often improve the quality of the data and increase the certainty of the interpretation. Furthermore, if the sample size is small, GCMS operation offers the most efficient method of sample introduction.

3. Field Ionization

The field ionization source is another technique that provides sample ionization at relatively low energy with resultant reduced fragmentation and increased relative abundance of the parent ion. As such, field ionization methods offer analytical information not always obtained from electron impact spectra and, in favorable circumstances, provide important auxiliary information.[61-65]

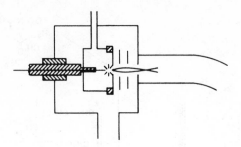

FIGURE 2.15. Schematic diagram of field ionization source. Insulated high-voltage blade or wire operates at 7,000–10,000 volts.

In the field ionization source (Figure 2.15), a very high positive electric field (in the range of 10^7–10^8 volts/cm) is produced by a sharp blade or thin wire held at a high positive voltage (7,000–10,000 volts) with respect to the first slit. (See also Damico, page 417.) The high electric field strength induces electron tunneling through a potential energy barrier in the molecule, and the resulting positive ion is accelerated out of the chamber and into the mass spectrometer analyzer.

The energy available for field ionization and subsequent excitation of a molecule is generally about 12–13 eV. Since organic molecular ionization potentials range from 7–13 eV, many molecules will have very little excess energy in the parent ion to cause fragmentation. A mass spectral pattern obtained from a field ion source is compared with the electron impact mass spectrum in Figure 2.16.

It is not possible to generalize with respect to the amount of fragmentation that will occur with field ionization. Values for the parent ion abundance may be as high as 99+% for hydrocarbons or as low as 10–20% for aldehydes. The extent of fragmentation is partly related to the ionization potential, particularly if this potential is close to the available energy. When excess energy is available, i.e., for molecules of lower ionization potential, the extent of fragmentation does not correlate with the ionization potential but rather depends on the unimolecular chemical reactivity of the parent ion. This difference is clearly indicated in Figure 2.17 which compares the field ionization spectra of n-heptanal and 3-heptanone. Both molecules have ionization potentials that permit the parent ion to acquire a few electron volts of excess energy in the ionization process. However, fragmentation is minimal

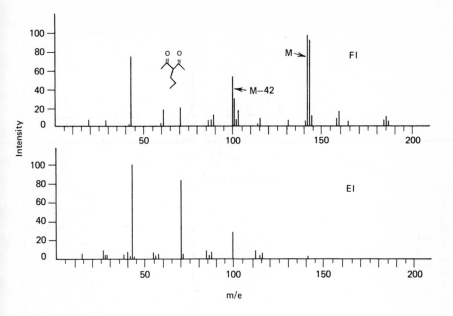

FIGURE 2.16. Comparison of field ionization spectrum with electron impact spectrum.[67]

for the ketone but significant for the aldehyde. Current studies are in progress to correlate this anomaly with known chemical characteristics and reactivities.[66, 67]

For analytical studies, the principal advantage of field ionization methods is the large abundance of the parent ion. As shown above, this feature can generally be expected but it is not reliably observed for all molecules. Frequently, a quasi-molecular ion $(M + H)^+$ is observed due to a surface reaction of the sample with adsorbed water on the blade or emitter. Thus, in Figure 2.17, a significant ion peak is observed in the heptanal spectrum at mass 113 $(C_7H_{14}OH^+)$. This feature may have some diagnostic value in analytical studies, but at the present, the available information is scattered and considerable additional study is required.

Field ionization methods have found several important analytical applications and the use of a GCMS mode has proven of value.[63] The reasons for GCMS in connection with field ionization analyses are essentially the same as for the electron impact or chemiionization methods. Tandem GCMS tech-

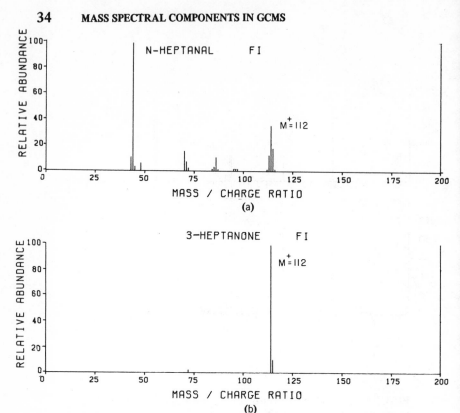

FIGURE 2.17. Comparison of field ionization spectra of n-Heptanal and 3-heptanone.[66]

niques provide a convenient method of introducing small amounts of reasonably pure samples to the mass spectrometer. The vacuum and gas flow considerations are the same using a field ionization source as with an electron impact source.

Sensitivity of the FI source is undoubtedly the most serious drawback for current analytical applications. In addition, the blade or emitter surface is unstable and severely affected by past history. Generally, the expected current sensitivity is 10–100x lower than that attained by electron bombardment which means that even though the parent ion peak is of higher relative abundance, the absolute ion signal may not be greater than that obtained from an electron impact study. In many cases, the fact that the spectrum is less cluttered with fragment peaks, particularly minor impurity peaks, is of sufficient advantage that the field ionization spectrum provides the important data necessary for an identification. This value can be enhanced by operating in a GCMS mode. However, most researchers agree that the state-of-the-art for field ionization is not developed to a point that permits easy routine analysis, and applications are still in the realm of the specialist. However, the field ionization method does offer very significant advantages for fundamental studies of ionic fragmentation,[59, 62, 63] and field desorption mass spectrometry promises to be extremely valuable for analysis of nonvolatile compounds.[64]

C. MASS ANALYZERS

One of the more characteristic features of a mass spectrometer is the method used to obtain mass dispersion of the ions produced in the ionization chamber. In general, when one asks the question "what type of mass spectrometer do you have?" the inquiry is directed to the method of mass analysis, not to the ion chamber, inlet system, or collector/computer assembly. There are at least 15 to 20 different types of mass analyzers, but for the most part, only a few have had extensive applications for analysis of organic compounds and only four are used in GCMS.

The most important parameter determined by the analyzer is the mass resolution. For most organic chemistry applications, the resolution R is defined by the equation

$$R = M/\Delta M \qquad (2.5)$$

where M is the mass of the first peak in a doublet and ΔM is the difference in the masses of the two peaks. The loose jargon "resolution of 10% valley" is commonly used and means that two recorded peaks of approximately equal intensity are resolved or separated when the height of the valley between them is 10% of the peak height.

This practical definition of resolution is illustrated in Figure 2.18 which

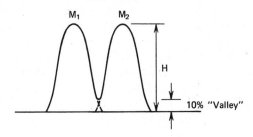

FIGURE 2.18. Simulated mass peaks illustrating resolution M_1/M_1-M_2 with a 10% valley.

shows two peaks of masses M_1 and M_2 resolved with a 10% valley. The choice of 10% valley is arbitrary and occasionally reference is also made to resolution of 1% or 50% valley. In the example, if M_1 = 500 (e.g., $C_{34}H_{60}O_2$) and M = 501, then the resolution would be 500. However, suppose both ions have the same nominal mass of 99. If M_1 is $C_6H_{11}O$, mass = 99.08098 atomic mass units (amu), and M_2 is C_7H_{15}, mass = 99.11737 amu, then R = $M/\Delta M$ would be about 2800.

Different analytical problems are served by different values of resolution, and the question arises "What resolution is necessary?" There is always a tendency to desire a mass spectrometer with maximum available resolution since it is generally difficult to know in advance what might be needed for future problems. Actually, this temptation should be tempered with careful appraisal of the job on hand and with a careful realization of the fact that increased resolution will mean higher "cost." In part, this cost will be the initial capital investment, but quite as important, increased resolution will mean an increase in the amount of sample used for the analysis. In a magnetic sector instrument, higher resolution is attained by reducing the source and collector slits with a corresponding decrease in sensitivity. Also, with higher resolution, the time taken to scan across the ion peak is decreased. If the amplifier and recording system have been optimized to work at a lower resolution, the total scan time must be correspondingly increased. The various factors that must be considered in balancing the resolution, scan time, and sensitivity are discussed further in Chapter 6.

The other feature of mass analyzers that influences GCMS performance is the length of the ion flight-path. In conventional mass spectrometry, the pressure in the analyzer section is usually 10^{-6} Torr or lower and at that pressure, the mean free path of the ions will be greater than 10^3 cm. (See Chapter 4.) The probability of collisions between the residual gas and the ions is thus quite low even for flight paths of 1 or 2 meters. On the other hand, in GCMS studies the pressure in the analyzer will frequently determine the amount of effluent gas that can be sampled. If the flight path is around 150 cm, the analyzer pressure would have to be one order of magnitude lower than in a small mass spectrometer with a 15-cm flight path.

Use of differential pumping between the source housing and analyzer section reduces the importance of the length of the flight path. In general, one to two orders of magnitude of pressure differential is obtained in the analyzer section using differential pumping (see page 162).

1. The Single-Focusing Magnetic Deflection Mass Spectrometer

The single-focusing magnetic mass spectrometer is the most common type used in organic analysis. The popularity arises from various factors. The system permits relative simplicity in construction and operation. Resolution of 300–1000 is easily obtained in small single-focusing mass spectrometers and resolution of 300–7000 is attainable in larger systems. The operator can select lower and intermediate resolution values by adjustment of the source and collector slits. In addition, the magnetic deflection mass spectrometer gains some popularity from the prejudice of historical precedence. The fact that many of the earlier catalogued mass spectra were obtained from this type of mass spectrometer was important in the past, but as more spectra are submitted from a wide variety of mass analyzers, this consideration has become less important.

The mechanism of a typical magnetic-sector analyzer is illustrated in Figure 2.19. Ions formed in the source are accelerated through the source slit S_1 toward a homogeneous magnetic field. For ions with an electronic charge e and mass m, the kinetic energy will be related to the accelerating voltage V, by the equation

$$Ve = 1/2\ mv^2 \tag{2.6}$$

where v = ion velocity. As the ions enter the magnetic field H, they experience a force orthogonal to the field which results in a curvature of the ion path. This accelerating force, Hev, is balanced by the centripetal force so

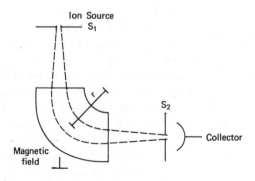

FIGURE 2.19. Schematic diagram of 90° magnetic sector showing direction focusing of divergent ion beam.

that

$$Hev = \frac{mv^2}{r} \tag{2.7}$$

where r is the radius of curvature. Elimination of the velocity term gives the equation

$$m/e = \frac{H^2 r^2}{2V} . \tag{2.8}$$

Thus, at a fixed radius r, and for a singly charged ion, the mass focused at S_2 and collected by the detector is proportional to the square of the magnetic field and inversely proportional to the accelerating voltage. By varying either of these two parameters, ions of different mass-to-charge ratio can be deflected to the collector and in this fashion the mass spectrum is scanned.

For most applications it is preferable to vary the magnetic field and maintain a constant accelerating voltage. When the voltage is varied over the course of a mass scan (with constant magnetic field), the efficiency of transmitting ions of low mass is much greater than that for ions of high mass. This mass discrimination is due to the fact that an ion of mass 400 will have one-tenth the accelerating voltage of an ion of mass 40. Since the higher mass region is the more important part of the spectrum, voltage scanning is used only for special cases where magnetic scanning is impractical.

Restriction to magnetic scanning is of no consequence for conventional mass spectrometry. In GCMS, there are certain circumstances in which a fast reproducible scan is important. GCMS monitoring is facilitated by a good oscilloscope display (page 264). Computer acquisition of GCMS data requires a precise scan function that is more easily attained by electric scan (page 247). Multiple ion detection scan methods can be applied only by using a fast electric jump (page 252). These special techniques all impose restrictions on the use of magnetic scan mass spectrometry. The various means of overcoming these obstacles are discussed in the later sections indicated above.

The resolution of a sector-field mass spectrometer is determined principally by the radius of curvature and by the width of the source and collector slits. The radius determines the mass dispersion, i.e., the distance by which two mass peaks are separated. The source slit defines the width of the ion beam. The collector slit determines the distance the focused image or peak must traverse in the course of scanning. A number of other factors such as energy dispersion, second-order directional divergence, magnetic fringe-field aberrations, space charge effects, and effective line of focus must be considered for optimal design of a system, but these are not parameters that can be controlled by the GCMS operator, either before or after purchase. For most mass spectrometers, the dispersion will be proportional to the radius, and the resolution can be defined as

$$R = \frac{kr}{S_1 + S_2} \cdot$$ (2.9)

Resolution is thus seen to be inversely proportional to the slit width, but decreasing the slits affects a corresponding decrease in beam intensity. To assure maximum flexibility, a magnetic mass spectrometer should be equipped with variable slits to provide a choice of operating resolution. Continuously variable slits are most common, but a selection of 3 or 4 different slit widhts is sufficient. For most GCMS work, a minimum resolution of 300–400 should be available to assure maximum sensitivity and scan convenience (see page 239).

The modern single-focusing mass spectrometer can be designed for convenient simple operation. A typical instrument is shown in Figure 2.20. Most of these mass spectrometers are quite small and sufficiently rugged to permit easy transport and assembly. Routine operational techniques can usually be mastered within a week.

2. The Double-Focusing Mass Spectrometer

In Figure 2.19, the beam from the ion source is shown to diverge as it exits through slit S_1 and then, after passing through the magnetic field, to converge and focus at S_2. This direction focus is a property of the magnetic field and if it did not occur, it would be necessary to use a collimated ion beam with a resultant loss of beam intensity. The magnetic sector instrument is thus termed "single-focusing."

In addition to the angular divergence illustrated in Figure 2.19, the beam also possesses an energy divergence. This energy divergence results from the difference in the position at which various ions are formed in the chamber and from the natural kinetic motion of the molecules. Thus, even though accelerated to an energy of a few thousand electron volts, all ions do not have exactly the same energy to within a few tenths of an electron volt. If equation 2.7 is rewritten in the form

$$r = \frac{mv}{He}$$

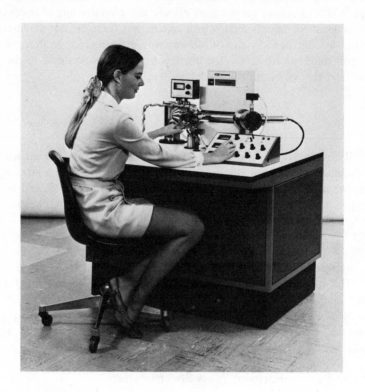

FIGURE 2.20. A single-focusing magnetic mass spectrometer (DuPont Instruments, Model 21–490). Compactness and ease of operation are typical design features of small modern mass spectrometers.

it is seen that the magnetic field does not produce true mass separation as is commonly suggested, but rather momentum separation. The velocity dispersion will thus broaden each individual peak and limit the attainable resolution.

To counteract the effect of velocity dispersion, an additional ion lens, generally a radial electric field, must be used. If two sectors of cylindrical electrodes are centered on a common axis as shown in Figure 2.21, an ion

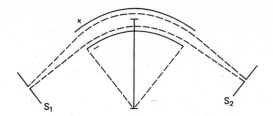

FIGURE 2.21. Schematic diagram of 90° electric sector showing direction focusing of divergent ion beam.

beam entering the field experiences an electrostatic force and describes a circular path. The centripetal force on the ion is Ee, where E is the field strength in volts/cm. This force must be balanced by the centrifugal force so that

$$Ee = \frac{mv^2}{r} . \tag{2.10}$$

For a given electrostatic field, the radius of curvature of the ion depends only on the energy, and ions of the same mass but with different velocity will be separated. Because there is no mass dispersion, the radial electric field cannot be used by itself as a mass spectrometer, but it can be used in combination with a magnetic field to obtain energy dispersion selection.

With proper design of the components, a mass spectrometer containing both an electrostatic sector and a magnetic sector can give focus with respect to both velocity divergence and energy divergence. For most electric and magnetic sector configurations, ions of different mass achieve velocity focusing on one locus and direction focusing on another locus. The intersection of these two lines is the point of double focus for a given mass. One specific geometry, the Mattauch-Herzog mass spectrometer, accomplishes double-focusing for all masses in a plane.[68, 69]

The principle of double-focusing is illustrated in Figure 2.22 for the popular Nier-Johnson configuration.[70, 71] A divergent ion beam from the source S, enters the electric field E where it experiences an accelerating force according to equation 2.10. Ions of a given energy are directionally focused at B_1; those of a lesser energy at B_2. The slit at this focal point, called the β stop, can be adjusted to accept a wider or narrower energy

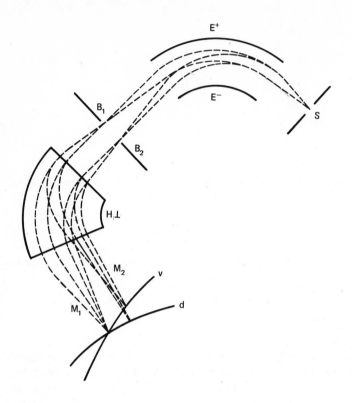

FIGURE 2.22. Arrangement of electrostatic and magnetic sectors in Nier-Johnson mass spectrometer. Double focus is achieved only at the intersection of the velocity and direction focus loci.

spread in the beam. The beam continues into the magnetic field where the ions are separated according to their momentum (equation 2.7). Directional focusing occurs for all ions on the locus d, and velocity focusing occurs on the locus v. The point of intersection of these lines is the point of double focus and defines the position of the collector slit.

Another double-focusing mass spectrometer configuration commonly used in organic chemical studies is due to Mattauch and Herzog.[68, 69] The significant difference between this geometry and that of Nier and Johnston is that all ions experience double-focusing in a straight-line plane. As illustrated in Figure 2.23, this feature has the advantage that a photographic plate P can be used for ion detection (see page 68). Electrical detection is

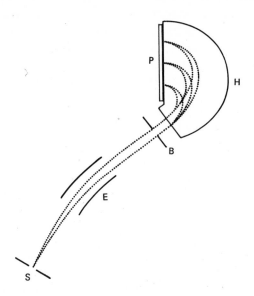

FIGURE 2.23. Arrangement of electrostatic and magnetic sectors in Mattauch–Herzog mass spectrometer. Double focus is achieved in a plane at the border of the magnetic field.

also an allowed option. Because of the severe restrictions placed on a scanning system at high resolution (see page 239), photoplate detection offers a practical method to obtain high resolution mass spectra of chromatographic peaks.[72, 73] (See also Habfast et.al., page 408.) Since no scan is actually made, there are no time constant restrictions in the amplifier or recorder. The maximum resolution usable for GCMS is determined only by the limitations of the mass spectrometer and not by the restrictions of the scan rate/recorder interaction.

The purpose of a double-focusing mass spectrometer in organic studies is to achieve sufficient resolving power to separate doublet or multiplet peaks for accurate mass determination. A resolution of 10,000–18,000 is sufficient to give mass accuracy of a few parts per million. Occasionally, resolution as high as 40,000–70,000 is needed to separate certain mass doublets. In GCMS studies, this can be accomplished only by using photoplate detection.

Fast scanning at high resolution necessitates fast response in the amplifier/recorder system, but in addition, the sensitivity is severely reduced due to two factors (see also page 288). First, to attain high resolution, the slit width is narrowed (equation 2.9), causing a proportional decrease in beam intensity. This factor is present using either photo detection or electrical detection. Second, when a peak is scanned very rapidly, the number of ions in a specified time window is significantly reduced. For example, an ion beam of 1.6×10^{-14} amps corresponds to 10^5 ions/sec. If the ion peak is scanned in 1/5000 sec, then the detector will collect an average of less than 20 ions. This amount is barely sufficient to obtain a reasonable statistical distribution of the ions to establish the peak profile for accurate mass measurement.

These restrictions of amplifier/recorder frequency and of sensitivity have necessitated a compromise for high-resolution GCMS scans. As a rule, 8–10 sec/decade is used at 10,000 resolution (see Table 6.2). Unfortunately, this rate is 2–3 times too slow for optimum scanning. It has been shown, however, that suitable mass measurement can be attained at resolution as low as 1000, provided that the mass peaks are resolved.[74] Unfortunately, at this resolution sample peaks are not resolved from the reference compound peaks (most commonly, perfluorokerosene). To circumvent this objection, a double-beam mass spectrometer has been proposed.[75] The double-beam system consists of two indepedent inlet and ion source assemblies, one electrostatic and magnetic sector, and two indepedent amplifiers and recorders (Figure 2.24). In this way, the sample and reference beams are produced and detected indepedent of each other, but because they pass through the same electric and magnetic field, the mass scale for each is precisely matched. Thus, sample-reference doublets do not occur, and accurate mass measurement can be attained at resolution 1000–2000 with the advantages of increased sensitivity and scan rate. Mass measurement accuracy of 15 ppm has been reported for major peaks in the spectrum with 2×10^{-9} gm of total sample eluted from the column.[75]

The double-beam technique is useful for many high-resolution applications involving very small samples, but it is limited to compounds in which the mass peaks are not partially separated doublets. In addition to accurate mass assignment, the double-beam system serves as an excellent mass marker for low resolution GCMS spectra. The commercial instrument is shown in Figure 2.25.

For most GCMS studies, a low-resolution run should precede any attempt to obtain accurate mass measurement. Because of the rapid introduction of chromatographic peaks using GCMS, several hundred spectra may be obtained in a day and many of these can be identified from the low-resolution spectra. Only a few spectra need the additional effort of a high-reso-

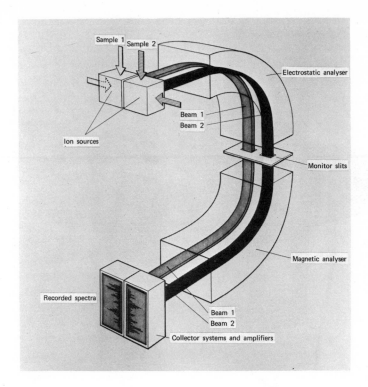

FIGURE 2.24. Principle of the double-beam mass spectrometer.[75]

lution study, and because of the extensive tabulation of data that results from a high-resolution run, the effort should be restricted accordingly. Large high-resolution double-focusing systems will most likely continue to be used mainly in the realm of the specialist.

In some cases, particularly biological studies involving alkaloids, antibiotics, or drugs, it can be advantageous to obtain the high-resolution data without a low-resolution screening run.[72, 73] However, unless the need for the more sophisticated data is established, low-resolution spectra should be used to obtain preliminary knowledge of the sample. This rule is applied to advantage for both batch and GCMS samples.

High-resolution mass spectral data should be handled by a computer system if the data is to be thoroughly processed.[76] The alternative manual methods are so unwieldy that the information is only partially reduced and interpreted. As little as 4 or 5 high-resolution spectra per week warrants the cost of a computer.

FIGURE 2.25. Double-beam, double-focusing mass spectrometer of Nier-Johnson geometry (Associated Electrical Industries, Model MS30).

3. The Time-of-Flight Mass Spectrometer

Various mass spectrometers have been devised that select or disperse ions according to their velocity.[77-82] For ions that are accelerated through a definite voltage drop, the velocity is characteristic of the mass as given by equation 2.6, and use of a pulsed voltage system or some form of radio frequency field permits an accurate determination of the time the ion takes to traverse a given flight path. The flight path can be circular or helical[79-81] as in a cyclotron resonance spectrometer, but the only system to have significant use for organic GCMS analysis is the linear time-of-flight mass spectrometer.[82]

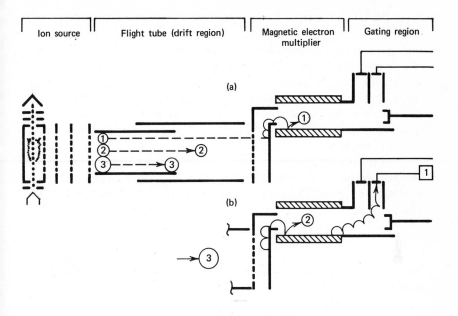

| Ion source | Flight tube (drift region) | Magnetic electron multiplier | Gating region |

FIGURE 2.26. Schematic diagram of time-of-flight mass spectrometer. (a) Shows hypothetical position of ions after 10 μsec. Ion 1 has arrived at collector. (b) Shows position of ions after 14 μsec. Ion 1 has been recorded; ion 2 has just arrived at collector.

The principal of the linear time-of-flight mass spectrometer is illustrated in Figure 2.26. Ions are formed in the ionizing region either by a continuous or pulsed electron beam. A repeller voltage pulse of a few nanoseconds duration is applied to the repeller electrode and a bundle of ions is ejected out of the ion chamber through the grid system. This ion bundle is accelerated by a voltage of a few thousand volts applied to the final accelerating grid. The ions then enter into the field-free drift region where they continue with a constant velocity to the collector. The velocity of each individual mass is determined by the accelerating voltages. During the flight, ions of different masses are thus separated as illustrated in Figure 2.26, and the arrival for each ion group at the detector is correlated with the time of the original voltage pulse. A typical flight time for a 1-meter drift tube is in the range of 1 to 30 μsec. The repetition rate of the process may be 10,000 to 50,000 times per second, but 20,000 cps is commonly used.

The ions are seen to traverse three regions: the ion chamber, the accelerating region, and the drift tube. The total time of flight is the sum of the time spent in each section. Ideally, if the ions are formed in an infinitely narrow plane with zero thermal velocity, the velocity acquired from the accelerating voltages can be accurately calculated. In practice, the ions are formed in a region of finite thickness and have a natural kinetic velocity in all directions. An excellent account of the motion of an ion in this real system is given by Wiley and McLaren[82] and optimization of design parameters is discussed.

In the present discussion, it will be sufficient to consider that the ion velocity is determined primarily by the large accelerating voltage. The approximate time of flight can then be easily calculated from equation 2.6. Thus,

$$v = \sqrt{2eV/m} \qquad (2.11)$$

The flight time t is given by the path length, l, divided by the velocity, hence

$$t = \sqrt{ml^2/2eV} \quad . \qquad (2.12)$$

For an accelerating voltage of 2000 volts and a flight path of 1 meter, a hydrogen ion would have an approximate flight time of 1.7 μsec. An ion of mass 2500 would have a flight time of more than 50 μsec. For a typical pulse repetition rate of 20,000 cps, ions greater than 2500 amu would therefore arrive at the collector after the beginning of the next cycle. This imposes a mass range limitation, but for obvious reasons, the limitation has not yet been of any consequence in GCMS analyses.

Resolution of a time-of-flight mass spectrometer depends upon the initial space distribution. Because ions cannot all be formed in an infinitely narrow plane with zero thermal velocity, these factors contribute to the time spread of the ion bundle. In addition, the sharpness and duration of the ion pulse, and mode of electron bombardment (i.e., pulsed or continuous)[83] contribute to the time spread of the beam. Broadening due to natural spatial and velocity variations can be reduced by optimizing electrode voltages and distances so that space focusing and time lag energy focusing occur.[82] Resolution in the range 500–600 is attained, which is sufficient for most GCMS studies. Occasionally, compounds such as highly substituted trimethyl silyl derivatives may be encountered at higher molecular weights.

The difference in the flight time for ions of different masses can be calculated using equation 2.12 in the form

$$t_2 - t_1 = \sqrt{l^2/2eV} \; (\sqrt{m_2} - \sqrt{m_1}). \qquad (2.13)$$

For the previous example in which l = 100 cm and V = 2000 volts, the allowable time difference between the arrival of masses 500 and 501 to give resolution 500 would be 37 nsec. If the spread of adjacent ion bundles exceeded 80 nsec, the resolution would be only about 100. Quite obviously the resolution of a time-of-flight spectrometer depends upon having a fast sharp accelerating pulse and a fast response amplifier.

Because the ion bundles arrive at the collector with only 10^{-7} to 10^{-8} sec separation, the amplifier and detector must have a correspondingly fast response and an electron multiplier must be used for amplification. In most applications prior to 1961, an oscilloscope was used for observation and for photo recording of the mass spectra. Suitable amplifier and recorder time constants were thus easily attained. Since 1961, a gating system has been used whereby the current from the multiplier is pulsed towards a collector anode during that brief period in which a specific ion arrives at the collector. This anode signal is input to an analog amplifier which acts as a short-term storage device and outputs a signal to an oscillographic recorder. By continuously varying the delay time of the deflecting pulse, the mass spectrum is scanned at rates compatible with most oscillographic recording systems and an output is obtained comparable to the mass spectral output from other types of spectrometers. These unique features of the time-of-flight scanning/recording system permit several operational advantages in GCMS which are discussed in detail in Chapter 6 in the sections on total ion monitors and multiple ion detection methods.

The mass spectral pattern of the time-of-flight mass spectrometer does not differ significantly from that obtained on deflection instruments (see page 54). One feature of time-of-flight mass spectra that offers an advantage is that a higher abundance of the molecular ion is usually observed. This difference arises from the fact that most time-of-flight mass spectrometers operate with the ion source at a lower effective temperature (about 100–150°C) than the temperature conventionally used for the ion source of a magnetic spectrometer (about 200–250°C). However, ion source temperatures are independent of the type of analyzer. If both the time-of-flight and magnetic sector instruments are operated at the same source temperature, the mass spectral patterns will be quite similar.

The time-of-flight mass spectrometer was the first system to receive widespread acceptance as a versatile instrument for GCMS. This well-deserved popularity was a result of one singular feature, namely that the instrument was supplied with an electron multiplier. Thus, the fast response needed for GCMS scanning could be attained without sacrifice of sensitivity. In contrast, prior to 1963, most magnetic mass spectrometers did not have an electron multiplier as standard equipment, and their use in GCMS was restricted to specialists who could modify the hardware and methods to meet the unfavorable operating requirements. Today, electron multipliers are standard equipment in all organic analytical mass spectrometers, and this overriding advantage of the time-of-flight no longer exists.

Figure 2.27 shows a recent model of a time-of-flight mass spectrometer

FIGURE 2.27. Time-of-flight mass spectrometer (Bendix Corporation, Model MA–2).

equipped with a large differential pumping system.

4. The Quadrupole Mass Spectrometer

One rather unique mass analyzer that has gained recent popularity is the quadrupole mass filter.[84-89] This spectrometer consists of a quadrant of four parallel hyperbolic or circular rods which provide a specific radio frequency field (Figure 2.28). Opposite rods are electrically connected. The applied voltage consists of a constant DC component U and a radio frequency component V_0 (cos ωt). ($\omega = 2 \pi f$ where f is the radio frequency.) The potential difference (P. D.) between the two sets of rods is thus $U \pm V_0$ cos ωt. In practice, the ratio U/V_0 must be less than one or all ions would be collected on the negative poles.

As a consequence of this unique oscillating field, a positive ion injected into the quadrupole region will oscillate between the adjacent electrodes of opposite polarity. At a specified radio frequency, ions of a given mass undergo stable oscillation between the electrodes. Ions of lower or higher mass

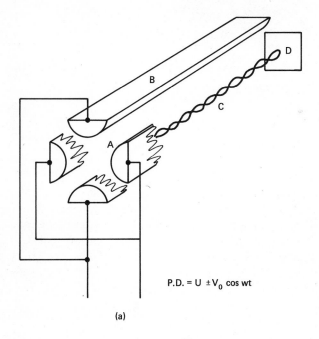

P.D. = U $\pm V_0$ cos wt

(a)

FIGURE 2.28. (a) Schematic diagram of quadrupole mass filter. A = ion injection, B = quadrupole rods, C = oscillating ion beam, D = collector.

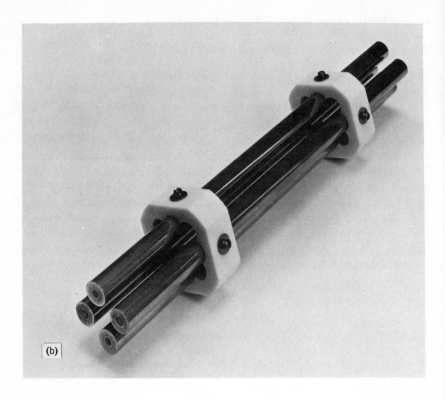

(b)

FIGURE 2.28. (b) Quadrupole rod assembly (Finnigan Corporation).

undergo oscillation of increasing amplitude until they are collected on the quadrupole electrodes. Within the quadrupole field, there is no force in the longitudinal direction so that an ion with a stable oscillation continues at its original velocity down the flight path to the collector. The accelerating energy is thus of secondary importance and there is no need for a high voltage in the range of several thousand volts. With 5–30 volts of ion accelerating potential, the ions undergo a sufficient number of oscillations during the flight period to provide reasonable mass separation.

The equations of motion that describe the ion oscillation in the x and y directions define the relation between mass, the dimensional parameters of the quadrupoles (i.e., radius and spacing), and the operational parameters U, V_o, and ω. Solution of these equations shows that for

defined values of these parameters there is a region of oscillatory stability
and a region of instability.[84, 86, 87]

A hypothetical example of a stability diagram for the quadrupole
system is given in Figure 2.29[85] which shows the nature of the plot of U/V_O
vs. $\kappa V_O/m$ (where κ is a constant containing the frequency and radius of
the poles). Any values of the quadrupole parameters and mass that lie
within the stable region result in passage of those masses to the collector.
For lower values of the ratio U/V_O, the range of mass stability is extended
and resolution is decreased. As indicated in Figure 2.29, for $U/V_O = a$,
masses in the range of 399.5–400.5 undergo bound oscillation and are
collected. All other masses are filtered out. At $U/V_O = b$, all masses in the
range 398.5–402 are collected.

Quadrupole spectrometers are conventionally mass scanned by varying
the dc voltage U and the radio frequency voltage V_O in unison to maintain
a constant U/V_O ratio. Since the abscissa contains the term V_O/m, any
change in V_O necessitates a corresponding change in m to keep the ratio
V_O/m in the stable region. The voltage/mass scan function is thus linear
with any change in voltage, i.e., $m = m_O + k V_O$. A mass scan can also be
performed by varying the frequency with U and V_O kept constant, but this
mode is considered less convenient. The frequency/mass function is
inversely proportional to the square of the frequency which leads to a more
difficult mass calibration.

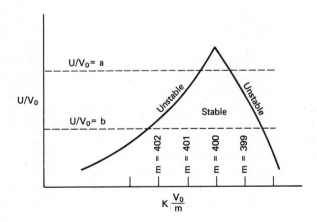

FIGURE 2.29. Stability diagram for quadrupole mass filter.

Theoretically, the resolution of a quadrupole mass filter can be increased to a high value by selecting the U/V_0 ratio close to the apex of the stability region. In practice, a significant percentage of the selected ions oscillates with a sufficient amplitude to strike the quadrupole electrodes and thus reduce the transmission efficiency. This errant motion depends upon the thermal velocity component in the x or y direction and also upon the position at which the ion enters the electrode cavity. In addition, the alignment of the electrodes must be very precise and the electrodes must be free from any nonconducting film (such as pump oil creep, column bleed, or excess sample condensation) that would distort the symmetric field. The practical resolution limit for most commercial quadrupole mass spectrometers appears to be 500–700 at the present time, although a few quadrupoles have demonstrated resolution of 1000 and better.

The ability to select a wide range of stable masses by using a low value of the U/V_0 ratio provides a convenient method of monitoring the GC run. With a low value of U/V_0, most ions are transmitted down the tube, and the output can be recorded as a chromatographic trace. When the scan switch is activated, a different U/V_0 value is set, and the spectrum is then scanned at the desired resolution. The brief discontinuity in the chromatogram conveniently marks the point of scan. This method of recording the chromatographic progress will be discussed further in Chapter 6 and compared with the alternative processes.

The quality of the mass spectral pattern obtained from a quadrupole mass spectrometer is strongly dependent upon the pole alignment and the tuning parameters. By modest adjustment of the injection voltage and the U/V_0 ratio, the operator can tune the quadrupole system so that the mass spectral pattern has a fairly close resemblance to that obtained from other mass analyzers. There is, however, a considerable mass discrimination against ions of higher mass and above mass 250, the ion intensities will have lower relative abundance. This characteristic is detrimental for observing low-intensity, high-mass molecular ions. It is not usually significant for comparing spectral patterns of compounds below molecular weight 250. Figure 2.30 shows a comparison of the spectrum of n-propyl benzoate obtained on the four different types of mass analyzer used in GCMS. Although each pattern is slightly different from the others, no important intensity variations are observed in this limited mass range. In the higher mass ranges (above 250), more serious mass discrimination occurs. The extent varies with different instruments and with the precise adjustment of the quadrupole field. In a recent comparison, the intensity of the parent ion of methyl eicosanoate was 18% from a magnetic mass spectrometer but only 8% from a quadrupole.[89]

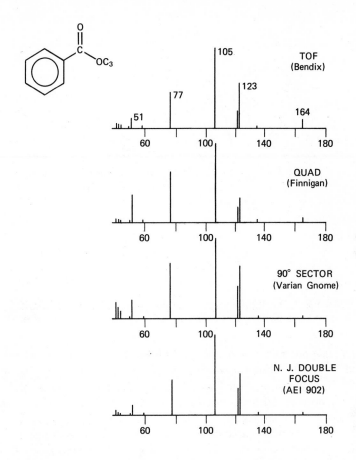

FIGURE 2.30. Comparison of the mass spectrum of n-propyl benzoate as obtained on four different commercial mass spectrometers. (a) Bendix time-of-flight. (b) Finnigan quadrupole. (c) Varian-MAT 90° sector (Gnome). (d) AEI Nier-Johnson double focus (MS902).

Two variations of the quadrupole mass spectrometer encountered in GCMS work are the monopole and the duodecapole mass spectrometers.[90, 91] The monopole mass spectrometer is designed as one quadrant of a quadrupole system. Because the equipotential lines in a quadrupole are symmetrical, a planar pole surface can be used to simulate parts of the total quadrupole configuration. As is illustrated in Figure 2.31, a monopole mass analyzer

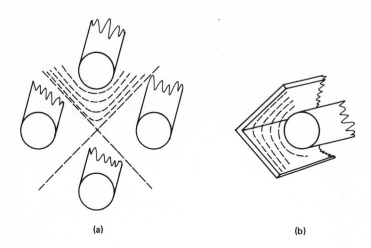

(a)   (b)

FIGURE 2.31. Equipotential line plot of (a) quadrupole and (b) monopole mass filters. Equipotential lines are drawn in dashes.

uses the same potential well as the quadrupole except that one quadrant is isolated by a right-angled bar which serves as the other electrode (Figure 2.31b). The constant and oscillating voltages are applied between this bar and the monopole. Such a configuration has the advantage that only two parts must be carefully aligned. It has the disadvantage that the angle and position of ion entrance are more restrictive, and for comparable geometric size, the ion transmission is reduced. A few operational features are different, but such a small number of monopoles have been used in GCMS that there is not sufficient data to indicate a preference for either the monopole or quadrupole.

The duodecapole mass analyzer is not really a twelve-pole mass filter but actually a quadrupole system with two "trimmer" electrodes in each quadrant. These trimmer poles, shown in Figure 2.32, permit easier adjustment of the equipotential field and thus overcome irregularities due to pole misalignment or non-conductive coatings. The performance appears to be about the same as is obtained from a conventional quadrupole. With future refinement and improved manufacturing precision, this variation may lead to improvement in resolution and/or sensitivity.

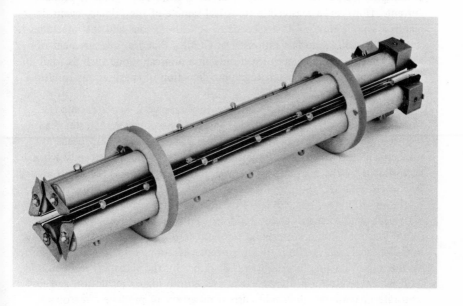

FIGURE 2.32. Rod assembly of duodecapole system showing location of trimmer electrodes relative to the four quadrupoles (Hewlett Packard, 5930A MS).

D. AMPLIFICATION AND RECORDING

Ion current values encountered and measured in modern mass spectrometry range from as high as 10^{-10} amp down to minimum values determined by the application and analytical method (10^{-16} to 10^{-19} amp). Typical minimum values may be 10^{-15} to 10^{-18}, and at these threshold levels, it is extremely important to bear in mind the significance of a small current in terms of the number of charged particles. Since the total charge in 1 Faraday (96,500 coulombs) is equal to 1 mole equivalent, the current corresponding to 1 singly charged ion/second is 1.6×10^{-19} amp. Thus, a current of 10^{-18} amp corresponds to about 6 ion particles/second.

Such a very low current value of 10^{-18} to 10^{-19} amp can be counted and amplified by an electron multiplier. With a typical gain of around 10^6, a multiplier output of 10^{-12} amp is obtained from 10^{-18} amp of primary current. However, a reasonable statistical sample of the primary event demands a sample time of 3–4 seconds to get 20–25 ions and this precludes scanning a peak at the fast rate used in GCMS. Practical measurement of a very low current can be performed only in a nonscan mode such as total ion monitoring or selective chromatographic detection of specific ions (multiple ion detection, page 252).

For most GCMS runs, a scan time in the range 2–6 sec/decade (i.e., from mass 30 to 300 in 2–6 sec) is considered acceptable. A faster scan is uncommon even for exceptional situations such as very sharp peaks at the beginning of a chromatographic run. A slower scan is used only when necessitated by the restrictions imposed in high-resolution mass spectrometry. In conventional GCMS, a typical scan rate of 5 sec/decade is often used at resolution 1000 (see page 242). This scan rate and resolution mean that a mass peak is traversed in about 10^{-3} sec and an ion current of 10^{-15} amp (6000 ions/sec) would correspond to about 6 ions in each mass peak. Thus, in GCMS analyses, the limit of ion current detection is not usually established by the noise limit of the amplifier, but rather by the statistical ion counting limit which results from the fast scan rate. Both higher resolution and faster scan rate increase the threshold current necessary to produce a reasonable statistical number of ions. In GCMS, these operational parameters should be kept as low as is practical to achieve high sensitivity on small samples. The statistical concepts of measuring low ion currents are discussed further in Chapter 7, page 288.

1. The Electrometer Amplifier

Whenever high precision and accuracy of measurement of ion beam intensity are more important than speed and sensitivity, the electrometer amplifier is the preferred method of current amplification. This device is used in most precision quantitative mass spectrometry and is also used for current/voltage conversion of the electron multiplier output. It is thus an integral part of any mass spectrometer system. (Distinction is made here with the mass spectrograph which, strictly speaking, uses photo detection and not electrical detection.)

A typical electrometer amplifier circuit is shown schematically in Figure 2.33. For GCMS work, only two design parameters need to be considered, the input resistance R_i, and the output resistance R_o. The value of the input resistance is important in controlling the allowed scan rate of the mass spectrometer. The value of the output resistance must be considered when the mass spectrometer is coupled to fast oscillographic recorders or to

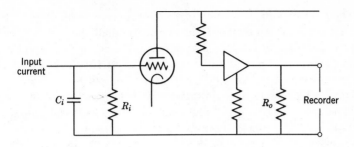

FIGURE 2.33. Schematic diagram of electrometer amplifier circuit.

an analog-to-digital converter for on-line computer applications.

Electrometer vacuum tubes are designed to have a high ground impedance and low ground capacity, thus permitting an input resistance at least as high as 10^{11} ohms. An ion signal of 10^{-14} amp develops an input voltage of 10^{-3} volt (V = 10^{-14} amp x 10^{11} ohms). In a unity gain amplifier, the output voltage across R_O is also 10^{-3} volt with a current gain of about 10^8 (for R_O = 1000 ohms). The output signal is thus compatible with a conventional potentiometric recorder.

Such a recorder is far too slow for GCMS recording. The characteristic decay constant τ of an input ion signal is given by the product of the input resistor and the input ground capacity. Thus, $\tau = R_i C_i$. The ground capacity of an electrometer circuit will be in the range of 2–5 pF so that for an input resistance of 10^{11} ohms, τ will be around 0.2 sec. This is at least 2 orders of magnitude too slow for GCMS scanning. To meet even the least demanding GCMS needs (i.e., low resolution, modest scan rate), the input resistance must be significantly reduced with a corresponding reduction in the amplifier output voltage. In addition, the recorder must be a faster, less sensitive galvanometer recorder. In GCMS, loss of sensitivity can be accepted only out of dire necessity, and whenever possible, an electron multiplier must be used to enhance the input current.

In spite of this limitation in fast scan systems, the electrometer amplifier is still the most convenient method of measuring the total ion current or some fraction of the unseparated beam. This measurement is frequently used for monitoring the progress of the chromatographic run. (See page 257). Since no mass scan is involved, the electrometer can have an input resistance of 10^{11} ohms which gives a current sensitivity of around 10^{-15} amp when used with a 1-millivolt full-scale potentiometric recorder.

The most important function of the electrometer amplifier in GCMS is to convert the amplified current from an electron multiplier into a signal that is compatible with fast recording devices. The attainable current gain from an electron multiplier is in the range of 10^5 to 10^7, so that a statistically limited ion current of 10^{-15} amp gives an output signal greater than 10^{-9} amp. With such high amplification, the input resistance of the amplifier can be as low as 10^7 to 10^8 ohms and still give a sufficient output voltage ($10^{-2}-10^{-1}$ volt) to drive a suitable fast recorder. The time constant of the amplifier ($R_i C_i$) is around 10^{-4} sec, so that a GCMS peak can be scanned in about 10^{-3} sec. With the very fast sampling rates used in computerized high-resolution GCMS (analog-to-digital conversion up to 25 kHz), the time constant would have to be reduced even further. (On-line computer operation is discussed in Chapter 7.)

2. The Electron Multiplier

The electron multiplier[92, 93] is now universally used in qualitative organic mass spectrometry. This remarkable device produces current amplification of 10^3 to 10^8 with virtually no noise or dark current, and with negligible time constant or signal broadening. Such an excellent gain is accompanied by disadvantages.[94] First, the electron multiplier lacks the stability of an electrometer, and the gain varies depending upon previous operating conditions. Second, the sensitivity depends on the mass of the impinging ion, the molecular structure, and the atomic composition. Third, overloading and saturation effects occur when the output current exceeds 10^{-8} amp. Fourth, the sensitivity must be determined by calibration, but the exact value differs from time to time. In spite of these disadvantages, the high-gain/high-speed feature is so important in organic analysis that an electron multiplier is considered to be absolutely essential.

A photograph and schematic diagram of an electron multiplier are shown in Figures 2.34 and 2.35. Most multipliers in common use contain 12–20 dynodes electrically connected through a resistive network. The ion beam emerges from the analyzer and strikes the first dynode, the so-called conversion dynode. This impact causes a shower of electrons to be emitted (approximately 1–2 electrons per positive ion), and these electrons in turn are accelerated toward the second dynode where their impact causes additional electron emission. The process is repeated down each stage of the multiplier, and the final dynode or collector is connected to an electrometer for amplification and recording.

The number of electrons given off for each incident ion, and for each subsequent electron impact at the dynodes, is a complex function of the

FIGURE 2.34. Photograph of commercial electron multiplier. (Associated Electrical Industries, MS 902.)

energy and composition of the ion, accelerating voltage across each dynode, and the nature of the dynode material. A typical 16-stage copper-beryllium (2% Be) multiplier has a gain of about 10^7 at an accelerating voltage of 3200 volts (200 volts/stage). Assuming the ion conversion stage has the same efficiency as the fifteen subsequent electron multiplication stages, this corresponds to about 3 secondary electrons/primary impact ($3^{16} \approx 4 \times 10^7$). For most GCMS analyses, the multiplier is run at a gain of about 10^6 and additional gain is of value only in a few isolated circumstances. Higher multiplier voltages often produce unwanted noise spikes.

Another type of electron multiplier used in GCMS work is the continuous dynode magnetic electron multiplier[95] (Bendix time-of-flight mass spectrometer). In this device, the copper-beryllium dynodes are replaced by two parallel glass plates which have been coated with a high-resistance

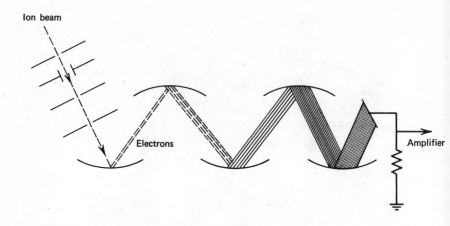

FIGURE 2.35. Schematic diagram of electron multiplier (voltage/resistance network omitted).

coating (Figures 2.26, 2.36). An electric field is produced by applying a high voltage ($\sim$ 3000 volts) down the length of the plates which are located in a magnetic field of a few hundred gauss.

Impact of the positive ions causes conversion to electrons. In the crossed electric/magnetic field configuration, the motion of the electrons is a cycloidal or toroidal path so that the electrons move in a "looping" fashion toward a position of lower electric field. As the electrons proceed down this path, they continually strike the surface of the glass plate and cause additional electrons to be given off. The resulting gain in electron current can be as high as 10^6 to 10^7.

The long-term stability of the Be-Cu dynodes is greater than that of the glass field strips, and the gain of the latter decreases rapidly if used at high current levels. Fortunately, the resistive coating on the glass is easily cleaned by soft abrasion, but it is annoying to have a shutdown every few months to correct low multiplier gain. On the other hand, if the gain of the Be-Cu dynodes is lost (e.g., by accidental air exposure at high voltage), it is generally necessary to send the multiplier to the factory for reconditioning.

Measurement of the gain of an electron multiplier should be a simple straightforward process which compares a given ion signal measured first on the multiplier and then directly on the electrometer amplifier. Lamentably, appropriate hardware for this simple, very important measurement is not

FIGURE 2.36. Continuous dynode magnetic electron multiplier (Bendix Instrument Corporation).

commonly provided, and frequently the multiplier gain is not known. In some cases, it is possible to make circuit changes and wire the electrometer directly to the conversion dynode. This gives a satisfactory measurement, but the process of wiring and rewiring is an unnecessary loss of effort.

An alternative method for measuring multiplier gain involves comparison of the output from the total ion beam monitor with that from the multiplier. This comparison is feasible provided one knows (1) the percent of the total ion current due to the selected peak, (2) the percent of the total beam intercepted by the beam monitor, and (3) the percent transmission of the beam to the electron multiplier. Factors (1) and (2) are easily obtained, but (3) is difficult to measure unless the multiplier gain is already known. Hence, this method is not always practical.

Another method for determining multiplier gain utilizes the fact that single ion peaks or spikes can be recognized and hence used as a measure of the multiplier output. The mass spectrometer is focused just off the base of a peak so that very few ions are collected. The multiplier is set at high gain and the single ion spikes that are put out by the amplifier are observed on an oscilloscope or recorded at a fast chart speed. The product of the height in amps (output voltage divided by amplifier input resistance) and the width in seconds gives the output charge in coulombs due to one ion. Since the charge on one ion is 1.6×10^{-19} coulomb, the gain of the multiplier can be calculated. A typical single spike might be 10^{-10} amp in height and 10^{-3} sec in width, so that the charge equals 10^{-13} coulomb. The multiplier gain is thus $10^{-13}/1.6 \times 10^{-19}$ or 6.3×10^{5}.

This particular method of measuring multiplier gain is limited to a system in which the gain is already quite high. If the gain is being measured to check the reason for low overall sensitivity, single ion spikes might not be observed and the method could not be used.

These various methods for determining multiplier gain have inherent errors, but there are few situations in GCMS analysis in which an accurate value of the gain is necessary. It is important to know the gain on a relative basis and this should be measured periodically, or whenever the system sensitivity is in question.

3. The Oscillographic Recorder

A recorder for GCMS analysis must have a frequency response compatible with the resolution and scan rate used in the mass spectrometry. This condition can be satisfied by a modern oscillographic recorder illustrated schematically in Figure 2.37. Several small galvanometer assemblies, each with its own input terminal, are inserted into a magnetic block. A current passing through the small coil causes deflection of a miniaturized mirror which reflects an intense light beam from a mercury arc lamp onto ultraviolet-sensitive paper. The physical dimensions of the coil and mirror are as small as possible, and a typical galvanometer ensemble, including the enclosing case, is only about 6 cm long and 0.3 cm in diameter. Consequently, the frequency response is quite fast and galvanometers are available with up to 5000 cps

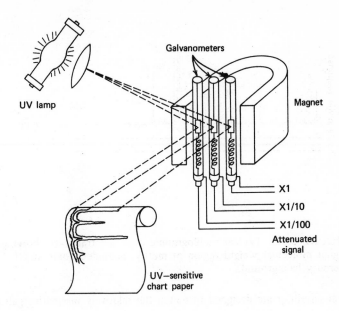

FIGURE 2.37. Schematic diagram illustrating the operation of an oscillographic recorder.

natural frequency. A common choice for GCMS is a 1670 cps galvanometer with a sensitivity of 20 mv/mm and a coil resistance of 70 ohms. Higher-frequency galvanometers are necessary at higher resolution.

In mass spectrometry, three to five galvanometers are commonly used. Each galvanometer receives the same mass spectral signal but in different attenuations and the mass spectrum thus can be scanned in a matter of seconds over a wide dynamic range. In GCMS, three galvanometer systems are quite popular with attenuations of X1, X1/10, and X1/100. This provides a range of close to 1000 between the minimum detectable signal on the X1 galvanometer and the maximum off-scale signal on the X1/100 galvanometer. A typical mass spectral trace is shown in Figure 2.38.

As noted above, galvanometer coil resistance is quite low, generally in the range of 30–100 ohms. Such a low resistance cannot be coupled directly to a mass spectrometer amplifier with an output resistance of 1000–10,000 ohms, and some form of impedance match is necessary. A power amplifier is used with a high input resistance to match the mass spectrometer amplifier output resistance R_o. The output power and output resistance of

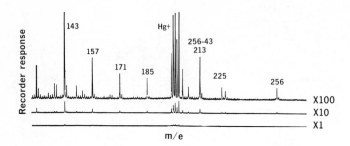

FIGURE 2.38. Typical oscillographic mass spectral trace showing higher molecular weight region of methyl pentadecanoate and mercury background.

the power amplifier are designed to match the relatively insensitive galvanometers. In the most common configuration, a separate power amplifier is used for each galvanometer, and the desired signal level to each galvanometer is provided by appropriate attenuation of the output of the power amplifier.

Although oscillographic galvanometers are relatively insensitive, the high current gain of an electron multiplier more than makes up for this loss. Consider for example, an ion beam of 10^{-15} amps that is scanned in 10^{-3} sec. As was previously shown, the electron multiplier receives about 6 ions in this time period (a statistical minimum). The amplification and recording of these 6 ions can be considered in three stages.

First stage: multiplication.

ion beam, 10^{-15} amp $\rightarrow$ multiplier, 10^6 gain $\rightarrow$ output, 10^{-9} amp
of electrons

Second stage: amplification.

multiplier output, 10^{-9} amp $\rightarrow$ amplifier, $R_i = 10^8$ ohms $\rightarrow$
output voltage = 0.1 volt

Third stage: recording.

amplifier output, 0.1 volt $\rightarrow$ power amplifier, X1 $\rightarrow$
recorder, 20 mv/mm $\rightarrow$ signal peak, 5 mm

It is thus seen that a signal in which only 6 ions are collected can give a recorded peak of 5 mm deflection with an expected signal-to-noise ratio of about 5:1.

One disadvantage of the oscillographic recorder is that the mass spectral chart must be protected from excess light. Exposure to sunlight or continuous use in fluorescent light causes irreversible fading. In some laboratories, yellow fluorescent lights or a mask of yellow cellophane are used to prolong chart life. With care, the mass spectra can be studied as long as is necessary, but they must be covered when not in use. Mass spectral records that have been properly handled can be stored for 10 years or more without significant loss of image. A wet chemical fixing process is available, but because of the large number of spectra encountered in GCMS work, the procedure is considered unwieldy and is unnecessary if the charts are carefully handled.

4. The Magnetic Tape Recorder

There are many GCMS applications in which the use of an oscillographic recorder is inconvenient or impractical. If excessive amounts of data are anticipated or the scan speed exceeds 1 sec/decade, a faster recording mode is needed. For subsequent off-line computer data reduction and interpretation, the primary data record must be taken in a form that permits further processing. One convenient system for this is the analog magnetic tape recorder.

Recording of mass spectral data on magnetic tape was first proposed for GCMS data obtained from high-resolution mass spectrometers.[96-99] Because of the very narrow peak width obtained with high-resolution mass spectra, use of an oscillographic recorder is impractical, particularly with fast scan. A frequency response of 5000–10,000 cps is necessary for scan rates in the range 10–20 sec/decade. A typical oscillographic scan of 300–400 mass units would have to be 25–40 ft long to give a clear usable trace. In contrast, the same data can be conveniently recorded on a few inches of magnetic tape, and hundreds of spectra are contained on a single roll.

The principal application of tape recording for low-resolution GCMS is with spectra that are to be processed by a computer. Many GCMS laboratories generate several hundred spectra per week, and the cost of mass marking and general data reduction is considerable. Ideally, this large volume of spectra should be input directly to an on-line computer, but for many laboratories, the cost of such a sophisticated system is excessive. On the other hand, a small analog tape unit can acquire the data at any reasonable scan rate and resolution, and the data can be processed either by an in-house computer facility or by a commercial data processing service. (Applications of computers are considered in Chapter 7.)

Several analog tape recorders are available for GCMS work in the price range $10,000–$15,000. The sensitivity and frequency response exceed those of the oscillographic recorder, and several recording channels can be used to attain the desired dynamic range of more than 1000. Prior to computer processing, the output can also be played into a conventional strip chart oscillographic recorder for general monitoring purposes.

5. Photoplate Recording of Mass Spectra

Certain mass spectrometer geometries are amenable to the use of a photoplate for recording mass spectra (page 42). The method has advantages for some GCMS applications, but it is certainly the most unpopular recording technique and is avoided whenever possible. The many disadvantages include: (1) the record is not immediately available, (2) photo development of the plate requires careful and inconvenient wet processing, (3) when developed the record cannot be examined without a magnifying projection unit, (4) very accurate mass measurement (2–3 ppm) for the determination of elemental composition requires a comparator which costs $25,000 or more, although less accurate mass measurement (20–30 ppm) can be made on a comparator that costs about $3000, and (5) measurement of relative intensity is quite inaccurate and the dynamic range on one spectrum is limited to about 100.

The advantages of photoplate ion detection in GCMS are derived essentially from one feature. A scan of the mass spectrum is not required.[100] Basically, the photoplate is very much less sensitive than an electron multiplier which detects as little as 5–6 ions. The photoplate requires 1000–10,000 ions (10^{-16} to 10^{-15} coulomb) for a detectable trace. This would appear to be an overwhelming disadvantage, but the limitation is offset by the fact that the photoplate is an integrating device that works on all ions during the complete period the sample is in the ion chamber. It has been shown that an ion peak of 3×10^{-15} amp would yield an average of 9 ions if scanned in 1/1000 sec—which is essentially a statistical detection limit. (See also Figure 7.3.) On the other hand, in a no-scan mode with photoplate detection, this same ion peak of 3×10^{-15} amp would yield 9,000 ions/sec and could be integrated over 10–30 sec during the evolution of the GC peak. A sensitivity gain of one or two orders of magnitude is thus possible.

Figure 2.39 shows a photoplate obtained from a Mattauch-Herzog mass spectrometer.[101] Thirty or more individual spectra can be repetitively obtained on a single plate by manually shifting the plate to a new position in the plane of the ion beam. Proper exposure for each line is determined from the beam monitor. The spectral detail available on such a plate is

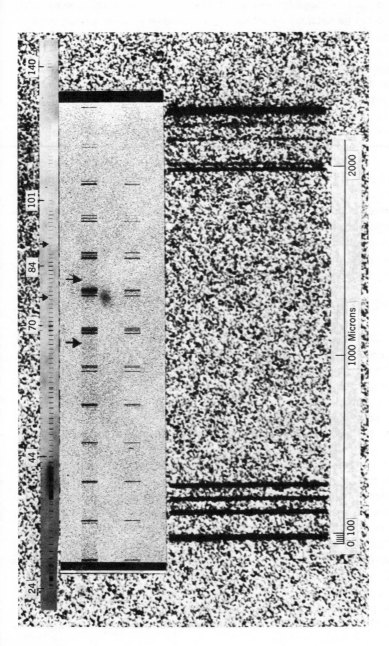

FIGURE 2.39. Photoplate record of mass spectra obtained on a Mattauch-Herzog mass spectrometer.[101]

shown in the two magnified sections of Figure 2.39. An accurate mass determination can be obtained by measuring the distance between an unknown
peak and known reference lines. In addition, the comparator output can be
plotted on a strip chart recorder to give a conventional mass spectrum.

The various disadvantages of photoplate recording are a sufficient
deterent that the technique is almost never used for low-resolution GCMS.
For high-resolution GCMS studies the choice between photoplate or magnetic
tapes is generally made on the basis of the available hardware. In laboratories
that have a Mattauch-Herzog mass spectrometer equipped with both a photoplate system and a magnetic tape system (or an on-line computer), tape
recording will be the chosen method except for cases where higher sensitivity
is essential.[102-104] Because of the severe demands imposed by high-resolution
GCMS techniques, a complete low-resolution GCMS study generally precedes
any consideration for the high-resolution measurement. If a study of a chromatographic peak at resolution 30–40,000 is indicated, photoplate detection is
the only practical procedure.

An excellent paper by Desiderio reviews some of the critical aspects of
mass spectral photoplate recording.[105]

E. OVERALL MASS SPECTROMETER SENSITIVITY

Sensitivity is possibly the most important parameter used in the evaluation of
mass spectral performance. In spite of this importance, it is the most ill-
defined and misunderstood of all mass spectral characteristics. The reasons
for such confusion are not immediately apparent but may be a result of the
fact that many different types of analyses are performed by mass spectrometry, and as a consequence, different definitions of sensitivity have come
into use: parts per million, amps per Torr, grams/second, grams total in
effluent, etc. Each of the many definitions has a specific use, but the overall
effect has been a deplorable lack of true understanding.

In essence, there are two aspects of mass spectrometry that must be
considered in defining sensitivity. One is the basic instrument sensitivity
that gives the amount of current arriving at the collector for a specified
sample consumption rate. The second is the method of handling the sample
which thus determines the efficiency of sample utilization. Fundamentally,
the first of these is a physics problem dealing with the ion optical system from
the point of ion production to the point at which the ion beam is collected,
amplified, and recorded. The second aspect is a chemical problem dealing
with the mass transfer of small quantities of chemicals.

In the previous sections, frequent implication was made regarding the effect of a particular part of the mass spectrometer on the sensitivity. It should now be quite apparent that the overall sensitivity depends on the efficiency of each component and their interplay with each other. Because of the very high gain made possible by the electron multiplier, it is often quite practical to trade off a loss in one function with available gain in another. The net operating result is that modern mass spectrometers have sufficient sensitivity to meet most analytical needs and in many instances, the ultimate detection limit is determined more by the background level than by instrument parameters. The limitation is particularly critical in high-temperature GCMS studies and will be discussed further in Chapter 3 in the section on column bleed.

1. Basic Mass Spectrometer Sensitivity

This parameter should unequivocally describe the output ion current received at the collector for a specified rate of sample consumption. The definition should state: (1) the chemical substance, (2) the mass peak that is measured, (3) the mass spectrometer resolution, (4) the quantity of sample used per second, (5) the electron beam intensity, and (6) the ion current arriving at the conversion dynode of the electron multiplier. Practical units would be amps of ion beam/nanogram/sec/amps of electron current at resolution R.

It is very desireable to be able to compare the sensitivity of different mass spectrometers. Yet, in a comparison of sensitivity specifications given for fifteen different GCMS mass spectrometers manufactured by eight different instrument companies, not a single pair of sensitivity claims used the same units.[106] Some manufacturers specified a result with methyl stearate; some with cholesterol; many did not specify a compound. Some referred to signal-to-noise of the parent ion; some a recognizable pattern; some that the compound could be identified. Some specified grams/second; some grams total; some grams in inlet. Consequently, it was not possible to compare the sensitivity of the various instruments.

The compound chosen for sensitivity determination is too often a material of interest that may thermally decompose prior to electron bombardment. Cholesterol is a common example. However, even though thermal decomposition is a critical consideration, this aspect of sensitivity should not interfere with the measurement of the basic mass spectrometer sensitivity, but rather should be considered separately as a part of the sample utilization.

The compound used to establish the basic sensitivity should meet certain criteria: (1) unlike the compound used to establish the inlet efficiency, it should be highly stable, (2) it should be easily available, (3) preferably, it should be a liquid to facilitate handling and have sufficient volatility to permit introduction via a 150°C inlet, (4) a fairly high molecular weight in the range 200–400 is preferred, and (5) the mass spectrum should show a significant molecular ion (5–30% of base peak).

It is not the function of this volume to establish any industry-wide standards, but it is important to emphasize the need for uniformity. Conditions (3) and (4) can be mutually restrictive, and if all the other desired properties are met, the choice of compounds is relatively narrow. Methyl laurate (MW 214) or methy myristate (MW 242) are two possibilities or, if the need for higher molecular weight is more important, benzyl myristate (MW 318) could be considered. A large number of halogens, particularly fluorocarbons, might be used, but condition (5) would not always be met. The simple saturated hydrocarbon hexadecane (MW 226) meets most of the features and may be a suitable compromise.

Measurement of the base peak (e.g., mass 74 in a methyl ester spectrum) gives the most uniform sensitivity comparison between different instruments. Measurement of the parent ion peak often indicates different spectral characteristics due to different instruments (page 54). Ideally, the base peak should be used to specify the basic instrument sensitivity and the ratio, base peak/parent peak, should be given to indicate instrument variations.

Several methods can be used to determine the rate of sample consumption, but the one described on page 11 is the most convenient. The use of grams/second rather than moles/second may be open to criticism from a fundamental point of view, but since one term can be easily converted to the other, the choice is not critical. Grams/second is more descriptive of the amount of sample that must be handled.

The term amps/Torr is particularly useful for residual gas analysis. It should not be used in organic mass spectrometry because it implies a static condition and fails to describe the rate of sample consumption. Furthermore, the pressure in the ion chamber would have to be calculated from the conductance of the chamber and the sample consumption rate using equation 2.2 (Q = PC). Thus, to an organic chemist, amps/Torr is a secondary standard, not a primary one.

The intensity of the electron beam should be specified as a parameter in defining sensitivity. However, in modern qualitative organic mass spectrometry, sensitivity is optimized at the expense of quantitative precision. Consequently, high-intensity electron current is commonly used (in the range 100–500 μamps) with a resulting uncertainty in the efficiency of ion extraction. This practice can make an unrealistic comparison of electron ionizing current in different instruments, and it may be sufficient to compromise by stating the sensitivity for "optimum operating conditions of electron current."

When basic sensitivity is defined as described above, the organic chemist has a definition that can be used to compare various instruments and one that can also be related to the total amount of sample. Uncertain characteristics such as percent ion extraction from the ion chambers or efficiency of transmission of the ion beam through the analyzer are inherent in the definition, and a more explicit specification involving these factors is unnecessary.

2. Efficiency of Sample Utilization

The amount of sample consumed per unit time is a critical factor in evaluating mass spectrometer performance. Ideally, sample should enter the ion chamber in a square-wave fashion for a period of time that is only slightly longer than the scan time. Sample utilization, i.e., the ratio of sample consumed during the scan to the total sample used in the sample introduction, would thus be maximized. In practice, this is difficult, but as will be shown, GCMS introduction is remarkably close to this idealized behavior.

Sample introduction profiles for the various methods of introducing sample to a mass spectrometer are shown in Figure 2.40. Examples (a) and (b) represent typical curves for a batch inlet. In (a), if the reservoir volume is 1000 cc and the leak rate around 0.2–0.3 cc/sec, the half-life would be several thousand seconds (calculated from equation 2.3). The drop-off of intensity in the ion signal is barely noticeable in the 6-min time period shown in the diagram. For a scan of 5–10 seconds (as indicated by the dotted lines), an extremely small amount of total sample is used in data collection, namely about 0.01%.

If the reservoir volume is reduced to 100 cc, or the leak rate is increased to 2–3 cc/sec, the amount of sample required in the inlet to give the same initial pressure in the ion chamber is reduced by a factor of 10. The half-life is also reduced to a few hundred seconds as shown in (b) trace 1, and the sample should be scanned within the first minute or two to minimize the decrease in signal level. Utilization of sample for data collection would be about 1%. Sample consumption at this rate represents a reasonable compromise when a batch inlet system must be used with small sample amounts of material.

In (b) trace 2, a sample profile is shown in which the sample consumption is 100 times faster than in (a). This could be accomplished using a variable leak, a smaller inlet volume, or a combination of both. The amount of sample in the inlet for the same initial signal is 100 times less than in (a), but since the half-life is about 30 sec, the mass spectrum must be run very quickly in order not to lose the sample altogether. This rate of sample effusion is inconvenient and would be used only when necessary, but

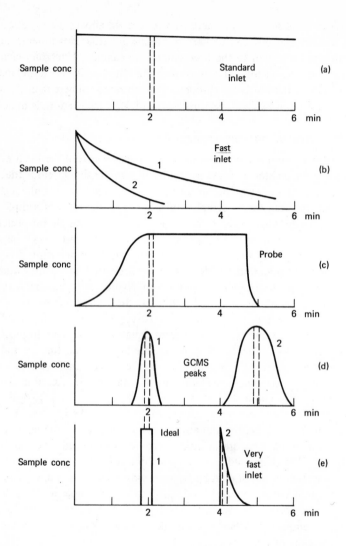

FIGURE 2.40. Sample introduction profiles for different methods of introducing sample into the mass spectrometer.

sample utilization could be as high as 5–10%.

The sample profile shown in (c) is obtained when a direct introduction probe is used. The sample is heated to increase the vapor pressure to a reasonable rate of evaporation and the spectrum is then run. A constant sample evaporation rate is maintained until all the sample is used. For the case shown in (c), the quantity was sufficient for about 300 sec, so that actual sample utilization during the 5–10 sec scan would only be 1–2%. If the sample in (c) had been only 1/5 as large, utilization would be 5–10%, but the signal would drop to zero after about 90 sec. For this reason, scans of probe samples must be taken at frequent intervals to assure that some record is obtained even if the sample is suddenly depleted.

In many ways the combined GCMS system offers the closest approach to ideal sample introduction. In (d), two possible sample profiles are shown for chromatographic peaks. The first profile represents a relatively narrow peak with half-width about 15–20 sec. Such a peak might occur early in a run or from a capillary chromatographic column. For this peak, a 5-sec scan taken close to the top of the peak would give sample utilization of 25–35%. The second peak profile represents a broader peak with half-width of 50–60 sec. For a 5-sec scan taken near the top of the peak, sample utilization would be 5–10%. Of course, if either trace 1 or trace 2 is scanned early or late on the peak, the utilization factor is correspondingly reduced.

The idealized situation is illustrated in (e) trace 1 which shows a "square wave" sample profile that is only 1 or 2 sec wider than the scan time. Sample utilization of 80–90% is achieved, but there is no realistic method of obtaining this sample profile.

One possible approach, aside from a sharp chromatographic peak, is to use a high-conductance batch inlet with a half-life of 10 sec (Figure 2.40e, trace 2). Sample utilization could be close to 30–40% for a 5-sec scan performed within 1 sec of sample introduction, but obvious experimental problems would make it difficult to obtain a satisfactory scan using this technique. In practice, (b) trace 2 or (e) trace 2 are encountered in some forms of stop-start GCMS or in-line trapping methods.[107, 108] . In these techniques, the sample is trapped in a small volume of about 100 cc, the carrier gas is pumped away, and a valve is opened to give fast sample effusion.

It must be emphasized that the previous discussion referred to the amount of GCMS peak that entered the mass spectrometer. Actually, the percent of total gas chromatographic effluent that can be put into the mass spectrometer may vary from 0.1–100% depending on the carrier gas flow rate, the interface enrichment, and the mass spectrometer pumping system. The true sample utilization is thus reduced by the sample introduction split or yield factor.

It is not intended to gloss over this consideration but rather, because of the tremendous importance of sample transfer in the GCMS interface, the subject is treated in depth in Chapter 5. The previous discussion is intended to show why different inlet methods need different amounts of sample.

3. The Meaning of Sensitivity in Mass Spectrometry

To assess the significance of a statement of sensitivity, one must recognize the difference between detection and identification. In gas chromatography, the output signal is usually only a detection, and identification on the basis of retention time uses considerable background information of the sample. In mass spectrometry, it is seldom sufficient to obtain a minimum signal at the base peak, and a signal 30–100 times above the detection limit may be needed to have enough information for spectral interpretation. This does not eliminate a fortuitous identification on the basis of one or two prominent peaks, but such luck is limited to a small number of compounds.

Consider the mass spectral patterns shown in Figure 2.41. Each of these compounds has close to 50% of the ion current in the parent peak which fortunately occurs at a mass number not commonly observed at high intensity in the mass spectral background. These unique ions quickly identify the compound, and the identification limit is the same as the detection limit. For example, an estimated 10^{-11} gm total of CH_3Cl was separated in a peak 8–10 sec wide from an orange juice extract.[109]   The mass 50 ion peak was only 5–6 mm high, but ion current at this mass is so uncommon that the compound was readily identified.

This type of example clearly implies that a minimum sample requirement for mass spectral identification of an unknown cannot be specified. A sensible maximum of 10^{-9} gm/sec can be suggested for most compounds encountered in GCMS, but the minimum quantity depends upon the amount of information needed. A methyl ester may be class identified at a low sample level from a very small signal at mass 74, but the molecular weight and possible branching in the chain would be unknown. A $C_{10}H_{16}$ terpene can reasonably be suggested from small peaks observed at masses 93, 121, and 136, but a choice between the many possible isomers requires more intense and more precise spectral data. Indeed, as is lamentably well known, it is never possible to say a priori that a given sample size will permit identification. Abundant sample guarantees only that there will be a satisfactory mass spectral pattern.

To appreciate the extremely high sensitivity available from a mass spectrometer, one should look at the capabilities of operating in a no-scan mode with an electron multiplier.[110]   Suppose it was necessary to establish the existence of methyl esters to a very low sample level. For a chromatographic peak which contained 10^{-10} gm total of methyl undecanoate (MW 200), the

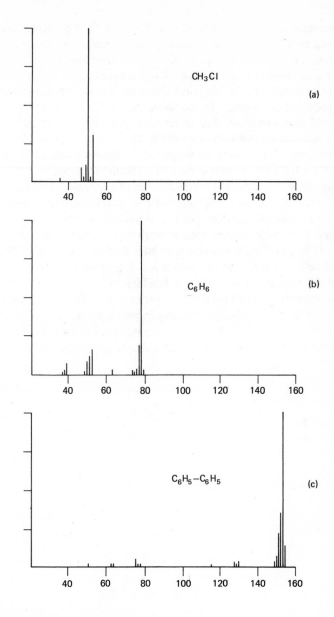

FIGURE 2.41. Examples of simple mass spectral patterns that can be identified at the detection limit. (a) Methyl chloride. (b) Benzene. (c) Biphenyl.

most abundant ion at mass 74 would be about 10^{-15} amp average current corresponding to 6×10^3 ions/sec. If the complete mass spectrum is obtained
with a fast scan, only an average of about 6 of these ions are collected. If
the mass spectrometer is set at mass 74 in a no-scan mode, the ions are continuously collected and the statistical limitation due to ion counting is reduced
several orders of magnitude. In the above example, 10^{-13} gm total of the
ester would give 6 ions/sec. For an electron multiplier gain of 10^6 this would
mean an electron current output of about 10^{-12} amp. With an electrometer
input resistance of 10^9 ohms, an output of 1 millivolt would be obtained,
enough for a full-scale deflection on a potentiometric recorder.

Specific ion detection is extremely valuable in many biological systems
where a trace of a known compound must be established. In a more sophisticated form (multiple ion detection), 2–8 ions are selected in sequence and
the output from each selected mass is obtained as a chromatographic trace.
Details and applications are discussed in later sections (pages 252, 396), but
it should be noted now that for most research problems, the minimum sample level is determined more often by handling techniques than by mass spectrometer sensitivity. Use of a no-scan operational mode to attain ultimate
sensitivity is limited to chemical systems in which the identity of the subject
compound is known. For all intents and purposes, the mass spectrometer
is being used as a specific detector, and not as a qualitative analytical instrument.

3

Fundamentals of Gas Chromatography Pertinent to GCMS Operation

A. INTRODUCTION

In the past two decades, the gas chromatograph has become a familiar instrument in most chemical laboratories. There are many fine reference books on the subject[111-119] and every chemist has some occasion to use gas chromatography. The techniques are applicable to a wide variety of chemicals ranging from light gases to derivatized nonvolatile compounds, and the degree of separation, the speed and simplicity of operation, and the high sensitivity are superior to many alternative processes. As a consequence, gas chromatography has been applied in virtually every area of chemistry and many related fields of science.

Gas chromatography is an excellent tool for the separation, detection, and quantitation of the components of a complex mixture. It is not a good tool for qualitative identification. The retention time can always be used to deny the possible existence of a compound in an unknown mixture, but the accuracy of retention measurements is not sufficient to eliminate the many thousands of compounds that might elute within the observed time period. Use of two or more different columns to obtain different retention indices is practical only for simple mixtures or pure compounds. For good qualitative analysis, gas chromatography should be combined with some other analytical system, and often, confirmation by two or more methods is desireable. Mass spectrometry is frequently a preferred choice because of its high sensitivity and relatively specific spectral information.

The basic components of a gas chromatograph are shown in Figure 3.1 and for the most part, these components function the same way in GCMS as

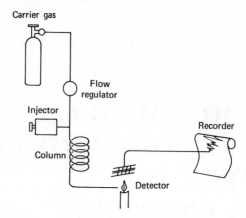

FIGURE 3.1. Schematic diagram of the components of a gas chromatographic system.

in conventional gas chromatography. However, for good GCMS operation, it is essential to optimize the gas chromatographic method, the mass spectrometric techniques, and the interface conditions. To fail in any one of these operations will result in an unsatisfactory experiment. Although the interface problems are generally the least understood, it is not uncommon to observe GCMS failure because no effort was made to adapt the chromatographic method to the more rigorous requirements of GCMS.

Optimization of the combined system often necessitates modification of operational parameters such as carrier gas flow and temperatures. Careful consideration should be given to the type of column and stationary phase material to assure the best chances of a successful GCMS run. The allowable temperature, the best flow rate, and the sample size limitations should all be determined in advance. The necessity or possible deleterious effects of a separator interface must be carefully weighed. Various chromatographic detectors can be chosen which provide important auxiliary data for certain types of mixtures. These and other aspects of the GCMS combination will be discussed in this chapter.

B. TYPES OF COLUMNS

1. Packed Columns

Packed columns are the most common type used in conventional gas chromatographic analysis and offer the maximum advantages in preparation, operational simplicity, versatility, and choice of substrate.

For most analytical purposes, the common 0.32 cm OD column (nominally called 1/8 in., average ID = 0.18 cm) gives maximum utility. Sample size can vary from the subnanogram minimum detection level to a heavy overload of 10–50 mg. Separation of fairly complex mixtures is attainable, although the separation power of the packed column is not as great as that obtained from open tubular columns. For sample collection, the 0.32 cm column can be loaded to yield suitable quantities of separated components.

The flow rate used in a 0.32-cm column may be as low as 15 cm^3 atm/ min* or as high as 75–80 cm^3 atm/min. For convenient analysis, the flow rate is chosen as a compromise between the separating power and the time of analysis. The best separations are usually obtained with 20–25 cm^3 atm/min, but if the mixture is one that requires temperature programming over an extensive range, the analysis time is sometimes reduced by increasing the flow rate.

If a gas chromatographic method is being developed for subsequent GCMS applications, careful consideration must be given to the effect of various flow conditions on the GCMS system sensitivity or sample utilization. The quantity of effluent gas that can be taken into the mass spectrometer depends primarily on the pumping rate of the vacuum system and for a first estimation, a differentially pumped mass spectrometer can receive 1 cm^3 atm/ min. (1 cm^3 atm/min = 1.3 x 10^{-2} liter Torr/sec.) If the interface consists of a simple flow splitter which directs 1 cm^3 atm/min to the mass spectrometer, a total flow of 20 cm^3 atm/min means that only 5% of the effluent is utilized in the GCMS analysis; for a total flow of 80 cm^3 atm/min, the percent of effluent that goes to the mass spectrometer will only be 1.2%. Whenever a low sample level is anticipated, higher flow rates should obviously be avoided.

The various enrichment processes that are used to improve this sample utilization will be discussed in Chapter 5. It will be shown that yields of 20–40% can be expected with a 0.32-cm column, but in any event, increasing

---

*For GCMS, one must specify the quantity of gas, and hence the pressure dimension is important. Volume alone does not specify quantity of gas.

the total gas flow makes it more difficult to achieve high sample utilization.

Because of the need for high sample utilization, 0.64-cm OD packed columns (ID = 0.46 cm) are seldom used for GCMS studies. Flow rates of $60-150$ cm^3 atm/min impose severe restraint on the percent of sample that enters the mass spectrometer. In general, the chromatographic separation is not as efficient with larger diameter columns. Their principal advantage is the higher load factor, but in GCMS studies, the increased sample load is partially offset by the increased carrier gas. However, there is no reason to avoid 0.64-cm OD columns if high sample utilization is not an important factor in the analysis.

Occasionally, 0.16-cm OD packed columns (micropacked columns) are used for special applications.[120] These columns operate at low flow rates and with a reasonably high sample load. However, many users have reported trouble obtaining reproducibility, presumably because the $0.07-0.1$-cm ID does not permit smooth packing of the column material during fabrication. As a consequence, the 0.16-cm packed column is not widely used and the various open tubular columns are preferred for low-flow GCMS applications.

GCMS operation is not affected by the length of the column, so that this parameter can be decided primarily by the demands of the chromatographic separation. Lengths of $1.5-3.0$ meters are most commonly used, but any length consistent with acceptable chromatographic practice will be compatible for GCMS. For applications needing greater chromatographic resolution, column lengths of $6-9$ meters are suitable. A further extension of column length is limited by the physical size of the columns, and also by the relatively low gas permeability of the 0.32-cm column, which necessitates a considerable increase in the carrier gas pressure with longer columns. Therefore, even though resolution is related approximately to the square root of column length, this method of increasing separations gives a diminished yield when long columns are extended.

2. Open Tubular Columns

For many gas chromatographic applications, increased separation can be obtained using open tubular columns. Since GCMS studies are often performed on very complex mixtures, the maximum chromatographic resolution is needed to get good representative spectra. The open tubular columns offer several advantages when directed toward this objective.[115, 116]

Most open tubular columns, often referred to as capillary columns, are made from long stainless steel tubing with 0.025, 0.05, or 0.075-cm internal diameter. To prepare an effective column, tubing walls must be cleaned with excessive care prior to the process of coating the walls with stationary phase. Several recipes are now commonly used which assure a reasonably well-coated column.[116, 121-124]

The most significant advantage of the open tubular columns is their high permeability (i.e., the low resistance to gas flow) relative to packed columns. Practical open tubular columns can be made in lengths up to 100 meters for 0.025-cm ID tubing, and up to 300 meters for 0.075-cm ID tubing. Furthermore, the small diameter permits coiling the column to a size that is convenient for most gas chromatograph ovens (Figure 3.2). This considerable increase in column length results in much higher separating efficiency. Figure 3.3 shows typical chromatograms that compare the resolving power of a 150-meter x 0.075-cm capillary column with that of a 3-meter x 0.32-cm OD packed column.[122, 124]

Another special type of column often used in GCMS work is the support-coated open tubular column (SCOT).[125-127] Instead of being coated simply with a thin film of the stationary phase, the coating in SCOT columns also contains a porous layer of very fine support material. This material may be a colloidal silica or diatomaceous earth of about 0.01-cm diameter which has been slurried into the initial coating solution. The presence of the support material results in a greater amount of stationary phase coating per unit length which permits a larger sample load. In addition, with the support layer the stationary phase is distributed thinner and more evenly so that the chromatographic performance is generally better than that of a conventional capillary column of the same length.

For GCMS studies, the choice between liquid-coated and support-coated columns is made on the basis of the chromatography. In some applications, SCOT columns are less effective at high temperatures, and higher molecular weight compounds are often not eluted. The SCOT column is less stable and must be prepared or recoated by the commercial supplier, whereas many laboratories can prepare satisfactory capillary columns in their own facilities. The principal advantage of the SCOT columns, particularly in GCMS, is the larger permissible sample load. (See Table 3.1.)

The degree of separation per unit length is different for the various diameter capillary columns and in general, one expects to attain close to the same separation for 0.025-cm, 0.05-cm, and 0.075-cm ID columns in lengths of 30 meters, 75 meters, and 150 meters, respectively.[122] Equivalent performance from a 0.05-cm SCOT column has been obtained with a 15-meter length. For additional resolution, two columns can be joined together with no serious loss of efficiency due to eddy currents at the join. The convenience of these small bore columns is emphasized by the fact that 300 meters of 0.075-cm capillary tubing can be coiled on a mandril small enough for most gas chromatographic ovens.

FIGURE 3.2. Photograph of a capillary dual column assembly connected to apparatus for cleaning or coating columns. The columns shown are 300-meter, 0.075-cm ID.[122]

The rate of flow of carrier gas through capillary columns must be very carefully considered in setting up a GCMS method. From a strict chromatographic point of view, the average linear velocity should be close to the minimum in the plot of HETP (height equivalent theoretical plate) vs. flow (the

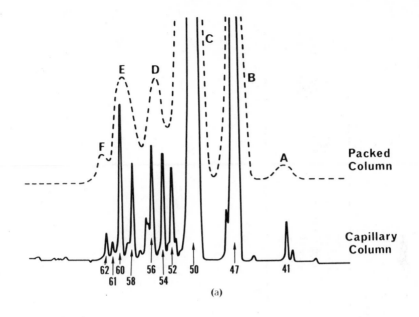

(a)

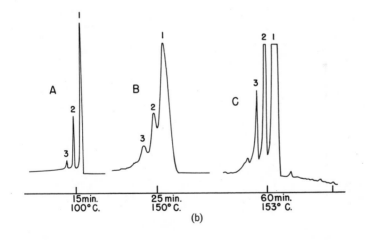

(b)

FIGURE 3.3. Comparison of typical chromatograms obtained from a packed column with those obtained from an open tubular column. (a) Separation of hop sesquiterpene hydrocarbons on 3-meter, 0.625-cm OD packed column and on 45-meter, 0.025-cm ID capillary column.[124] (b) Separation of isopulegol isomers on: A, 22-meter, 0.025-cm ID capillary column; B, 9-meter, 1.25-cm OD packed column; C, 300-meter, 0.075-cm ID capillary column.[122]

TABLE 3.1. Operating characteristics of different-size columns in GCMS analysis.

Column Size, cm	Helium Flow, cm³ atm/min	Max. Allowed Sample Load of Main Component, gm	% Helium to MS[1]	Amount Main Component to MS[2], gm	Dynamic Range[3]	Amount of Min. Sample Needed, gm
0.025	1	5×10^{-6}	100	5×10^{-6}	5×10^{3}	10^{-9}
0.05	5	5×10^{-5}	20	1×10^{-5}	10^{4}	5×10^{-9}
0.05 SCOT	5	2×10^{-4}	20	4×10^{-5}	4×10^{4}	5×10^{-9}
0.075	10	2×10^{-4}	10	2×10^{-5}	2×10^{4}	10^{-8}
0.32 (OD) Packed	30	10^{-2}	3	3×10^{-4}	3×10^{5}	3×10^{-7}

[1] Assumes differential pumped MS; no enrichment device (Chapter 5).
[2] These quantities are all in excess.
[3] Assumes 10^{-9} gm total is minimum quantity that gives a usable mass spectrum.

so-called van Deempter plot).[112] The optimum average linear velocity will
usually be around 20 cm/sec. At this rate, an air peak retention time or
"front" time would be 12.5 min for a 150-meter column.

Depending on the complexity of the sample, analysis times of more
than one hour may be encountered. If only a few runs are anticipated, this
time is of no consequence, but for many repeat runs of similar mixtures, or
of authentic compounds for retention time measurement, the feasibility of
decreasing the column length or increasing the flow rate should be investigated.

Volume of carrier gas flow is also an important parameter in GCMS
analysis. With a 0.05-cm 150-meter capillary column, 10–20 cm/sec velocity
corresponds to a volume flow of 5–10 cm^3 atm/min (assuming $P_i/P_o = 2$
and a negligible thickness of stationary phase coating). To minimize the
total amount of carrier gas in GCMS, a lower flow rate and a higher temperature are often used. This compromise is not always possible if high-boiling,
thermally unstable compounds are being analyzed, but for most GCMS
studies, it is practical to use a volume flow of 1, 5, and 10 cm^3 atm/min
corresponding to the three common column diameters of 0.025, 0.05, and
0.075-cm ID.

The lower volume flow used in capillary columns as compared to that
used in packed columns is often mistaken to be an important factor in
choosing these columns for GCMS analyses. Actually, any advantage is
strictly a result of the chemical system being studied, and the choice should
be decided primarily on the basis of the separations. If a packed column
will perform the chromatographic function, then a capillary column is seldom
used to advantage in GCMS.

The reason a lower flow rate is not automatically advantageous in
GCMS is that the maximum sample loading factor, i.e. the ratio of sample
to carrier gas, increases with the size of the column. For comparable performance, the optimum carrier gas volume increases with the square of the
column diameter, but the maximum allowed sample load increases in an in-
determinate fashion that is greater than the square of the diameter. Thus, on
the average, the volume flow through a 0.075-cm column will be nine times
greater than through a 0.025-cm column, but the allowable loading factor is
20–50 times greater. Consequently, the larger column may give better
GCMS results in terms of sample dynamic range, even though the percent
utilization of total sample may be as much as 7–9 times lower. The choice
of column type must be based on such factors as the total amount of sample
available, the concentration of sample in solvent, the dynamic range desired
in the analysis, the significance of trace components, and the complexity of
the sample. In other words, the type of column cannot be chosen on a

column characteristic such as flow rate without consideration for the nature of the sample and objectives of the analysis.

The term maximum sample load is not always clearly defined and in GCMS studies the meaning may be quite different than in conventional gas chromatography. In some cases, maximum sample load defines the amount of sample that can be used without serious loss of column separation. However, for trace analysis a column is often overloaded to obtain measurable amounts of the trace compounds of interest. This situation is commonly encountered in GCMS studies, particularly if the extract mixture is diluted by a large volume of solvent. The maximum sample load is not decided on the basis of column efficiency but rather by the total amount of sample and solvent that can be injected without destroying the column. Frequently, two separate GCMS determinations are needed, one in which the column efficiency is maintained with conventional sample load, and a second which uses a heavy overload to analyze the trace components.

For trace analysis studies, the maximum load factor cannot be definitively stated, and the quantity injected on the column is based primarily upon personal experience or prior art. It varies with the type of column or type of compound, and can be larger for a volatile solvent that remains mainly in the gas phase. To a great extent, load factor depends more upon the risk that the operator is willing to take with the column than upon sound scientific knowledge.

The significance of different columns, different sample loads, and different flow rates is summarized in Table 3.1. Conventional values of carrier gas flow are presented and the load per component is suggested as the maximum amount that a typical operator will allow on the column. It is assumed that a differentially pumped mass spectrometer system is used that can handle 1 cm^3 atm/min. A minimum sample of 10^{-9} gm total is assumed to give a satisfactory mass spectrum.

Even though the percent of total sample to the mass spectrometer decreases with increasing column size, increased loading in larger columns results in more sample being utilized, and one or two orders of magnitude in dynamic range are gained by using the 0.32-cm packed column. This is especially important for mixtures in which the solvent is 95–99% of the total volume. For such a situation, the dynamic range, exclusive of solvent, may be less than 1 part per 100 for the 0.025-cm capillary column but over 1 part per 1000 for the 0.32-cm packed column.

These figures are representative of an overload situation, and the dynamic range given in Table 3.1 refers only to the ratio of total sample injected into the gas chromatograph to the minimum amount needed for a reasonable mass spectrum. In practice, a separate run would be necessary to obtain mass spectral data for major components since any component in excess of 10^{-7} gm/sec would also overload the mass spectrometer. In some cases, use of partial bypass valves in the interface results in good data for the overloaded components.

It can be concluded that, as a first consideration, choice of column must be made on the basis of the desired separation. If chromatographic efficiency is of secondary importance and the sample has a large percent of solvent, the 0.32-cm packed column is the most suitable. If the total amount of available sample is only a few micrograms, then the higher sample utilization of the 0.025-cm capillary will be an advantage. For general purpose analyses, many laboratories find that the 0.05-cm SCOT column or the 0.075-cm open tubular column gives the best all-around utility for separation of complex mixtures.

C. CHOICE OF STATIONARY PHASE

1. Liquid Phase Characteristics

One of the important functions of the gas chromatographer is selection of the proper stationary phase. Certain properties such as low volatility, thermal stability, and wetting and solubility characteristics are important[111, 112, 119] but basically, the choice must be decided by the chromatographic separation. Hundreds of different liquids have been used for a wide variety of applications, but probably more than 90% of GCMS analyses are performed with about 10 or 12 different stationary liquids.

Most of these phases can be classed in the three categories polar, nonpolar, and intermediate polarity. In addition, some analyses are performed better with hydrogen-bonded liquid phases, and others use specific chemical properties of the stationary phase. Various theoretical and empirical techniques are used to determine the selectivity of a particular chemical.[128-132] These techniques are valuable for establishing a chromatographic method that will be continuously used on a well-defined chemical mixture. For separation of complex mixtures, especially of mixed and unknown functionalities, the most satisfactory technique is to try the mixture with three or four phases of different polarity to find the system that gives the most peaks. If the need for increased separating power is still indicated, it is often easier to use a more efficient column (i.e., change to a capillary or increase the length) than to seek endlessly for a better stationary phase.

Very little consideration has been given to the idea of selecting a stationary phase specifically suited to GCMS analyses. The features that can be optimized are virtually the same as those desired for good gas chromatography. The choice of stationary phase is thus left almost entirely in the realm of gas chromatography, and the needs of the combined GCMS operation are frequently ignored. It is important to recognize that some characteristics of the immobile liquid can have a considerable influence on the performance of a GCMS run. Of these, column bleed at higher temperatures is the most detrimental, but in the final analysis, even severe mass spectral interference from the stationary liquid must sometimes be tolerated to achieve the necessary chromatographic separations.

Column bleed is frequently the limiting factor that determines the GCMS sample utilization, particularly in programmed high-temperature runs. In gas chromatography, the adverse effects of column bleed can be reduced by using a dual column system that balances out the simultaneous bleed signal from a control column and the analytical column. The effects of the bleed are minimized and except for minor variations in the columns or operation at high amplifier sensitivity, a flat stable base line is obtained. Even if the dual column technique is impractical, good analysis can be obtained from a chromatogram with a steady, increasing baseline.

Unfortunately, in GCMS work the dual column technique is impractical. (It would require two complete mass spectrometers and a complex computer analysis of data.) The simple steadily increasing baseline of the chromatogram becomes a complex mass spectrometric background pattern which is considerably influenced by minor variations in the bleed intensity and by statistical fluctuations resulting from the scan process.[100, 103, 133] As a consequence, when the quantity of column bleed exceeds the minimum sample level of the mass spectrometer, the background mass spectral pattern interferes with the interpretation of the spectra, and small chromatographic peaks cannot be mass analyzed.

The problem is seldom serious for chromatographic runs that end before 200°C. A wide choice of stationary phases can be used up to that temperature without significant decomposition or vaporization. However, in analysis of less volatile mixtures such as long chain esters, drug metabolites, steroids, or high-boiling petroleum fractions, temperatures above 250°C are common, occasionally even above 300°C. At these limits, the stability of the column becomes the most important parameter in column selection, and the extent of column bleed often determines the mass spectrometric sensitivity.

Table 3.2 suggests temperature limits for a few popular stationary phases.[41, 44, 134−136] Safe temperature limits as recommended by manufacturers must have a reasonable safety factor, but some chromatographers find that operation above the safe temperature limit is permissible for a limited

TABLE 3.2. Temperature limits of common stationary phases.

Stationary Phase	Recommended Max. Temp., °C	Polarity
Apiezon L	300	nonpolar
Carbowax 20	250	polar
Dexsil 300	500	intermediate
Diethylene glycol succinate (DEGS)	190	polar
Poly M-phenyl ether (PPE−20)	450	intermediate
Silicone gum rubber (SE 30)	320	nonpolar
Silicone SF−96	300	nonpolar
OV−1	320	nonpolar
Versamide 900	250	polar
Chromosorb 100 series	250−300	several choices

number of runs. The success of operation at the excessive temperature varies with different laboratories and depends on the previous history of the column, particularly the conditioning process. Although excess temperatures may lead to a reduced column life, there are occasions when this sin is necessary to obtain the data. Fortunately, higher temperatures do not mean immediate column destruction and the effect of being above the safe limit for a few runs may not be noticed. Programming to a high temperature is preferred because any air spike introduced on injection passes through a relatively cold column. Isothermal operation at a high temperature level is more harmful.

To reduce column bleed to an acceptable GCMS level, the column should be conditioned above the anticipated operating temperature. Several acceptable recipes are available[137-139] and the choice of method is often a result of personal experience. One common process[138] holds the column at an intermediate operating temperature with an average carrier gas flow for 15–24 hours and then increases the temperature to 25–30°C above the maximum operating temperature for another 24 hours. If a GCMS run shows that the bleed level is too high, the column may be further conditioned at an additional 20°C higher. In some laboratories[137] an excessive flow rate (approximately fivefold higher) is used during this more extreme conditioning period. When possible, severe processes should be avoided as there is always a danger that the column will be destroyed.

Sometimes it appears that at the high conditioning temperature, the stationary phase is transported or redistributed by the flow of carrier gas. To avoid this problem, a static conditioning process is often used in which the column is thoroughly flushed with carrier gas at a low temperature and then the end is capped to stop the carrier gas flow.[139] The column remains in this static flow condition while the temperature is raised to the maximum level. After conditioning for 24 hours the column is cooled, and carrier gas flow is continued while the column is slowly cycled two or three times to the maximum temperature.

The effect of column bleed can be reduced without extreme temperature conditioning by use of a bleed-absorbing column between the analytical column and the GCMS interface. In this technique, a short column of very low-bleed material is joined to the end of the analytical column. The analytical stationary phase is used with the desired temperature program, but the excessive bleed from that column is stopped by the sort length of the low-bleed material. Levy et.al. have shown that this technique will greatly reduce the mass spectrometric background and permit trace analysis at the upper temperature limit of the column.[140] With Tergitol NP–35 (Nonyl phenyl polyethylene glycol ether) as the analytical phase and Carbowax 20M–TPA (polyethylene glycol-terephthalic acid ester) as the bleed absorbing column, the mass spectral patterns (Figure 3.4a and b) indicated the useful temperature for GCMS could be extended from about 150°C to over 190°C. Figure 3.4c shows the dramatic reduction achieved in the total ion monitor current using the bleed absorbing column. In recycle, the bleed absorbing column is slightly less efficient but can be discarded and replaced with no significant cost.

The bleed absorbing technique is not widely used but should be considered whenever column bleed is troublesome. The effectiveness varies with the nature of the analytical column, the bleed absorbing column, and the necessary temperature, but some improvement can always be expected. With the current trend toward high-stability columns such as Dexsil 300 and poly-

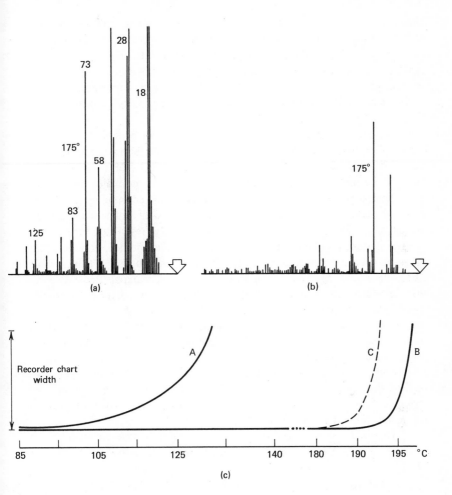

FIGURE 3.4. Effect of bleed absorbing column in reducing column bleed in GCMS. (a) Mass spectral background of Tergitol NP—35 without bleed absorbing column. (b) Mass spectral background using bleed absorbing column of Carbowax 20M—TPA. (c) Chromatographic baseline increase using: A, Tergitol NP—35; B, Tergitol NP—35 with bleed absorbing column of Carbowax 20M—TPA; and C, recycle of B.[140]

phenol ethers, a greater flexibility of column choice for GCMS may be possible by using these materials for bleed absorption. The minor variation in retention time due to the short length of the different column is insignificant but can be calculated as a linear combination of the retention times for the two phases. For some stationary phases, it may be advantageous to mix 3–5% Dexsil 300 or some other stable liquid directly into the coating solution.

Sometimes circumstances arise in which it is not possible to control the extent of column bleed and when this occurs, it may be necessary to attempt the GCMS analysis in spite of the adversity. As a first rule, it can be expected that the minimum sample limit will be increased, and major background mass spectral peaks must be subtracted. This is time consuming, and if the sample mass spectral pattern is not at least of the same order of magnitude as that of the background, intensity fluctuations will result in a very poor spectral pattern for the unknown. It should not be assumed, however, that excessive column bleed automatically eliminates any hope of getting useful GCMS data. The spectral interference of unwanted materials depends on their chemical nature, and knowledge of column bleed spectra can be used to select a more suitable column for specific types of unknowns. For example, the bleed mass spectral pattern from an Apiezon column will be primarily that of high-molecular-weight saturated hydrocarbons. If the unknown is a geochemical sample or petroleum fraction, use of Apiezon at high temperatures would result in maximum bleed interference. On the other hand, many of the principal mass peaks from the bleed of a silicone column do not seriously interfere with the important spectral peaks from traces of high-molecular-weight hydrocarbons.

The mass spectral bleed pattern from three stationary phases is shown in Figure 3.5[133] Most of the ion current from all phases occurs at masses 18, 28, and 44 (H_2O, CO, CO_2), and mass 73 is also prominent in many of the patterns ($Si(CH_3)_3$, C_4H_9O, $C_3H_5O_2$). Other characteristic peaks tend to be unique to a specific substrate, and the distribution of ion peaks throughout the higher mass ranges is often reduced and scattered. Since the more important spectral information is in the higher-molecular-weight region, useful data can often be obtained even in the presence of excessive column bleed. Figure 3.6 shows the mass spectral pattern (probe samples) of the polyphenyl ether Convalex 10 and the silicone oil SF96–50. The patterns are predominated by 8–10 peaks and there is a good chance that these will not interfere with the sample peaks. Similar spectra are obtained with other silicone oils.[4, 141] The important point is that column bleed does not blot out all mass peaks and often the run can be performed even when the overall picture appears to be a complete mess.

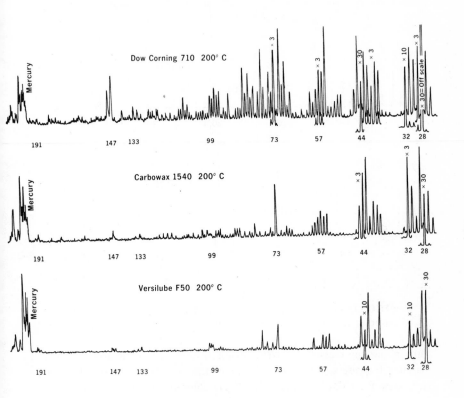

FIGURE 3.5. Mass spectral pattern of column bleed at 200°C for (a) Dow Corning 710, (b) Carbowax 1540, and (c) Versilube F50. Pattern not shown above mass 207.[133]

It is very difficult to quantitate the total ion current increase due to column bleed as a function of temperature.[140, 142] The extent of column bleed is very dependent on the conditioning process and the history of the column. It may vary considerably from column to column even though the apparent history is the same. Furthermore, comparison of the increased total ion current in a mass spectrometer with an increase in a hydrogen flame detector is misleading due to the very low flame sensitivity of CO, CO_2, and H_2O. In GCMS, the important factor is not the absolute amount of ion current due to bleed but rather the distribution of this ion current

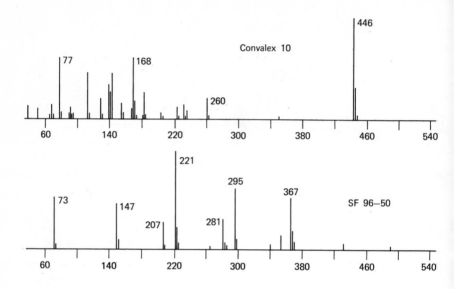

FIGURE 3.6. Mass spectral pattern of Convalex 10 and SF96−50. Samples introduced by direct probe.

throughout the mass scale. Nevertheless, a qualitative comparison of the baseline increase as a function of temperature aids the selection of a stationary phase for GCMS (Figure 3.7). If the column can be operated below the temperature at which the baseline increases rapidly, the mass spectral background due to column bleed will not be excessive.

Concern is sometimes given regarding the effect of high molecular weight bleed material to the GCMS interface and to the mass spectrometer ion chamber. If the temperature of the transfer lines, interface, and ion chamber are within ± 20°C of the upper column temperature, no adverse condensation or thermal polymerization is likely to occur. In extreme cases of column bleed, polymerization might occur in the ion source and necessitate disassembly for thorough source cleaning. Occasionally, the mass spectrometer filament is harmed due to high-temperature reactions with excess bleed and hence must be replaced. Unpleasant as these consequences may be, the damages are not irreversible and if unusual and important research projects are at stake, the risk may be necessary.

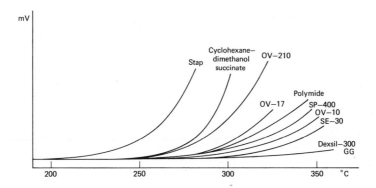

FIGURE 3.7. Qualitative comparison of baseline drift for several high-temperature stationary phases.[141]

2. Role of Separation Efficiency in GCMS

Qualitative analysis by any spectral or chemical method is most effective when performed on pure chemicals. Mass spectrometry is no exception and one of the important reasons for using GCMS is to obtain mass analysis on samples of maximum purity. High column efficiency is important in attaining this goal, but it is not always necessary to have complete separation or even a noticeable valley between two components to obtain good, easily interpreted mass spectra. The ability to interpret the data is dependent upon both the type of chemicals separated and the exact GCMS scanning technique. The GCMS literature contains many examples in which components have been identified without complete separation.[143-149] Unfortunately, the literature fails to emphasize the cases in which mass spectral identification might have been possible with better chromatographic conditions.

The possibility of easy GCMS identification without complete component separation was demonstrated early in the development of GCMS techniques during the analysis of a complex strawberry extract.[143] The chromatogram, shown in Figure 3.8, was obtained using a 60-meter x 0.025-cm capillary column coated with Tween 20 (Atlas Powder Co.). More than 150 components were indicated down to the 0.05% level. When GCMS analysis was performed, it was apparent that many of the symmetrical peaks were composed of two or more components. In some cases, not all of the compounds could be identified. For example, the three components 109, 110, and 111, were all eluted under one peak but only component 109 could be

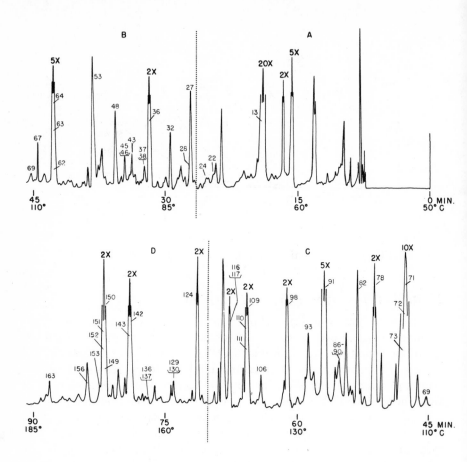

FIGURE 3.8. Chromatogram of a complex strawberry oil. Separations attained using a 60-meter x 0.025-cm capillary column coated with Tween 20.[143]

identified from the mass spectra. Similarly, components 116 and 117 came off as one sharp peak, but only component 117 was identified.

In some cases, the nature of the mass spectra allowed more favorable interpretation. One peak in the later part of the chromatogram was shown to contain the five components 149–153 (including a small shoulder peak). The four mass spectra obtained during the elution of this peak are presented in Figure 3.9. The first spectrum, taken about half way up the peak, showed ion signals at masses 143, 99, 82, and 67 characteristic of a hexenyl hexan-

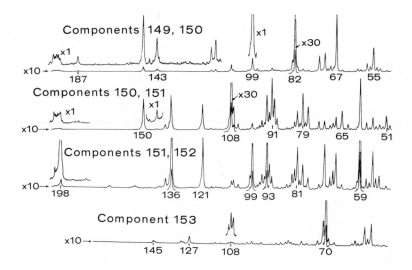

FIGURE 3.9. Mass spectra obtained on the chromatographic peak of Figure 3.7 eluted at 175°C after 82 min. Data taken with a Bendix time-of-flight mass spectrometer operating at a mass resolution of 150.[143]

oate. This spectrum also contained peaks at masses 150, 108, 91, and 79 indicative of benzyl acetate. In the second spectrum, taken at the top of the chromatographic peak, the ion signals due to the hexenyl hexanoate are essentially absent. Those due to benzyl acetate are enhanced and in addition other peaks have appeared at masses 136, 121, 93, and 59 signifying another component, α-terpineol. The third mass spectrum was scanned about halfway down the peak and gave the same type of favorable information, namely disappearance of benzyl acetate spectrum, enhancement of the α-terpineol spectrum, and the appearance of new mass peaks, in this case due to another hexenyl hexanoate isomer. Finally, the small shoulder peak was scanned and yielded a spectrum that contained essentially only the characteristics of a pentyl octanoate.

This mass spectral magic is possible only because the compounds appearing in each scan are of different chemical classes and are all common enough to be catalogued in mass spectral compilations (see page 272). Complete identification of closely eulted components is not generally possible if the materials are chemically similar. For example, most $C_{10}H_{16}$ terpenes are characterized by the intensity of the mass peaks at 136, 121, and 93 and, with only a few exceptions, the lower mass peaks are of minimal diagnostic value.[150] Thus, a series of closely eluted $C_{10}H_{16}$ terpenes could not be identified in the same way as the examples of Figure 3.9. The characteristic mass peaks would all fall upon each other, and the varying intensities would be a noninterpretable function of concentration, rate of scan, position in chromatographic peak, and detail of mass spectra. In a favorable case, the GCMS analysis might give identity to the major component. On the other hand, it might not even be possible to say how many similar compounds are present in the chromatographic peak.

These examples are typical of situations encountered in interpretation of GCMS data. It is obvious that maximum chromatographic separation should be attained whenever possible. For example, the spectra of the two hexenyl hexanoate isomers in the strawberry oil were taken about 15 sec apart. If these two compounds had been separated by only 6–8 sec, one could not establish the existence of both. However, prior to the GCMS run, it is not possible to know how much separation may be needed. Excessive time spent on this endeavor could be wasted, and it is often more profitable to perform a preliminary test on three or four columns, choose the best one, and proceed with the GCMS study.

One of the common gas chromatographic problems that can be troublesome in GCMS is the tailing of polar compounds due to adsorption on uncoated sites of the support material or column walls. This tailing results in a poorer degree of separation and will interfere with mass spectral interpretation if similar compounds are being eluted. Polar compounds, particularly alcohols and aldehydes, are the most troublesome, and if they are present in significant quantity, tailing may continue for several minutes. During this time frequent background scans must be taken so that a suitable correction can be made to the scans of small chromatographic peaks. If the small peak is chemically different from the tailing peak, the corrected spectrum is interpreted, but if it is chemically similar, identification may be impossible.

For example, large quantities of hexanol occur in many natural mixtures. Important characteristic mass peaks occur at masses 84, 69, 56, 55, 43, 42, 41, and 31. Even if these peaks were quite intense in the tailing background spectrum, they would not interfere with identification of most esters, aromatic compounds, terpenes, and many heterocyclic compounds. However, if a branched heptanol or octanol were eluted on this tail, the spectral characteristics of the new component would be completely swallowed up by the hexanol spectrum. The spectra of unsaturated aldehydes or ketones would be difficult to pick out of the considerable background at masses 69, 56, 55, 42, and 41. Thus, even though the tailing can be treated as a steady background, it may interfere with important identifications and should be eliminated whenever possible.

Many chromatographers anticipate this problem and add a small amount of surfactant to the stationary phase prior to coating the column packing or the tube walls.[151-152] A variety of additives have proven successful, although the exact mechanism of their function is not always clear. With anionic or cationic reagents, the result can be attributed to chemical reaction with solid adsorption sites or traces of reactive compounds in the stationary phase. The nonionic additives may function primarily as wetting agents. Another technique sometimes applied with open tubular columns is to include a solution of sodium hydroxide in the cycle of cleaning reagents.

Igepal CO880, Alkaterge T, and Span 20 are a few of the more common surfactant additives. In use, their concentration may vary from 1–5% in the stationary phase. The effectiveness of these additives in reducing tailing is illustrated in Figure 3.10.[153] In (a), n-hexanol elutes with a very broad tail extending for about 90 sec. With 1% Igepal the peak is sharper and the tailing is reduced to about 40 sec. With 5% Igepal, tailing is essentially absent.

If an efficient gas chromatographic separation is to be utilized in GCMS, the interface must not cause hold-up or adsorptive peak broadening, and the sample must be rapidly pumped out of the mass spectrometer. Rapid sample removal is accomplished by maintaining interface, connecting lines, and mass spectrometer inlet at or near the column temperature. In some cases, silylation is also necessary. These techniques are discussed further in Chapter 6.

The method used for the mass spectral scan also influences the effectiveness of the chromatographic separation. If the operator has exact control over the time at which the scan is taken, especially if there is a good oscilloscope display, maximum benefit is obtained from even the slightest separation of components. On the other hand, if specific features of the run necessitate a slow scan or a cyclic scan with relatively slow cycle time (6–8 sec), explicit mass spectral data such as was illustrated in Figure 3.9 may be sacrificed.

D. THE CARRIER GAS

Four rules determine the choice of carrier gas for standard GCMS. (1) It must be chemically inert. (2) It must not interfere with the mass spectral pattern. (3) It should have some property that will enable an enrichment of the sample in the carrier gas stream. (4) It should not interfere with total ion

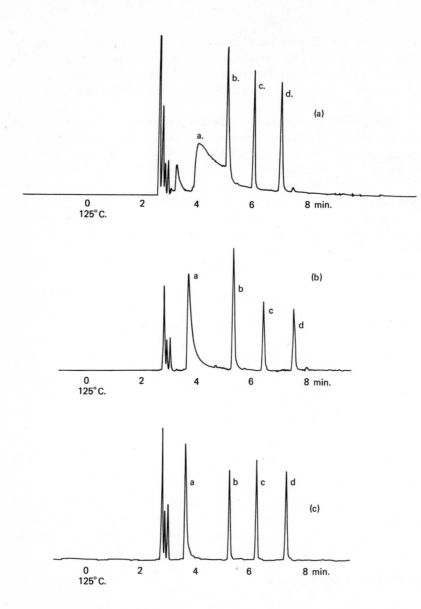

FIGURE 3.10. Comparison of chromatograms of polar test mixture on Apiezon C with varying amounts of surfactant. (a) 100% Apiezon C. (b) 99% Apiezon C + 1% Igepal. (c) 95% Apiezon C + 5% Igepal. Chromatograms obtained from a 22-meter x 0.025-cm capillary column.[154]

detection. The first condition is also desired for conventional gas chromatography. The other three are important for most GCMS runs, and carrier gas properties such as cost, availability, and viscosity are of very little importance.

The three most common carrier gases in gas chromatography are helium, hydrogen, and nitrogen. Of these gases, only helium meets all four conditions for GCMS. Hydrogen meets all conditions except (4) and in many applications is used interchangeably with helium. Nitrogen does not meet condition (3) or (4) and interferes with the mass spectral pattern in the low-mass range (at masses 28 and 29) in violation of condition (2). This latter point is particularly pertinent when a computer system is used that sets the mass scale calibration by the mass 28 air (N_2) peak. Consequently, nitrogen is almost never used for GCMS work. More exotic gases such as CO_2, argon, methane, and steam have a few specialized applications in gas chromatography but are used in GCMS only under the unique condition of chemiionization (see below).[53, 58, 60]

The choice between helium and hydrogen is not critical. In the USA where helium is readily available, it is preferred because of the additional safety factor, but in most GCMS situations equivalent results can be obtained with either gas. Hydrogen has some advantages with certain types of separators. If the enrichment factor depends upon the square root of the molecular weight as in effusive or diffusive separators, the lower molecular weight of hydrogen improves the separation efficiency by the factor $\sqrt{2}$. (See pages 171, 189.) This is not sufficient to be important for most analyses, however, and hydrogen is seldom used for this reason alone. With the palladium/hydrogen diffusion separator, use of hydrogen is obviously essential.

The ionization voltage of the carrier gas is an important factor when a chromatogram is obtained using the mass spectrometer as a total ion monitor (see page 257). Table 3.3 shows that the ionization voltage of most possible carrier gases is below 20 eV except for helium.[155] If the electron ionization energy is set at 20 eV, use of helium eliminates a large background current due to carrier gas ions. Other gases can be used, but the voltage signal due to background current must be compensated with an external voltage source which reduces sensitivity of the mass spectrometer as a chromatographic detector.

A high ionization voltage of the carrier gas is also an advantage when a field ion source is used in the GCMS study.[63] However, since the energy available from this type of source is only 12–14 eV, several possible carrier gases could be used without contributing a large background ion current.

Various GCMS carrier gases have been proposed for use with chemiionization sources.[54, 58, 60] For these studies, ionization of the carrier/reac-

TABLE 3.3 Ionization potentials of various carrier gases.[154]

Carrier Gas	Ionization Potential, eV
Hydrogen	15.4
Nitrogen	15.6
Carbon Dioxide	17.3
Water (steam)	12.6
Methane	12.8
Ethane	11.6
Isobutane	10.6
Argon	15.7
Neon	21.6
Helium	24.6

tant gas is provided by 150–500 eV electrons, and the ionization potential of the gas is of secondary importance. The principal concern is the mode of reaction with the unknown. The excessive current due to the reactant gas makes it very difficult to use the mass spectrometer in a chromatographic detection mode.

Helium has been used as a carrier/reactant in GCMS chemiionization studies, but the reaction mechanisms are not yet clearly understood. Anomalous results have been obtained indicating a need for further study.[58, 60]

Methane, ethane, and isobutane are commonly used in chemiionization studies, and because their reactions are better understood than those of other gases, they are the ones most commonly chosen for GCMS.[53, 54, 56] The easiest technique is to use the reactant gas as the carrier, and no significant changes are reported to occur in the chromatogram when the hydrocarbon is substituted for helium.[54, 60]

In another technique, helium is used as a carrier gas in the chromatograph, and the reactant gas is added at the GCMS interface.[59, 60] This method appears to give results that are comparable to those obtained with pure reactant gas and because the method does not change the chromatographic conditions, it is preferred by several research laboratories.

E. THE INJECTOR

There is no difference between the sample introduction techniques used in GCMS and those used in gas chromatography. Nevertheless, certain injector systems influence the sample utilization of the total system and should be considered. For example, it was shown on page 86 that a 0.025-cm capillary column could give the highest percent yield of sample into the mass spectrometer, and if only very small quantities of sample are available, this type of column should be used. The injector system of chromatographs used with 0.025-cm columns will usually have a splitter so that less than 1% of the total sample goes to the capillary column. This provides a convenient way of introducing microgram quantities to the column using a conventional microliter syringe. However, if microgram quantities of a mixture must be analyzed for components at the part per thousand level (nanograms per component), the sample cannot be taken up in, say, 0.5 microliters of solvent and injected with the stream splitter open. This would result in loss of 99% of the sample and failure of the experiment. On the other hand, injection of all the solvent onto the column may be forbidden due to possible column damage.

Techniques for handling microquantities of material are always difficult and even using a carrier gas, solvent liquid, or coprecipitated solid, adsorption sites or reactive sites may cause loss or destruction of the sample. Handling methods depend upon the nature of the material and must be decided before the analytical run begins. (Silylation of equipment and special carrier techniques such as adding a deuterated homologue are discussed in Chapter 6.)

If the unknown is sufficiently nonvolatile, the solvent can be evaporated on a solid inert material such as powdered silica. This material now becomes the carrier for the sample instead of the solvent, and can be packed into a small tube and inserted into a sample injector loop. A schematic method of performing this insertion is shown in Figure 3.11, in which the sample is placed into the loop via a three-way stopcock. Superficially, this method of introducing a small sample appears simple, but construction materials are a problem at higher temperatures. Stainless steel parts often gall even when used with a discreet amount of graphite or molybdenum sulfide

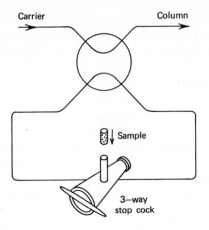

FIGURE 3.11. Schematic illustration of a method for sampling sub-microgram quantities of a mixture that has been solvent-evaporated on an inert powder.

lubricant. Teflon is a suitable plug material up to about 200°C at which point excessive outgassing may occur. The injector parts must be designed free from small pockets which could cause eddy currents and result in a slow, lingering sweep out of the vaporized sample. Since the flow to a 0.025-cm capillary will be in the range of 1 cm^3 atm/min, even a small eddy pocket could cause noticeable tailing.

When evaporation of solvent is not practical as described above, the total solution can be injected onto a 10–15-cm length of 0.32-cm packed column that precedes the capillary. In this way, the solvent can be eluted and vented through a bypass valve prior to the analytical column, and the sample then transferred at an elevated temperature by a low flow of carrier gas. This process can also be used for enrichment of various types of GCMS samples.

It is important to avoid significant eddy pockets in any part of the transfer line used in low flow rate chromatographic systems. Whenever the process of smooth laminar flow is broken by an irregular geometry, the sample swirls around in the cavities and if thoroughly mixed, the sweepout is exponential according to equation 2.3 ($P = P_0 e^{-C/V \, t}$, where P is the partial pressure of the sample). The characteristic time period (i.e., the time for the concentration to be reduced by the factor $1/e$) is thus V/C where the conductance C is equal to the volume flow rate. The initial concentration of sample in the cavity is thus reduced by 1000 in about 8 characteristic periods.

A small tube leading to a bypass valve or other functional part might be about 2–3 cm long and 0.12 cm ID. The volume of this tube is about 2.2×10^{-2} cm^3 and would be exponentially swept out in about 0.5 sec at a flow of 20 cm^3 atm/min. (V/C $\approx 6 \times 10^{-2}$ sec.) The peak broadening effect of the cavity would seldom be serious. On the other hand, if this tube is used with a flow of 1 cm^3 atm/min (1.7×10^{-2} cm^3 atm/sec) the clean-out time for an exponential sweep would be 10–15 sec, which would be completely unacceptable. Flow rates for 0.05-cm and 0.075-cm capillary columns are usually at least 5 cm^3 atm/min and 10 cm^3 atm/min so that the injector sweep-out time as calculated above would be reduced to a few seconds. This type of calculation should be performed whenever the smooth laminar flow through an open tube is disturbed with a splitter T, bypass valve, or larger diameter tube. The characteristic time given by the ratio of the eddy pocket volume and flow rate, V/C, should be close to 10^{-1} sec or less.

F. USE OF THE GAS CHROMATOGRAPH AS AN AUXILIARY DETECTOR IN GCMS

1. Retention Indices

There are degrees of uncertainty in any qualitative analysis and the chemist must establish an identification confidence level consistent with the importance of the unknown. Except for a few very simple examples, no important identification should be based on a single piece of datum, and whenever possible, the identification should be confirmed by at least one other analytical system. In GCMS studies, the automatic attainment of retention data provides one convenient check on the mass spectral results.[155]

Several useful compilations are available that list retention data for various stationary phases and at various temperatures.[156–158] Because retention times are so dependent upon the column condition and operating conditions, relative retention values are most commonly used. The retention time of the sample and the retention time of two standard compounds are expresses as a difference ratio so that the effects of minor uncontrollable variations in the operating conditions are minimized. Relative retention indices from the literature can thus be used in different laboratories. However, because precision depends so much on operating conditions such as minor temperature differences, exact flow measurement and control, sample size, and variations in column condition, many chemists prefer to obtain their own data. In any event, literature compilations are important for a preliminary check on the feasibility of a mass spectral identification. Ultimately, the authentic compound may be wanted for additional tests and for biological evaluation. If this means attempting an expensive synthesis, then a quick retention index denial of that compound or similar homologues saves a wasted effort.

One of the more certain methods of gas chromatography confirmation is the technique of "spiking" the unknown mixture with a small quantity of the suspected compound. This process assures that variations in the column characteristics due to excessive solvent or due to a major component are the same during the GCMS run as during the retention time test. Such variations are commonly noticed when a heavy sample load is employed for purposes of trace analysis. Similarly, the use of additives to reduce tailing also diminishes the reliability of catalogued retention data. Even a new column of the same stationary phase can give different relative retention values. Examples of retention time variations due to these effects are shown in Figures 3.12 and 3.13. In such cases, the "spiking" technique is often the most reliable method of obtaining good time correspondence.[159, 160]

Several disadvantages are associated with the "spiking" technique. First, if the unknown mixture is available only in small amounts, it is too valuable to use on repeated gas chromatographic runs. Second, the authentic compound must be available, whereas catalogued retention indices provide analytical data without the compound. Third, since the suspect compounds must be added to the complex mixture in rather small amounts, it is not convenient to use more than four or five compounds per retention time run. This necessitates long, laborious determinations if a large number of identifications must be confirmed. Consequently, "spiking" is most effective when used to obtain a very close retention time confirmation on a limited number of important identifications.

In a typical GCMS run, a large number of peaks are scanned and many compounds are suggested as possible components. The easiest way to obtain retention time confirmation is to compare a list of the retention indices for the unknown peaks with a list of the indices for the compounds suggested by mass spectrometry. If sufficient catalogued data is available, confirmation or denial of a hundred peaks can be made in 1−2 hours even allowing time for decision making when the index values do not match closely. Card catalogues or a computer data listing are commonly used for manual comparison. If computer time is available, automatic calculation and file search of retention indices provides a significant time saving.

Catalogued retention data can also be used as a prior guide to possible

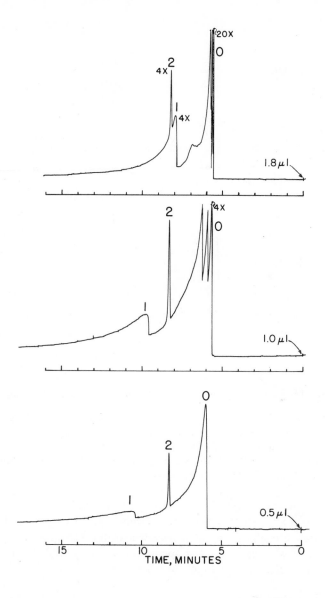

FIGURE 3.12. Effect of sample load on relative retention time. Solution 65% ethanol (0), 34% 1-propanol (1), and 1% 3-methyl-pentane (2) separated on 60-meter, 0.025-cm ID capillary column coated with SF96−50. Injector split, 200/1. No tailing reducer added.[160]

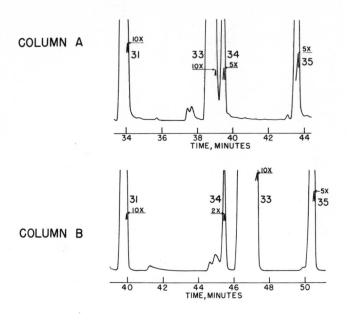

FIGURE 3.13. Example of retention time variations occurring between two columns of the same material. Column A, 150-meter, 0.05-cm ID capillary column coated with SF96−50 plus 5% Igepal CO−880 after rigorous cleaning. Column B, same as A, but coated once, washed with organic solvents only, and recoated. Peaks are (31) butyl acetate; (33) 2-hexenal, trans; (34) ethyl 2-methyl butyrate; and (35) 2-methyl butyl acetate.[160, 161]

mass spectral identification. Too frequently, mass spectral files (see page 275) do not contain a match of the GCMS spectrum, and the suggestions based on mass spectral correlations may be inconsistent with other information about the sample. In such cases, a list of compounds in the general retention region may suggest additional possibilities for mass spectral interpretation.

2. Specific Detectors

Several types of specific gas chromatography detectors provide valuable qualitative information as an auxiliary or confirmatory aid to mass spectral interpretations.[19] The most common of these are specific to certain elements such as sulfur, phosphorus, nitrogen, or the halides. In this capacity, these detectors partially serve the function performed by high-resolution mass spectrometry. The information is not as complete as that obtained by high-resolution mass spectrometry, but it can often be obtained more easily, at higher sensitivity, and at much less cost.

The flame photometric detector is one of the more popular systems used in conjunction with GCMS studies.[161] A photometric cell is placed in view of the hydrogen flame and a wavelength band is selected specific to phosphorus or sulfur. Sensitivity in the range of 10^{-10} gm/sec for sulfur and 10^{-12} gm/sec for phosphorus is easily attained. The complete system can be used in a dual fashion that gives simultaneous recording of the hydrogen flame current and the photometric current. This type of information is extremely valuable when obtained prior to the GCMS run.

Other methods of obtaining elemental analytical information are equally suitable. The coulometric detector has a versatility that permits determination of sulfur, nitrogen, or halogens, and can be used cooperatively in GCMS studies whenever the existence of these elements is uncertain.[162] Analysis can be performed on about one nanogram of material.

The microwave emission detector is specific to a large number of elements and has sensitivity in the range 10^{-7} gm/sec to 10^{-14} gm/sec depending upon the element.[163] This particular detector would be very attractive if it were commercially available. The cost of the components is in the range of $9,000 to $11,000, exclusive of installation expenses.

The thermionic detector is another highly sensitive system that can be used for qualitative analysis of phosphorus, halides, or nitrogen.[164-166] It has the advantage of being commercially available at a reasonable cost or if preferred, a conventional flame ionization detector can be easily modified to serve this purpose.[165] The sensitivity is in the range of 10^{-13} gm/sec for phosphorus, 10^{-11} gm/sec for halogens, and 10^{-10} gm/sec for nitrogen. The only operational difference between this system and the unconverted flame ionization detector is a change in the optimized air and hydrogen flow rates. Because of its simplicity, the modified system is often used in a simultaneous split mode with another detector such as the photometric detector.[167]

The electron capture detector is also very useful when used in collaboration with GCMS studies. Specificity is obtained for certain elements such as halogens and sulfur, but functional information is also given. The existence of conjugated carbonyls, nitriles, disulfides, trisulfides, and polycyclic aromatic compounds can be established if response ratios are determined.[168, 169] In some cases, such as analysis of complex mixture of sulfides, this detector is so specific that retention indices and response factors are sufficient to identify many components without a GCMS analysis.[168]   Conversely, the mass spectral pattern of alkyl sulfides and disulfides is sufficiently characteristic that auxiliary information is seldom necessary.

Establishing the existence of conjugated nitriles, carbonyls, etc., is one of the more important applications of the electron capture detector in GCMS. The electron bombardment process frequently scrambles the parent ion structure, particularly the location of double bonds, so that it is often not possible to establish whether the system is conjugated or not. Auxiliary data from another system is needed and although ultraviolet and infrared spectroscopy are often specific and of great value, the electron capture device is more sensitive and easier to use on gas chromatographic fractions.

The effectiveness of any of these specific detectors used with GCMS data depends upon careful appraisal of the chemical system being studied and it may be valuable to run two or three specific detectors during the preliminary gas chromatography method development. For example, any study on roasted proteins (nuts, meat, coffee, etc.) will have many nitrogen and sulfur compounds. The interpretation of the GCMS data is greatly facilitated if one knows a priori that a peak contains one or both of these elements. The value of this auxiliary data is described by Waldralt in Chapter 8 (page 112).

Sometimes there is not sufficient prior knowledge to anticipate which elements might be present. Fruit volatiles do not generally have significant quantities of nitrogen or sulfur, and the mass spectral pattern is often sufficient to establish the structure of the few such compounds that do occur. For such cases, it is more convenient to analyze the mass spectral data first and then perform a supplemental run for sulfur or nitrogen if necessary.

In the analysis of oxidation products such as those occurring from engine studies or autoxidation of foods, many saturated and unsaturated carbonyl compounds are encountered. The mixtures are always quite complex and an electron capture detector is of limited value by itself. The GCMS data of the saturated compounds found in oxidation studies are quite easily interpreted, but many of the unsaturated carbonyls cannot be identified beyond establishing the molecular weight. The use of an electron capture device can establish the existence or nonexistence of conjugation, and often this additonal information may permit mass spectral identification of the unsaturated carbonyl compounds.

4

Application of Elementary Vacuum Technology in GCMS Systems

A. MASS SPECTROMETRIC VACUUM REQUIREMENTS

It is well known that a mass spectrometer requires a good vacuum for efficient operation, and in general practice, a pressure of less than 10^{-6} Torr is considered desireable. Today, most mass spectrometer users obtain an instrument package with vacuum hardware designed by the manufacturer, and significant modifications of the pumping system are seldom requested in the purchase. Some commercial mass spectrometers have vacuum features specifically designed to increase GCMS compatibility, but neglect of certain details can restrict the flexibility of operation. It is important to understand the limitations that a particular vacuum configuration will impose on various GCMS analyses and to be able to appraise the capacity of a specific mass spectrometer to solve the problems at hand.

In combined GCMS operation, two separate systems are interfaced with a pressure differential of several orders of magnitude. As a result, it is necessary to operate the mass spectrometer close to a critical vacuum level where optimum mass spectral performance may be in jeopardy. Too frequently, the requirement of high vacuum is considered a severe incompatibility between mass spectrometry and gas chromatography; but actually, the main consideration should be the quantity of gas that can flow through the systems, and emphasis should be given to the methods available to maintain the vacuum level. It is also important to examine the reasons for a high vacuum, especially since they are not all valid if the offending gas is helium or another inert substance.

113

The principal reasons for high vacuum in mass spectrometry are:

(1) Significant amounts of oxygen burn out the mass spectrometer filament.

(2) High background pressure gives an interfering mass spectral pattern.

(3) Ion-molecule reactions occur at relatively high ion chamber pressures and change the fragmentation pattern.

(4) High pressure in the mass spectrometer ion source interferes with normal regulation of the electron beam.

(5) With high pressure in the ion chamber or ion source housing, the several thousand volts used for ion acceleration may cause an electrical discharge.

(6) The mean free path of the ions must be longer than the flight path from the ion source to the collector.

For good mass spectral operation, all these factors must be considered and if any necessary conditions are not met, the system will fail. Some of the reasons listed, such as reasons (1) and (2), depend upon the type of gas, but the effect of the others depends upon the total pressure. The following discussion will show that for most GCMS conditions, reason (6) is the only important factor in establishing a limiting pressure. If this mean free path condition is attained, the limit of the other conditions will seldom be exceeded

1. Effect of Oxygen on Filament Life

A high background pressure of air or oxygen will significantly reduce the life of the mass spectrometer filament. Under normal mass spectral operating conditions, the partial pressure of oxygen will be $1-2 \times 10^{-8}$ Torr. Filament life expectancy will be at least $5-10$ months and depend more upon the operating methods and types of compounds analyzed than upon the harmful effects of the background oxygen. In GCMS operation, the carrier gas pressure in the ion source housing may be 10^{-5} to 10^{-4} Torr, but since the carrier gas is always inert, it does no harm to the filament. However, small air leaks may be encountered in the GCMS interface which cause the partial pressure of air to be as high as 10^{-6} Torr. Filament life may thus be reduced $1-2$ months of continuous operation. Such leaks should be avoided when possible, but in any event, a high pressure of helium, hydrogen, argon, or nitrogen does not harm the filament.

2. The Carrier Gas Background Spectrum

In normal mass spectrometry, a low pressure level of background contamination is maintained to avoid interference with the sample spectrum. The amount of these background materials (oils and residues from previous samples) is kept low by using a fast vacuum system and an occasional mass spectrometer bake-out. In a clean vacuum, the partial pressure of contaminants will be less than 10^{-9} Torr.

Large amounts of helium, hydrogen, argon, or nitrogen used as a carrier gas do not contribute interfering ions to the overall mass spectral pattern so that this is not an interference that contributes to the need for a high vacuum. A large amount of nitrogen may obliterate the spectral pattern within one or two units of mass 28, but since nitrogen is seldom used as a GCMS carrier gas, this point is not important. The same statement can be given regarding the use of argon. This gas could give harmful effects in the region of mass 40, but argon is also seldom used for GCMS studies.

3. Effect of Ion-Molecule Reactions

Ion-molecule reactions or chemiionization will occur whenever the ion chamber pressure approaches 10^{-1} Torr or more (page 24). For planned chemiionization studies, a high-pressure condition is deliberately attained by use of a tight ion source and excess reactant gas. In GCMS studies, the ion chamber pressure should not greatly exceed 10^{-2} Torr to assure that the mass spectral pattern is uninfluenced by ion-molecule reactions.

The ion chamber pressure can be readily calculated from equation 2.2 if the quantity of helium (carrier gas) and the conductance of the source are known. A typical source will have a conductance of 1 liter/sec. If the overall mass spectrometer system can take 1 cm^3 atm/min of gas, the allowed quantity of helium is 1.3 x 10^{-2} liter Torr/sec. The pressure in the ion chamber is thus:

$$P = Q/C = \frac{1.3 \times 10^{-2} \text{ liter Torr/sec}}{1 \text{ liter/sec}} = 1.3 \times 10^{-2} \text{ Torr.}$$

This value is close to the suggested safe limit and if the pressure is significantly increased by taking in more carrier gas or reducing the conductance of the ion source, chemiionization may take place. This could make a significant change in the mass spectral pattern if hydrogen is used as a carrier gas. With helium, the chemiionization behavior is not well understood, but the chemiionization mass spectral pattern from helium reactant gas is very similar to that obtained by electron impact.[58, 60] Therefore, the consequences of excess helium are not serious provided the other operational limits are not exceeded.

4. Regulation of Electron Beam

In conventional mass spectrometry, quantitative precision is obtained by regulating the electron beam. The optimum amount of ionizing current is manually or automatically selected and current fluctuations are used as a feedback signal to regulate the voltage input to the filament. For qualitative analysis, such precision is not necessary and in fast GCMS scans, it cannot be attained due to other sources of variation. However, the automatic control is often retained because it is there.

When the pressure in the ion chamber starts to exceed 10^{-2} Torr, the passage of electrons through the chamber is impeded, and at gas concentration of 10^{-1} Torr or greater the electron beam is essentially stopped. The regulating circuit increases the filament current in an attempt to increase the electron beam, and thus may result in filament burn out. Chemiionization systems do not use filament current regulation, and greater electron beam penetration is attained by using a higher ionization voltage (150–500 volts). In GCMS, the problem can be avoided by maintaining an ion chamber pressure of about 10^{-2} Torr. If calculations of the source pressure indicate that this value might be exceeded, then the electron beam regulator should be bypassed.

5. Danger of Electric Discharge

Another consequence of high ion chamber pressure is the increased probability of electric discharge from the ion source to the ion source housing or the GCMS interface. In most deflection mass spectrometers, the ion source will be several thousand volts above the ground potential. The high voltage is isolated from the housing or inlet by using a glass tube or other nonconductive material as an inlet transfer line. However, if the inlet line pressure is greater than 10^{-1} Torr, there is a considerable probability that the gas will be conductive and cause an electrical discharge. This may damage the ion chamber parts or the electronic supply circuits. In chemiionization studies, electric discharge is avoided by having an insulated line back to a point where the gas has a high enough pressure to be nonconductive. Fortunately, in standard GCMS studies, the pressure in the inlet lines seldom exceeds 10^{-1} Torr so that a discharge does not occur. If unusually large quantities of gas are to be conducted into the mass spectrometer, a calculation of the ion source pressure should be made to assure that the region of electric discharge is not being approached.

6. Mean Free Path Requirements

In a GCMS system, the maximum allowable carrier gas flow to the mass spectrometer is usually determined by the mean free path of the ions through the spectrometer and the limiting pressure due to other factors is seldom reached. The mean free path of a gaseous particle represents the average distance an atom or molecule travels before it interacts with another particle. The interaction could be a gross collision of the billiard ball type resulting in a chemical reaction, or it may be a modest deflection due to the two particles approaching within an interacting distance. These small deflections are particularly important in ion optical systems such as exist in deflection mass spectrometers.

The mean free path of a particle depends upon the collision diameter and the concentration of the particles. In a neutral pure gas it does not depend on the velocity. From the kinetic molecular theory, the mean free path L, of a pure gas is given by

$$L = \frac{1}{\sqrt{2\pi} \, n\sigma^2} \qquad (4.1)$$

where n is the number of molecules/cm^3 and σ is the collision diameter.[170] If the collision diameter is known, the mean free path of a particle can be calculated as a function of pressure. Since the collision diameter of most gases does not differ sufficiently to affect order-of-magnitude calculations it is convenient to use the approximate rule of thumb given by

$$L = \frac{5 \times 10^{-3}}{P} \text{ cm} \qquad (4.2)$$

where P is the pressure in Torr. (1 Torr = 1 mm of Hg.)

Unfortunately, the available data for the mean free path of neutral atomic or neutral molecular systems do not accurately describe ion/atomic or ion/molecular collisions, especially in an ion optical system. There are three main differences. First, in the neutral molecule systems, the mean free path is usually measured by viscosity, low-angle scattering, or diffusion. The measurements give a good estimation of the distance between molecules and of their gross interactions, but diffusion or viscosity is not sensitive to the subtle deflections that would cause an ion beam to be deflected off course in a magnetic mass spectrometer. The operation of the mass spectrometer requires careful alignment of two slits that may be $2-3 \times 10^{-3}$ cm wide, and small deflections of the ion beam can cause noticeable loss of resolution and sensitivity. For a quadrupole or TOF mass spectrometer, this degree of pressure sensitivity is absent, so that these instruments can be operated at a slightly higher pressure for the same length of flight.

A second reason that the mean free path of neutral molecules is not strictly applicable in ion systems is that the collision diameter or cross section is greatly increased due to ion-molecule reactivity. The increase may be an order of magnitude or more, but unfortunately, very little exact information is available. It must be emphasized that the cross section under consideration is not the cross section for the occurrence of ion-molecule reactions but rather it is the cross section for an interaction that causes deflection of the ion so that it does not arrive at the collector. The uncertainty in the values of this parameter makes it difficult to estimate the mean free path in the mass spectrometer.

Fortunately, the increase in collision diameter or cross section is offset by the fact that the velocity of the accelerated ion is two orders of magnitude greater than the thermal velocity. Consequently, successful transmission of an ion from the source to collector is more probable. The subject can be treated in fairly exact form on the basis of kinetic theory, but for the present discussion it is sufficient to consider that the probability that the ion undergoes a collision by a neutral gas molecule is proportional to the time the ion spends in the flight path. Thus, the mean free path for the fast-moving ion is much longer than it would be for a particle moving at a thermal velocity.

Because of these three factors, a realistic calculation of mean free path in a mass spectrometer cannot be made. The ill-defined nature of the ion/molecule collision diameter and its significance in different mass spectrometer systems should be clarified but in the meantime, the ion/molecule mean free path must be established empirically for each GCMS apparatus. Fortuitously, the simple equation 4.2 gives a reasonable "ball park" estimation that can be used in a preliminary mass spectrometer calculation. This correspondence results from the fact that the adverse effect of increased collision diameter is offset by the longer mean free path of the high-velocity ion.

The importance of the mean free path can be understood by consideration of mass spectrometer pumping configurations. Figure 4.1 shows the two common pumping methods used in organic mass spectrometry. In (a), one vacuum pump is used to obtain the vacuum in the entire mass spectrometer housing. The source is assumed to be fairly tight and would have a conductance of about 3 liters/sec for helium. Thus, two distinct pressure regions exist (A and C). In the ion source region A, the pressure can be calculated

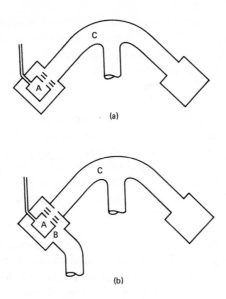

FIGURE 4.1. Vacuum system configurations commonly used in GCMS. (a) Single-pumped mass spectrometer. (b) Differentially pumped mass spectrometer. A, B, and C are regions of different pressure. A = ion source, B = ion source housing, C = analyzer.

using equation 2.2 (Q = PC). Although the partial pressure of a sample will seldom exceed 10^{-6} Torr, when a large quantity of carrier gas is used, the ion source helium pressure might be 10^{-3} Torr to 10^{-2} Torr.

An estimated mean free path of 5–0.5 cm in the ion source can be calculated from equation 4.2. Since the amount of travel from the point of ionization to the outside of the ion source is usually less than 0.1 cm, the mean free path is quite sufficient. Furthermore, in this region, the ion has not yet entered into the precise path through the analyzer and is still under the acceleration of the repeller potential or draw-out potential. As a result, collisions close to the exit slit of the ion source are of little consequence and the ion will continue and enter the accelerating/focusing region.

The pressure in the rest of the single-pumped mass spectrometer, region C, is for all intents and purposes equal to the pressure achieved at the entrance of the pumping line. This pressure is usually the parameter that limits the quantity of carrier gas allowed in a single-pumped mass spectrometer. For most mass spectrometers, the flight path is about 100 cm long (within a factor of 2) and from equation 4.2, a mean free path of 100 cm is obtained at a pressure of 5 x 10^{-5} Torr. A reasonable margin of error must be allowed, and successful operation is usually possible at 2 x 10^{-5} Torr pressure. For short mass spectrometers with only a 10-cm flight path (e.g. a quadrupole mass filter), the maximum allowable pressure is a factor of 10 higher, i.e. about 2 x 10^{-4} Torr.

Methods for determining the pumping speed at the entrance to the vacuum line are given in Section 4.B. If, for example, a pumping rate of 50 liters/sec is assumed, the quantity of gas that gives P = 2 x 10^{-5} Torr is 10^{-3} liter Torr/sec or 0.08 cm^3 atm/sec. This corresponds to only 0.4% of the total effluent from a column operating at a flow of 20 cm^3 atm/min, and such a severe loss of sample is often critical. As a consequence, an enrichment device (Chapter 5) is almost always necessary when the mass spectrometer is not differentially pumped.

Incidentally, the carrier gas pressure in the ion source would be only 3 x 10^{-4} Torr with the quantity of helium calculated above, which is one to two orders of magnitude lower than the known safe pressure level.

Figure 4.1b illustrates a differentially pumped mass spectrometer. The significant difference between configurations (a) and (b) is that vacuum is provided in (b) by two separate pumping systems. The opening that connects regions B and C is kept small, and should be no larger than is necessary to permit passage of the ion beam. The distribution of gas to the two regions B and C is proportional to the conductance of this small hole and that at the top of the source pumping line. Methods of calculating these conductances for specified geometrical dimensions are discussed in Section 4.B, but as a first approximation, it will be assumed that 10% or less of the total gas goes into the analyzer section C.

The consequence of splitting the gas flow in this manner is that for a given pumping speed and a required mean free path, ten times more total effluent can enter the mass spectrometer, at least in so far as the analyzer section is effected. Thus, with the specified pumping speed of 50 liters/sec, 10^{-2} liter Torr/sec of total effluent can enter the mass spectrometer, corresponding to about 0.8 cm^3 atm/min.

Region B will have a pressure nine times greater than that in C, assuming both vacuum systems have the same conductance. This pressure, about 2 x 10^{-4} Torr, corresponds to a mean free path of 25 cm. Since the length of travel from the exit of the ion source to the entrance of the analyzer is usually 1–6 cm, a pressure of 2 x 10^{-4} Torr provides an adequate safety margin.

In many differentially pumped systems, the opening to the analyzer (region C) may be quite small so that only 1% of the total gas enters this section. The quantity of effluent gas to the mass spectrometer can therefore be increased another factor of ten to 8 cm^3 atm/min without affecting the performance of the analyzer. However, at the assumed pumping rate of 50 liters/sec, the pressure in region B will now be about 2 x 10^{-3} Torr and the mean free path would be only about 2 cm. The distance of travel in the accelerating/focusing region between the source and analyzer is often 2–6 cm, so that the ion beam would be limited in region B. Therefore, the safe allowable quantity of effluent would still be restricted to about 2 cm^3 atm/min rather than the 8 cm^3 atm/min allowed by the limitation in the analyzer.

To overcome the limitation in region B, the pumping speed of the source can be significantly increased to, say, 200 liters/sec, but the ion source pressure would then be close to $0.3-1$ x 10^{-1} Torr and the limitation will occur in region A.

Calculations of this type should be performed on every GCMS setup. The results establish the maximum safe operating level and predict the efficiency of sample utilization exclusive of an enrichment interface. The principal obstacle to the calculations is obtaining the true speed of the vacuum system or at least the geometric measurements that permit its calculation. The details of vacuum pump evaluation, measurement of gas flow, and measurement of vacuum are elaborated in Section 4.B.

The calculations performed above are summarized in Table 4.1. It is apparent that with a minimum vacuum system, the quantity of effluent gas is limited in the analyzer section. With a differential-pumped vacuum system, the limitation occurs in the source housing section. Ultimately, increased vacuum power leads to the quantity of effluent being limited by the pressure in the ion source and an interface enrichment device is required for further improvement.

B. EVALUATION OF VACUUM PUMPING EFFICIENCY

1. Conductance of the Total Pumping System

It was shown in the previous section that the quantity of carrier gas that can be taken into the mass spectrometer depends upon the size and design of the pumping system. Unfortunately, many mass spectrometer vacuum systems have been designed without full consideration for the more rigorous conditions of GCMS analyses, and many mass spectrometer users are not aware of the influence of each vacuum component (pump, trap, valves, lines, etc.) on the overall pumping efficiency. It is meaningless to state that a diffusion pump has a speed of 300 liters/sec when the true speed of the pumping

TABLE 4.1. Comparison of expected GCMS efficiencies for the systems illustrated in Figure 4.1.[1]

	Total quantity effluent gas, cm^3 atm/min	Pressure, Torr			Limiting Region
		A	B	C	
Single-pumped mass spectrometer	0.08	3×10^{-4}	–	2×10^{-5}	C
Differential-pumped, ratio 9/1	0.8	3×10^{-3}	1.8×10^{-4}	2×10^{-5}	C
Differential-pumped 100/1	2	0.8×10^{-2}	5×10^{-4}	5×10^{-6}	B
Differential-pumped, ratio 100/1. Source conductance 200 liter/sec	8	3×10^{-2}	5×10^{-4}	2×10^{-5}	All close to limit. A may exceed.

[1]Ion source conductance, 3 liters/sec; vacuum conductance, 50 liters/sec except where specified. Analyzer 100 cm long.

system could be reduced to 20 liters/sec by a poor choice of a valve or trap. A meaningful value of the pumping speed that can be used to calculate the pressure in the various regions of mass spectrometer can be obtained only by taking into account the conductance of all the vacuum components.[171–173]

A typical vacuum pumping system is shown schematically in Figure 4.2. Each of the component parts has a conductance or pumping speed determined by its size and configuration. The diffusion pump provides a means of removing gas from the system so that additional gas can flow into the pumping region. The volume flow rate at which the pump will remove gas is defined as the pumping speed, S. The other parts have a resistance to the gas flow which increases with increasing length of the tube and with decreasing diameter or cross section. Since resistance is effectively the reciprocal of the conductance, a tube with a large diameter and short length has a high vacuum conductance. For the components illustrated in Figure 4.2, the conductances of the several parts (lines, valve, and trap) are designated C_L, C_V, and C_T.

The rules for calculating the total system gas conductance are the same as those used for calculating electrical conductance. Conductances of components in parallel are additive; conductances of components in series are reciprocally additive. Thus,

$$C = C_1 + C_2 + \ldots + C_n \tag{4.3}$$

for parallel connections, and

$$\frac{1}{C} = \frac{1}{C_1} + \frac{1}{C_2} + \ldots + \frac{1}{C_n} \tag{4.4}$$

for series connections.

For the example of Figure 4.2, the conductance at the mass spectrometer would be given by

$$\frac{1}{C} = \frac{1}{C_{L1}} + \frac{1}{C_V} + \frac{1}{C_{L2}} + \frac{1}{C_T} + \frac{1}{S} \, . \tag{4.5}$$

Numerical values typically encountered for these parameters might be C_{L1} = 60 liters/sec (He), C_V = 120 liters/sec (He), C_{L2} = 60 liters/sec (He), C_T = 150 liters/sec (He), and S = 180 liters/sec. (It will be shown later in this chapter that the conductance of a line is dependent on the gas, so that the type of gas must be specified. Pumping speed, S, varies only a small amount with different gases.) The total conductance would be given by

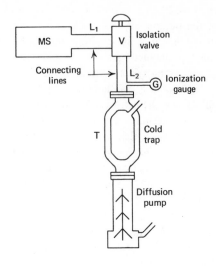

FIGURE 4.2. Schematic diagram of the components of a typical mass spectrometer vacuum system.

$$\frac{1}{C} = \frac{1}{60} + \frac{1}{120} + \frac{1}{60} + \frac{1}{150} + \frac{1}{180}$$

and C would thus be 18 liters/sec.

This surprisingly low value is typical of many GCMS pumping systems that have not been properly optimized. Each component appears to be quite suitable by itself and yet when all are joined in series the total conductance or pumping speed is so low as to severely restrict the efficiency of effluent utilization.

2. Conductance of Gas Through Tubes and Orifices.

The exact nature of the flow of gas through a tube or orifice depends upon the pressure or molecular density of the gas. If the pressure is fairly high, the molecules undergo many interactive collisions with each other and hence move down the tube in a bulk fashion whereby the molecules push one another. This type of flow is called viscous flow. On the other hand, at low pressures the molecular density is sparse and the number of intermolecular collisions is small compared to the number of wall collisions. The molecules then move down the tube independent of each other at a rate that is determined by the kinetic velocity. This type of gas flow is called molecular flow.

Viscous flow. For viscous flow the mean free path of the molecules is small compared with the diameter or other cross-sectional dimensions of the tube. In this condition, the gas flow behavior is similar to that of a fluid and the collisions between molecules cause them to flow in a viscous fashion governed by the Poiseuille equation.[172] With a pressure differential of $P_2 - P_1$, the quantity of gas Q is given by

$$Q = \frac{0.16}{\eta} \frac{d^4}{l} (P_2{}^2 - P_1{}^2) \text{ liter Torr/sec} \qquad (4.6)$$

(handwritten annotation: radius, pointing to d^4)

where

η = the viscosity (average viscosity for a gas mixture)
d = the tube ~~diameter~~ radius (handwritten: radius)
l = length of tube
temp. = 300°K

In GCMS studies, viscous flow is encountered in the gas chromatographic column, in the jet separator, in the entrance and exit of the effusive separator, and in forepump vacuum lines. It should never be encountered in any region of the mass spectrometer except in chemiionization studies where viscous flow may occur in the exits of the ionization chamber. Under vacuum conditions, $P_2{}^2$ is generally much greater than $P_1{}^2$, so that the exit pressure can be ignored in order-of-magnitude calculations. The fourth power dependence upon the diameter exerts a strong influence in controlling the flow, and variation of this parameter is used to control flow from an effusive separator into the ion chamber. Accurate calculations are difficult for small-diameter tubing (d < 0.2 mm) because aberrations in d are enhanced by the fourth power factor.

Equation 4.6 is not valid for short tubes. For square or rectangular pipes, it can be used for crude approximations, but preferably, reference should be made to more exact formulations.[172]

Molecular flow conductance through an orifice. Conditions for molecular flow are the opposite of those for viscous flow. In molecular flow, the mean free path of the molecules must be long compared to the cross-sectional dimensions of the tube. As a result, the molecules move completely independent of each other and most collisions occur on the walls of the vessel. This isolated behavior permits relatively simple calculations to be made on the molecular flow conductance of gases through tubes and orifices.

The rate of molecular flow through an orifice of unit area can be equated to the number of wall collisions per unit area. It is assumed that the back pressure on the outside of the orifice is insignificant, but a suitable correction for this factor can be made when necessary. From the kinetic theory of gases[170] the number of wall collisions ν per unit area is given by

$$\nu = 1/4\, n\,\overline{v} \qquad (4.7)$$

where n = number of molecules per cc and $\overline{v}$ is the mean molecular velocity. It follows that the net rate of volume flow out of a unit area orifice, i.e. the conductance, is $1/4\,\overline{v}$. Thus, the conductance for a specific area A will be

$$C_{or} = 1/4\, A\,\overline{v}\ \text{cm}^3/\text{sec.} \qquad (4.8)$$

The mean velocity can be expressed by the equation

$$\overline{v} = \sqrt{8RT/\pi M} \qquad (4.9)$$

where R = gas constant, T = absolute temperature, and M = molecular weight. Thus, the orifice conductance can be written

$$C_{or} = A\sqrt{RT/2\pi M}\ . \qquad (4.10)$$

At T = 300°K, and for R = 8.3 x 10^7 ergs/°K/mole,

$$C_{or} = 63\, A/\sqrt{M}\ \text{liters/sec.} \qquad (4.11)$$

The conductance of a 1 cm²-orifice will be 12 liters/sec for nitrogen and 31 liters/sec for helium. The equivalent conductances for a 1 in.²-orifice are 74 liters/sec and 190 liters/sec, respectively.

Temperature can be considered as a constant for most approximate calculations. Even a drastic change from 300°K to 600°K will increase the conductances only by a factor $\sqrt{2}$. Since most pumping lines are kept fairly close to room temperature, changes in this parameter can be overlooked. For calculations of effusion rate through a molecular separator or conductance of an ion source, it is necessary to bring temperature into account.

The conductance of an ion source can be calculated from these relationships provided the area of all exit holes is known. A schematic ion source block is shown in Figure 4.3. If the system is well designed, the ion chamber is tight and the only exit openings are due to the ion exit slit, the electron

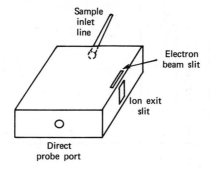

FIGURE 4.3. Schematic diagram of tight ion source showing the three openings that permit gas conductance out of source.

beam slit, and the direct introduction port. Typical values for the size of these openings (averaged from five commerical sources) are given in Table 4.2 along with calculated conductances. About 60% of the source conductance is due to the ion exit slit, so that modest changes in the dimensions of the other ports will not make a significant change in the total conductance. It is noted that the source conductance for helium may be an order of magnitude faster than that for a high-molecular-weight component. This means that in the ion source there is an enrichment of the component relative to the helium concentration but this must not be used to imply any increase in sensitivity. The total amount of sample into the source is completely independent of this enrichment, and once in the source, ionization of the sample is not influenced by the presence of helium, provided the pressure is not in the chemiionization region.

On the other hand, if the mean free path is smaller than the dimensions of the source openings, viscous flow will take place out of the ion source. This would mean that a high-molecular-weight compound at low concentration would be carried out of the source with the helium. The result could be as much as an order of magnitude loss in sensitivity. For the present example, a flow of 2 cm^3 atm/min would give a pressure of about 10^{-2} Torr (Table 4.1), and from equation 4.2, L would be about 0.5 cm. Since the length of the ion beam slit is 1 cm, the flow would be in the intermediate range between viscous and molecular flow and some loss of sample efficiency might be suffered. This calculation is necessary whenever a large source pump is used to obtain a high helium throughput. The ion source pressure should be kept below the level at which viscous flow will occur.

TABLE 4.2. Conductance of exit ports in typical ion source operating at 200°C.

| | Length, cm | Width, cm | Area, cm^2 | Conductance, liters/sec | | |
				Helium	Nitrogen	MW 200
Ion beam slit	1	0.05	0.05	1.9	0.75	0.27
Electron beam slit	0.25	0.075	0.02	0.77	0.30	0.10
Direct probe opening	0.125 dia.		0.013	0.50	0.19	0.07
TOTAL			0.083	3.2	1.2	0.44

Molecular flow conductance through a tube. There are only a few regions in the GCMS system in which conductance occurs through an orifice and in general, transfer of gas takes place through relatively long tubes (length $\gg$ diameter). Wall collisions are frequent, and because the molecules are reflected diffusely in random directions, these collisions cause back diffusion of the gas. The conductance of a tube C_t, is thus lower than the conductance of an equivalent orifice C_{or}, and is inversely proportional to the length of the tube. A derivation of C_t in terms of fundamental parameters[174] gives the expression

$$C_t = \left(\frac{\pi RT}{16M}\right)^{1/2} \frac{d^3}{l} \tag{4.12}$$

where d is the diameter and l is the length of the tube. The quantity of the gas flowing through the tube is obtained using equation 2.2, namely Q = PC. Hence,

$$Q_t = \left(\frac{\pi RT}{16M}\right)^{1/2} \frac{d^3}{l} P . \tag{4.13}$$

Equation 4.12 can be used to calculate C_t when $l \gg d$, but it is more convenient to use an approximation that expresses C_t in terms of C_{or} (from equations 4.10 and 4.12), namely

$$C_t = 1.4 \, C_{or} \frac{d}{l} . \tag{4.14}$$

The serious consequences of a long tube are immediately apparent. For example, an orifice of 5 cm ID has a conductance of 580 liters/sec for helium, but a pumping line of 5 cm ID and 40 cm length has a conductance of only about 100 liters/sec for helium. When this line is placed in series with a valve, a trap, and a pump, the overall conductance, as determined from equation 4.5, may be as low as 20 liters/sec.

The example is typical of many mass spectrometer pumping systems used for GCMS studies. Fortunately, many systems are now fabricated with short pumping lines of large-diameter tubing, and pumping speed in the range of 50–200 liters/sec for helium is frequently attained.

When equation 4.12 is used as suggested above, the calculated conductance is 10–20% high if d is only 5–10 times l. For a closer approximation, the orifice conductance must be taken into account so that the total conductance is given by

$$\frac{1}{C} = \frac{1}{C_{or}} + \frac{1}{C_t} .$$

(4.15)

A tube conductance calculated using equation 4.15 is accurate within a few percent provided $l \gtrsim 5d$.

Figure 4.4 gives the molecular conductance of helium as a function of length for several common different size tubes. It is noted that for $d > l$, the short tube has a conductance close to that of an orifice. For $l > 5d$, the conductance decreases rapidly according to equation 4.12 or 4.14.

Special formulae are available for calculating the conductance of trapezoidal pipes or awkward structures encountered in valves or traps.[171-173] However, in GCMS work, a considerable margin of error (up to a factor of 2) can usually be tolerated in applying estimated conductances, and satisfactory approximations are made by analogy with a straight pipe. A curve or elbow contributes a flow resistance equivalent to the length of the center line of the curve.

Comparison of molecular flow and viscous flow conductances. For most pumping applications, there is a significant difference between molecular flow and viscous flow. For molecular flow, the quantity of gas transferred per unit time is proportional to the first power of the pressure (equation 4.13). For viscous flow, the quantity of gas transferred per unit time is proportional to the square of the pressure (equation 4.6). As a consequence, conductance in the molecular flow region is independent of the pressure, but at higher gas densities, where viscous flow occurs, the conductance is a function of the pressure.

Figure 4.5 gives the helium conductance of a 1-meter pipe of various diameters as a function of pressure. When the mean free path is greater than the diameter, the conductance is invariant. As the pressure increases, a transition or intermediate flow condition is attained. At higher pressures, the flow becomes viscous and the conductance increases linearly with increasing pressure. The quantity of gas increases with the square of the pressure. This fact is important in GCMS work where large quantities of carrier gas are transported through a long tube to a forepump. Viscous flow will predominate and the resulting increase in tube conductance permits effective removal of the excess gas.

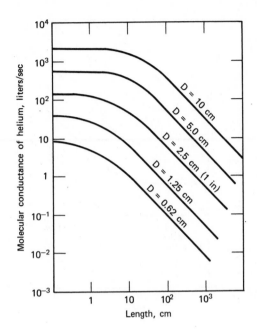

FIGURE 4.4. Molecular flow conductance of helium as a function of tube length. The parameter is tube diameter.

3. Vacuum Pumps

Forepumps. A wide selection of rugged mechanical forepumps is available which provide long-term, reliable operation with little maintenance beyond inspection of the drive belt or an occasional replenishment of oil.[171] Many of these pumps attain a vacuum level of 10^{-3} Torr in a tight system, but an ultimate pressure specification is of no importance for a continuous flow operation. The significant parameter is the rated pumping speed, and a considerable choice is available from 20 liters/min to as high as 1000 liters/min. (Conventionally, forepumps are usually rated as volume/min. The faster high-vacuum pumps are usually rated as volume/sec.)

For most GCMS studies, forepump capacity for either the analyzer section or the ion source housing are easily met, and a speed of 50–150 liters/min is quite sufficient. However, the higher quantities of helium that are pumped through a separator (sometimes 50 cm^3 atm/min or more) may

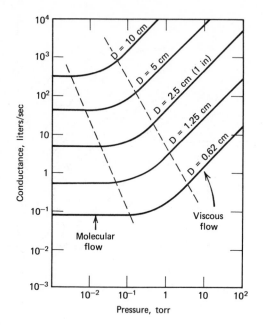

FIGURE 4.5. Flow conductance of helium in a 1-meter tube as a function of pressure. The parameter is tube diameter.

exceed the capacity of a small pumping system, and a calculation should always be made to assure that the pump will carry the load.

Consider the schematic forepump system shown in Figure 4.6a. If 18 cm^3 atm/min (Q = 13.7 liters Torr/min) of carrier gas is pumped through the separator, the pressure at the throat of a 150-liters/min forepump would be about 0.1 Torr (calculated from P = Q/S). The mean free path of 5 x 10^{-2} cm is considerably less than the internal diameter of the tube, and hence viscous flow would occur. An estimated conductance of the tube can be obtained from Figure 4.5 by assuming as a first approximation that the average pressure in the tube is 0.2 Torr. The average tube conductance would be close to 1 liter/sec or 60 liters/min. The combined conductance of the pump and pumping line is obtained from equation 4.2. Thus,

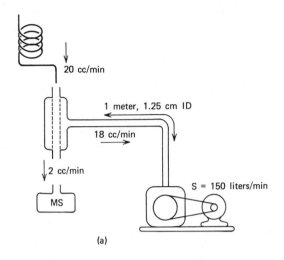

20 cc/min

1 meter, 1.25 cm ID

18 cc/min

2 cc/min

MS

S = 150 liters/min

(a)

FIGURE 4.6. Typical forepump configurations. (a) Used with effusion separator at a load of 18 cm^3 atm/min.

$$\frac{1}{C} = \frac{1}{C_{pump}} + \frac{1}{C_{line}} = \frac{1}{150} + \frac{1}{60}$$

and C = 43 liters/min. The calculated pressure at the outside of the separator would be 0.3 Torr which is close to the maximum that can be tolerated without having a serious amount of back effusion through the separator.

Figure 4.5 gives the tube conductance for the case where $P_2 \gg P_1$, so that as used in the above example, there is an error of about 15%. Similarly, calculation of P_2 from equation 4.6 is not very accurate when $P_2 \approx P_1$. Nevertheless, the approximation clearly shows that the components suggested for the example result in a marginal operational level. If the pressure inside the frit is 1 Torr, there would be 30% back diffusion. Any further restriction in the pumping speed or increase in gas flow would result in significant degradation of separator performance.

Most separator systems are equipped with forepumps rated at 150 to 200 liters/min. If larger pumps are used to obtain a more comfortable safety margin, it is important to match the conductance of the pumping line appropriately.

Suppose that the same forepump system is used with a diffusion pump

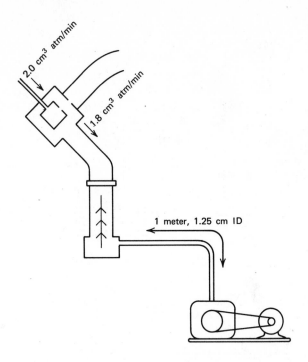

FIGURE 4.6. Typical forepump configurations. (b) Used with ion source diffusion pump at a load of 1.8 cm^3 atm/min.

on an ion source as shown in Figure 4.6b. Assume that 1.8 cm^3 atm/min of helium passes through the source pump (Q $\approx$ 1.4 liters Torr/min) and, as a first approximation, that the average line pressure is 0.1 Torr. From Figure 4.5, the conductance of the line is about 0.8 liter/sec ($\sim$ 50 liters/min) of viscous flow. The combined line and pump conductance is 38 liters/min and with the given value of Q, the estimated back pressure on the diffusion pump is 4 x 10^{-2} Torr.

For most diffusion pumps, the limiting forevac pressure is 0.3–0.5 Torr, so that the pumping system in the example provides an adequate safety margin. Nevertheless, for a GCMS system, a calculation of the forevac pressure is desireable to assure that this safety factor is indeed present.

High-vacuum pumps. Most mass spectrometer high-vacuum pumping systems utilize an oil diffusion pump or a mercury diffusion pump (Figure 4.7).

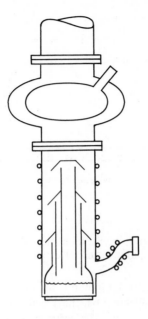

FIGURE 4.7. Structural schematic of an oil diffusion pump.

The pumping fluid is boiled up the chimney assembly and directed downward by the jets as a viscous flow, high-velocity stream of vapor. Momentum is transferred to the gas molecules which are removed at the bottom by a mechanical forepump. The pump vapors condense on the watercooled walls of the pump and return to the boiler. Three successive jets or stages of pumping are commonly used. Vacuum levels of 10^{-8} to 10^{-9} Torr are attained in tight systems if a liquid nitrogen cold trap is used to reduce the vapor pressure of the pump fluid. The pumping speed of a diffusion pump depends primarily upon the size, and may be anywhere from a few liters/sec to 1500 liters/sec or more. In mass spectrometry, the diffusion pump specifications are usually in the range of 150–500 liters/sec. Pumping efficiency is constant over most of the operating range, but at pressures of 10^{-3} Torr or higher, the speed is drastically reduced. The operating characteristics for a 1500-liter/sec diffusion pump are shown in Figure 4.8. For this pump, the maximum throughput Q_{max}, would occur at 10^{-3} Torr, and hence would be 1.5 liter Torr/sec or 110 cm^3 atm/min. ($Q_{max} = PS$.)

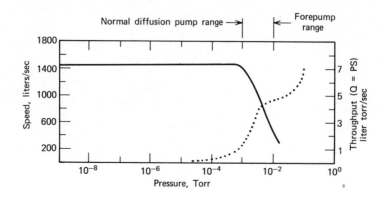

FIGURE 4.8. Operating characteristics of a fast oil diffusion pump. Solid line, pump speed. Dotted line, throughput Q.

If a 180-liter/sec diffusion pump is used to attain faster pumping for the separator system of Figure 4.6a, the 18 cm^3 atm/min of carrier gas would exceed the Q_{max} of the pump. The pressure would increase above the operating range and the diffusion pump would be ineffective. For safe operation, a minimum pump speed of about 250 liters/sec would be desired to handle the specified 18 cm^3 atm/min of carrier gas from the separator. Whenever a diffusion pump is used to handle large volumes of gas, care must be taken to avoid exceeding the Q_{max} specification.

Conversely, the amount of gas that goes through a mass spectrometer will never exceed the maximum diffusion pump throughput provided the necessary vacuum levels are maintained in the source housing ($\sim 10^{-4}$ Torr) or analyzer ($\sim 10^{-5}$ Torr). If the pressure at the throat of the pump is 10^{-4} Torr and S = 180 liters/sec, the throughput is equal to 1.8 x 10^{-2} liter Torr/sec or 1.4 cm^3 atm/min. This is an order of magnitude less than the allowed maximum.

Specifications given by vacuum equipment manufacturers refer to the pumping rate or conductance of the component for nitrogen. Lighter gases such as hydrogen or helium are pumped faster but the increase is not particularly significant. The magnitude of the increase may vary with the design of the pump but will seldom exceed 25% and for most calculations, this merely provides a modest safety factor. Of course, the conductance of a valve or trap will be considerably greater for helium than the specification given for nitrogen as was emphasized in Section B.2.

Diffusion pumps are usually operated with a liquid nitrogen trap to keep the pump vapors from entering the vacuum system. If the pumping fluid is mercury, a cold trap is absolutely essential since the vapor pressure of mercury at room temperature is greater than 10^{-3} Torr and back migration is considerable. Most diffusion pump oils have a vapor pressure at room temperature lower than 10^{-7} Torr and can be operated without a refrigerant if desired, especially for standby.

For GCMS analyses, the choice between oil and mercury as a pumping fluid is somewhat arbitrary. The background mass spectrum due to mercury consists of seven isotopic peaks in the mass range 196–204. (See Figure 2.38.) The intensity of these peaks can be kept at a fairly low level with liquid nitrogen so that they do not interfere with the mass spectral analysis. In practice, these mass peaks are often useful as mass markers and sometimes a "mercury inlet" is deliberately provided so that a small leak of mercury can be introduced for this purpose.

The background mass spectrum from a typical pump oil will be quite similar to the background spectrum from column bleed of many substrates (see Figures 3.5 and 3.6.) The intensity of pump oil background is extremely low when liquid nitrogen is used and seldom makes a significant contribution to the total background spectrum. In Chapter 3, it was pointed out that column bleed was often difficult to control and may be the factor that limits the minimum sample level. With many pump oils, even when used without liquid nitrogen, the contribution to the background is small compared to column bleed, and GCMS operation without the refrigerant is often acceptable.

With a mercury diffusion pump, a refrigerated baffle at $-30°C$ is placed just above the top jet of the pump. This cold trap condenses the mercury as a liquid so that it can drop back into the pump. If an intermediate refrigerated baffle is not used, mercury is frozen on the trap as a solid and the boiler charge must be frequently replenished. A refrigerated baffle over an oil diffusion pump provides a convenient compromise between dry trap operation and liquid nitrogen operation, and some operators now insert a cooling cartridge or cooling coil (e.g., Cryocool) into the conventional cold trap. A trap temperature of $-50°C$ is attained which is sufficient for most applications.

Sputter-ion pumps. One type of high-vacuum pumping system that is frequently considered for mass spectrometric work is the sputter-ion pump.[175] In these pumps, a fresh metal surface of titanium is continually sputtered from a cathode area at a potential of 3–5000 volts, and chemisorption of neutral gases occurs on this active film (Figure 4.9). These pumps can be

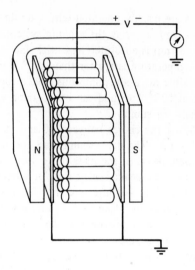

FIGURE 4.9. Schematic diagram of sputter-ion pump. Electrons spiral in the positive anode cavities causing ionization of gaseous molecules. The positive ions bombard the titanium cathodes and sputter off metallic titanium which deposits on the anode, trapping gaseous molecules.

purchased with almost any desired pumping speed from several manufacturers under the trade names "Vac-ion," "Drivac," "Orb-ion," "Magnetic ion," etc. They are widely used in standard mass spectrometry with great reliability and convenience, and because they operate without any pump fluid, mass spectral background from pump oils is absent. Sputter-ion pumps are not suitable for GCMS work for two reasons. First, the lifetime of the pumping element is directly proportional to the quantity of gas pumped, and second, ion pumps are not efficient for pumping rare gases. Even using a special rare gas pump, the pumping speed is low relative to the physical size of the pump.

If an ion pump is rated for 30,000 hours of operation at 10^{-6} Torr, only 1500 hours would be obtained at a typical GCMS analyzer pressure of 2×10^{-5} Torr. The element would have to be cleaned or replaced at nine-month intervals unless the system is used on a light work schedule. In a differentially pumped mass spectrometer, with a split ratio of 100/1 (third example, Table 4.1), the analyzer pressure of 5×10^{-6} Torr would be low enough to give 3–4 years of operation even with a heavy work load. Here, the ion pump would be quite satisfactory and would give a clean oil-free vacuum which is particularly advantageous in the electric sector of double-focusing mass spectrometers. However, a sputter-ion pump would not be acceptable as the pump for the ion-source housing. The operating lifetime of the pumping element would be only 150 hours at a pressure of 2×10^{-4} Torr. Furthermore, at this higher pressure the pumping speed is only 50% of the rated value.

Because of the chemisorption nature of ion pump action, rare gases are not efficiently occluded and specially designed pump elements are necessary. Even so, the pumping speed is noticeably reduced, and a rare gas ion pump rated at 140 liters/sec for nitrogen will have a speed of only 100 liters/sec for helium. This necessitates purchase of a physically larger pumping system than if a diffusion pump is used, but the additional size and extra cost are not usually important. If hydrogen is used as a carrier gas, the speed of the sputter-ion pump is increased but the lifetime is still short if used for ion source pumping. Thus, even for hydrogen, the ion pump would be suitable only on a differentially pumped analyzer tube.

Turbomolecular pumps. Another type of high-vacuum pump that is used in mass spectrometry is the Turbomolecular pump or molecular drag pump.[176] This remarkable mechanical device obtains pumping action by imparting a specific directional velocity to the gas molecules from high-speed rotors (Figure 4.10). The rotors are oblique slotted disks that rotate at 20,000 rpm between slotted stators. Each pair of rotors and stators forms a stage of pumping and with 15 to 20 stages, a vacuum level of 10^{-7} to 10^{-9} Torr can be achieved. Pumping speeds are available from 160 liters/sec up to 1600 liters/sec.

The turbomolecular pump offers many advantages. Since the system uses no pumping fluids, background contamination is absent. The pumped gas is removed from the system and cannot be released back into the vacuum chamber. (This is a frequent objection given for sputter-ion pumps.) All gases are pumped at a comparable rate. The pump is not damaged by accidental exposure to an atmosphere. Liquid nitrogen is not necessary to attain a high vacuum.

The only significant disadvantage of turbomolecular pumps is that the cost is 3–5 times higher than that of diffusion pumps of comparable performance. The pump will be suitable for most GCMS purposes except for use with high-resolution mass spectrometers which need vibration-free mounting.

The characteristics of the three pumps that can be used in GCMS are compared in Table 4.3. Although the initial cost of the turbomolecular pump is higher, the long-term operating cost may be lower, particularly if the

FIGURE 4.10. (a) Photograph of 260-liter/sec turbomolecular pump (Sargent-Welch Scientific Company Model 3103).

diffusion pumps are used continuously with liquid nitrogen.

4. Measurement of Vacuum Pressure

$P = 10^{-3}$ to 10^{-8} Torr.   Two types of gauge are commonly used for high-vacuum measurement in mass spectrometry.   These are the hot filament ion gauge[177, 178] and the cold cathode ion gauge.[179]

The hot filament ionization gauge produces a positive current of ionized gas molecules in essentially the same way as the electron bombardment source in a mass spectrometer.   Thermally emitted electrons are accelerated toward a loosely wound helical grid and pass through the grid into an ionization region (Figure 4.11).   Positive ions formed by ionizing collisions are collected on a cathode and the measured current is directly proportional to the pressure.

Either of the electrode configurations shown in Figure 4.11 is suitable for GCMS work, but the Bayard-Alpert gauge is used most often.   Actually, this is a mistake.   The function of the inverted electrode configuration is to reduce the photoelectric current emitted from the cathode and thus extend measurement to the 10^{-10} Torr region.   In GCMS work, measurement below 10^{-7} Torr is never needed.   On the other hand, the exact placement of the wire in the Bayard-Alpert configuration influences the gauge calibration.   Var-

TURBO-MOLECULAR PUMP

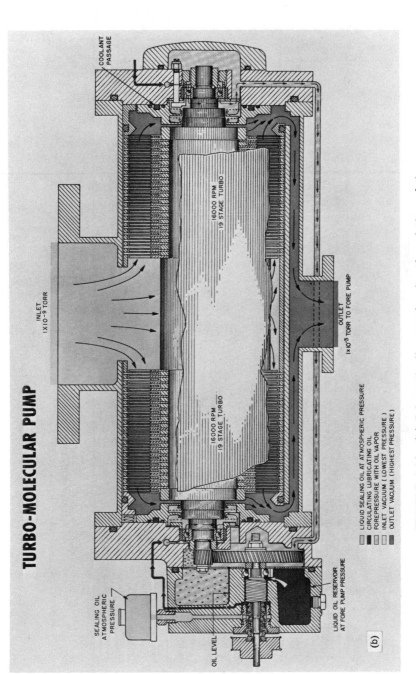

COOLANT
PASSAGE

INLET
1×10⁻⁹ TORR

16000 RPM
19 STAGE TURBO

16000 RPM
19 STAGE TURBO

OUTLET
1×10⁻³ TORR TO FORE PUMP

SEALING OIL
ATMOSPHERIC
PRESSURE

OIL LEVEL

LIQUID OIL RESERVOIR
AT FORE PUMP PRESSURE

LIQUID SEALING OIL AT ATMOSPHERIC PRESSURE
CIRCULATING LUBRICATING OIL
FOREPRESSURE WITH OIL VAPOR
INLET VACUUM (LOWEST PRESSURE)
OUTLET VACUUM (HIGHEST PRESSURE)

(b)

FIGURE 4.10. (b) Schematic diagram showing the relation of the rotors and stators in a turbomolecular pump.

TABLE 4.3. Comparison of operating characteristics of high-vacuum pumps.

	Possible pump fluid contamination	Cold Trap	Max. operating pressure, Torr	Normal lifetime at 10^{-5} Torr (He)	Pump, all gases equal	Cost, 160 liters/sec, $
Mercury diffusion	yes	essential	10^{-2}	Add Hg, 3 mos.; clean system, 15 mos.	yes	900[1]
Oil diffusion	yes	yes	10^{-3}	Clean pump, 12 mos.	yes	700[1]
Sputter-ion	no	no	10^{-4}	Change elements, 12 mos.	Noble gases, approx. 1/2 speed	2400[2]
Turbo-molecular	no	no	10^{-2}	No maintenance required	yes	3000

[1] Includes trap for oil diffusion pump and baffle and trap for mercury diffusion pump. Estimated liquid nitrogen cost, $1000/yr.
[2] Includes magnets and power supply.

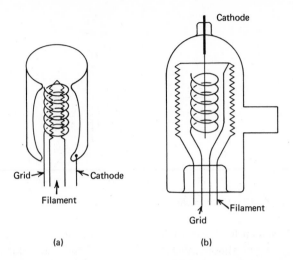

FIGURE 4.11. Hot filament ionization gauges. (a) Conventional electrode configuration.[177] (b) Inverted Bayard-Alpert design.[178]

iations occur due to lack of manufacturing tolerances and due to operation in nonvertical positions. Although the errors are not critical for day-to-day comparisons, the uncertainty can be avoided by using the conventional electrode system shown in Figure 4.11a.

The Penning or Phillips cold cathode ionization gauge attains electron emission by applying a high voltage between a cylindrical anode and two flat cathodes (Figure 4.12). A voltage of $2-4$ kV is used, but the total electron emission is considerably less than that given off from the hot filament. To increase the ionization efficiency, the electrode assembly is placed in a magnetic field of several thousand gauss which constrains the motion of the electrons to a helical path, thus increasing the probability of ionizing collisions. The sensitivity is comparable to that of the hot filament gauge.

The cold cathode gauge is considered to be more rugged than the hot filament gauge. Accidental exposure to a high pressure does no harm. The commercially available Penning gauges are all-metal and can be handled and operated with virtually no danger of damage. On the other hand, the hot filament ion gauge can be damaged by air or by an excessive exposure of column bleed. Damage from pump oil is encountered when cold traps are allowed to warm up for stand-by operation.

Unfortunately, cold cathode gauges tend to sputter and result in a short circuit which necessitates minor shutdown for gauge cleaning. More critically, cold cathode gauges often fail to start or fire at low pressures ($\sim 10^{-6}$ Torr), and the readings may be completely erroneous. This occurs most often in older gauges and has frequently resulted in the false belief

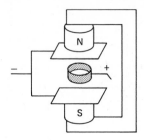

FIGURE 4.12. Cold cathode ionization gauge.

that a good vacuum level was attained when it was not. Generally, hot filament gauges give a more reliable measurement and many experienced mass spectrometrists insist on using them rather than a Penning gauge. Nevertheless, it is claimed that reliable operation is attained from a Penning gauge if it is cleaned every 3–4 months.

It is essential to remember that ionization efficiency depends upon the type of gas, and manufacturers' gauge calibrations are almost always given for air. When the principal gas in the system is helium, the meter reading will be 7–8 times lower than that obtained for air. When the gas is hydrogen, the reading will be about 2 times lower. These calibration factors vary slightly from gauge to gauge and if an accurate measurement is needed, calibration with the pure gas should be performed. For most GCMS work, a comparison with previously established operating conditions is sufficient. For reporting measurements with helium, the gauge reading should be multiplied by 7 to give a realistic estimation of the true pressure condition.

The location of the gauge in the vacuum line is important for a meaningful pressure measurement. The best results are obtained with the gauge located on a separate part of the ion source housing or analyzer tube, but it is also satisfactory for it to be right at the start of the pumping line. If the gauge is located further down the pumping line, the true system pressure should be calculated from the conductance of the line components. Since the quantity of gas Q is the same through all sections, the product $P_A C_A$ for a point A can be equated to the product $P_B C_B$ for any other point. For a well-designed system, the pressure at the pumping port will be about twice the pressure measured halfway down the line. If some component is poorly chosen, the true system pressure may differ significantly from the measured value.

Consider the system shown in Figure 4.2. If an ionization gauge is located just above the cold trap, the pressure in the mass spectrometer ion source housing P_H would be calculated from

$$P_H C_H = P_G C_{T,S}. \tag{4.13}$$

C_H is the total system conductance of 18 liters/sec as calculated from equation 4.5 (page 124). $C_{T,S}$ is the conductance for the trap/pump combination as calculated from equation 4.4 and for this example, $C_{T,S}$ would be 82 liters/sec. The pressure P_H in the ion source housing would thus be about 4.5 times greater than the pressure at the gauge, P_G. Because the system is pumping helium, a calibration factor of 7 must also be used, so that for this case, the gauge reading would be 32 times lower than the true pressure in the ion source housing. To obtain a mean free path in the ion source housing of 20 cm, the pressure should be 2.5×10^{-4} Torr as calculated from equation 4.2. The corresponding gauge reading at the trap would have to be 7×10^{-6} Torr on a gauge calibrated for air.

P = 10 to 10^3 Torr. Pressure measurement in this range is seldom important in conventional mass spectrometry. During an initial pump down, it is convenient to monitor the forevac pressure attained by the rotary pump, but it is not essential. The total quantity of gas that passes through the system is small, and most forepumps easily maintain the forevac level necessary for efficient diffusion pump operation. In some commercial mass spectrometers, the fore line gauge is omitted or replaced with an on/off discharge circuit. The manufacturers claim (erroneously) that this simplifies operation.

In GCMS systems, it is occasionally important to monitor the forevac pressure due to the larger quantities of gas that must be pumped. On page 133, it was shown that a forevac line and forevac pump could be operating at a marginal level. The pressure can be calculated from system parameters, but it is easier to measure it directly and thus assure efficient diffusion pump operation.

In addition, if a molecular separator is not operating efficiently (Chapter 5), careful attention must be given to the operating pressure. Although it is not conventional to equip these devices with pressure gauges, occasionally it may be necessary to put a gauge in the system if the separation efficiency appears to be low.

Pressure measurement in the range 10^{-3} to 10 Torr is simple and straightforward and either a Pirani gauge or a thermocouple gauge can be conveniently used. There are many reliable commercial suppliers. Both of these gauges operate on the principle of thermal conductance, so that the meter reading is not correct when helium or hydrogen is used. The calibration factor vaires considerably over the operating range, but fortunately it is seldom greater than 1.4–1.5 and can be ignored for most GCMS "ball park" measurements.

The Pirani gauge is generally more accurate and has a wider operating range. The thermocouple gauge is considered to be more rugged and is completely satisfactory for measurements in GCMS systems.

C. MISCELLANEOUS VACUUM HARDWARE

1. Vacuum Valves

High-conductance valves. During the course of standard mass spectrometric maintenance, it is frequently convenient to be able to isolate the pumping system from the mass spectrometer. Any valve used in this critical part of the pumping line must have a high conductance or the pumping efficiency will be severely reduced. Three types of valve that can be used are shown in Figure 4.13. For proper balance in a GCMS pumping system, the opening should be a minimum of 5 cm which would correspond to a valve helium conductance of 200–400 liters/sec, depending on the style of the valve.

The piston valve (Figure 4.13a) is most frequently used. It is available in various forms and sizes from almost all manufacturers of vacuum equipment. A welded bellows generally provides the stem vacuum seal. The seat may be soft metal or elastomer depending on the applications. These valves

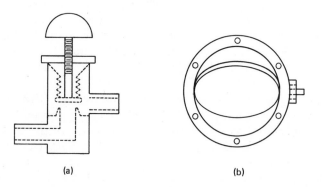

(a)   (b)

FIGURE 4.13. High-conductance isolation valves. (a) Piston valve. (b) Butterfly valve.

FIGURE 4.13.   High-conductance isolation valves. (c) Gate
valve. (Thermionics Laboratory, Inc., Hayward, California)

give reliable maintenance-free service, but because of the relatively long conductance path from flange to flange, a large valve is needed to assure minimum flow resistance. Typically, a 5 cm stainless steel piston valve would have
a conductance for helium of 200 liters/sec. The approximate cost would be
about $250.

The butterfly valve (Figure 4.13b) provides a convenient way to close
a vacuum line without introducing noticeable flow resistance. When this type
of valve is placed in a pumping line of the same diameter, there is virtually no
change in the overall conductance. However, butterfly valves are not available

in a wide selection of sizes. They also have the disadvantage that the compression seal at the valve stem can be an annoying source of vacuum leak. The cost of a 5 cm stainless steel butterfly valve is around $200.

Gate valves (Figure 4.13c) have the same advantage as the butterfly valve in that they do not significantly reduce the pumping line conductance. They can be purchased from several manufacturers with 5, 7.5, 10, and 15-cm openings so that the dimensions of the pumping system can be properly matched. If an aluminum cast body is acceptable, the cost of these valves is about the same as a stainless steel butterfly valve, but if stainless steel construction is demanded, the cost will be about three times higher. This style of valve is quite popular but, like the butterfly valve, it has the disadvantage of possible vacuum leaks at the stem seal. In the more expensive stainless steel gate valves, the seal motion is transmitted through a bellows that gives leak-free operation.

Although an isolation valve should always be provided to shut off the diffusion pump and cold trap, mass spectrometer manufacturers often omit this feature for the sake of convenience and economy. In the long run, the omission causes considerable operational inconvenience, particularly for ion source changes. Unfortunately, it is often difficult to add an isolation valve after the mass spectrometer is designed and assembled.

Flow control and metering valves. Selection of a metering valve for control of effluent gas is one of the more common problems encountered in setting up a GCMS interface. There are a considerable number of needle valves and metering valves available at prices that range from $25 to over $500. Because of the variety of applications that arise, price is not the important consideration, and indeed, the more expensive valves are generally for precision metering which is seldom needed.

The important considerations in selecting an interface control valve are (1) avoiding dead volumes that cause loss of chromatographic efficiency, (2) possible seizure or galling of the moving parts due to dry operation at elevated temperatures, and (3) the valve must be capable of providing the flow control needed for the application.

Needle valves are widely used in GCMS interface applications and, depending upon the stem point style, can serve most of the interface functions. Several common point styles are shown in Figure 4.14. In a typical GCMS interface, a fine metering valve (a) is often used at the exit of the column to control the flow of total effluent to the separator or mass spectrometer. Sometimes it is convenient to have a throttling or regulating valve (b, c) to isolate the separator and mass spectrometer when interface components are changed. A throttling valve is satisfactory but the vacuum sealing seat can be

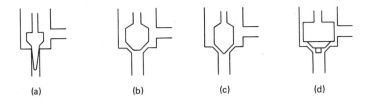

FIGURE 4.14. Needle valve stem point styles. (a) Fine metering.
(b) Throttling. (c) Regulating. (d) Elastomer soft seal.

damaged by excessive torque or cycling. The elastomer seat (d) is preferred
when the operating temperature is not excessive, but elastomer valves cannot
be used for flow regulation due to slow creep recovery of the seal deformation.

A bellows stem seal valve should be used whenever possible since frequent temperature cycling of GCMS interface lines and ovens causes leakage
of packed stem valves. If dead volume restrictions dictate the use of a packed
stem, the packing should always be suspected when interface leaks develop.

Several types of needle valves are shown in Figure 4.15. Although the
selection is taken from only a few sources, most manufacturers of miniature
valves or fine-metering valves have similar choices, and a valve designed for a
specific purpose can be duplicated by different suppliers. The three valves
designated a, b, and c in Figure 4.15 can be used for throttling or sealing on
the vacuum side of the GCMS interface. They would not be practical for
metering small quantities (0.2–2.0 cm^3 atm/min) of effluent gas from the
column exit at one atmosphere pressure. For one thing, it would be difficult
to get the necessary flow control but more important, as can be seen in the
diagrams, these valves have a dead volume in the proximity of the tip which
may be 0.05–1.0 cm^3. The exponential sweep-out time constant of this
dead space is the ratio of the volume to flow rate (V/C as in equation 2.4),
and 6–8 characteristic time periods are needed for acceptable sweep-out.

Suppose 18 cm^3 atm/min is to be metered from a column effluent
into a separator. A packed stem valve such as is shown in (a) would have
about 0.1 cm^3 dead volume and the characteristic time constant would be
0.1/18 min or 0.3 sec. A reasonable, clean sweep out would be expected in
about 2 sec. Thus, the metered flow of 18 cm^3 atm/sec is close to the minimum for this valve. A bellows or diaphram valve, such as (b) or (c), would
have dead volume 5–10 times larger and could not be used at atmospheric
pressure without serious loss of chromatographic efficiency.

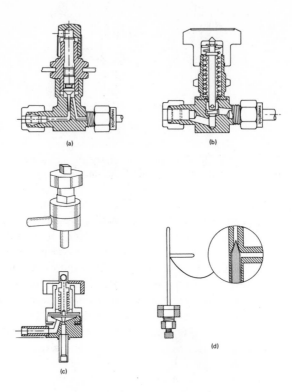

FIGURE 4.15. Examples of typical metering valves for GCMS interface applications. (a) Nupro 1/M fine-metering valve with 0-ring stem seal (cross section Nupro straight pattern). (b) Nupro 1/M throttling valve with bellows stem seal (stainless steel). (c) Hoke 1/M regulating valve with diaphram stem seal. (d) Precision Sampling Corporation microcapillary valve with packed stem seal.

The approximate cost of stainless steel valves in styles (a), (b), and (c) is $20–$50. Unless a large number is purchased, the bellows type (b) is preferred over the packed stem valve. For use in vacuum, the conductance C is much greater so that the dead volume time constant V/C is often insignificant. The diaphram valve (c) is quite acceptable, but tends to have a shorter lifetime than a bellows valve.

Several fine-metering valves are available that provide easy control of flow in the range 0.2–20 cm^3 atm/min. These valves can be obtained with

packed stems or with bellows. For many applications, the choice depends
upon the dead volume restrictions, and a simple valve such as is shown in (d)
is often preferred. The dead space in this valve is negligible. A packed stem
valve such as (a) may have about 0.1 cc dead space, but bellows valve often
has as much as 0.9 cc dead space.

A micrometer action can be obtained on most metering valves and may
be useful for close control. However, the fine tolerance of micrometer
screws often results in metal seizure at high temperatures, so that these valves
should be used only under circumstances in which the micrometer threads can
be kept cool.

All needle valves should be handled carefully, especially fine-metering
valves such as (a) and (d). If a needle valve fails to seal tightly, excessive
torque will only cause further damage to the seat, and it is better to tolerate
a minor leak than to have to replace the valve completely. A delicate valve,
such as (d), will give years of service if properly used, but untrained personnel can ruin it with one overly tight closure.

Valves with Viton O-rings or sealing tips can be operated up to 150°C.
Teflon gaskets, etc. are useful up to 230°C without excessive outgassing.
For higher temperatures, all-metal valves should be used. At high temperatures,
metal seizure is a common failure, but the risk can be minimized by judicious
use of Molykote or some similar high-temperature lubricant.

Other metering valves are available for specialized applications. Needle
valves with excellent precise flow control are made for use as a variable leak
in a batch inlet. In another style of valve, a plunger is moved down a close-
fitting barrel to attain precise delivery of very small quantities of gas. These
specialized valves are well made and can be used at high temperatures. However, they are not usually designed for metering small quantities of gas at
higher pressure and the dead volume is often prohibitive. The cost of such
specialized valves may be $200–$600.

High-temperature, inert material valves. One of the most difficult problems in GCMS is to avoid decomposition of high-boiling reactive materials
typically encountered in analysis of biological chemicals. Silylation
of active sites is commonly employed and in extreme cases, an all-glass
system is used. Unfortunately, there are very few all-glass valves for use in
the GCMS interface, and operational compromise is often necessary.

Metering an effluent flow through an inert interface is most commonly
performed using a short calibrated section of glass capillary tubing. Although
a shut-off valve before the capillary is convenient, it is usually omitted,
primarily because no suitable valve is available. The major inconvenience of
metering by a calibrated capillary is that the quantity of gas delivered is a

function of temperature and cannot be varied. Furthermore, it is difficult to join a fine capillary tube into a system and get the exact desired flow. When it is necessary to change the interface flow of, say, 18 cm^3 atm/min from a packed column to 5 cm^3 atm/min from a capillary column, the whole interface/column system must be shut down and refabricated.

An all-glass metering valve can be made by having a close-fitting glass rod slide down a piece of glass tubing. Because close tolerances are needed for good flow control, this type of valve is very difficult to construct and gives a lot of trouble during operation. A Teflon sleeve must be used as a motion seal and limits the operating temperature to about 250°C. Such a difficult solution would be chosen over the simple calibrated capillary only if frequent gas chromatography flow changes are anticipated.

Even though an isolation valve is not essential between the interface and the end of the column, it is very convenient to have one between the interface and the mass spectrometer. At this point, the gas is at a fairly high vacuum so that the volume flow rate is more than 100 times faster than at the interface entrance. Consequently, dead-space volume is of minimal concern. A solenoid-activated glass valve such as is shown in Figure 4.16 can be used. These valves hold a vacuum of about 10^{-5} Torr against one atmosphere. The solenoid activation is generally not strong enough to open the valve against one atmosphere, so that to start it up, it is necessary that a rough vacuum can be obtained on the interface side or the atmosphere will

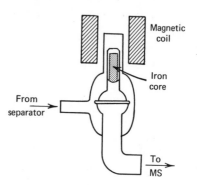

FIGURE 4.16. All-glass isolation vacuum valve. The valve can be used to isolate the mass spectrometer from the GCMS interface. Because of the large dead volume, it cannot be used between the chromatograph and the GCMS interface.

hold it closed. The large dead volume precludes applications anywhere that a vacuum is not used.

Teflon or Teflon/glass valves can be purchased or fabricated in several styles and, if Teflon parts can be accepted in the interface, this solution to high-temperature decomposition should be considered. Other techniques, such as applying a gold plate or ceramic coat to the valves and lines, are usually difficult to perform and commercial parts are available only in very limited design.

Interface bypass valves. Large quantities of solvent or major components of the sample are inevitable during GCMS trace analysis and frequently, the amount of organic material going into the mass spectrometer may exceed a tolerable maximum. Whenever more than 10^{-7} gm/sec of organic material passes through the ion chamber, some impairment of mass spectrometric performance can be expected. The ion gauge current may increase to trip the protective devices. The filament emission may decrease due to temporary carbonization. Adsorption of the solvent may result in a lingering excessive background.

If a metering valve is used at the column/interface junction, manual adjustment of the valve can reduce or eliminate the excess sample. If the effluent split to the separator is metered by a length of stainless steel capillary, a switching valve such as is shown in Figure 4.17 should be employed. A good selection of rotary shear valves of this type is commercially available and more complex switching patterns can be used if necessary. However, as a general rule, the GCMS interface should be kept as simple as possible.

A rotary shear valve or any valve used for bypass purposes at atmospheric pressure should have an effectively zero dead volume. When metal surfaces cannot be tolerated, a simple three-way Teflon barrel glass stopcock may suffice at temperatures up to 230°C.

2. Vacuum Couplings

The design or modification of a GCMS interface necessitates knowledge of available vacuum couplings. The familiar Swagelok or Gyrolok fittings (Figure 4.18a) are widely used for gas chromatography couplings and as a result, they are also applied frequently in the vacuum connections of the GCMS interface. When properly assembled, these couplings give a strong tight vacuum seal. The joints can be heated higher than 300°C, although Teflon ferrules will flow excessively and fail. No welding or brazing is necessary to form the union and a versatile selection of components facilitates coupling of all common sizes of tubing.

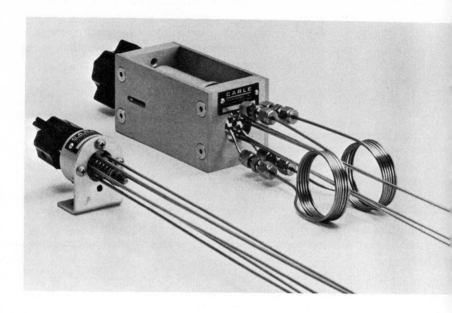

FIGURE 4.17. Rotary shear valves with zero dead volume for use in the GCMS interface at atmospheric pressure (Carle Instruments, Inc.).

Swagelok-type fittings have the disadvantage that after a few reassembly cycles, vacuum failure becomes probable unless great care has been taken in the process of connection. Excessive torque will, at best, give only a partial reduction of the vacuum leak and may lead to a broken part or a pressure weld of the mating flange and ferrule. In such an unhappy event, the component must be forced open and reconstructed, which involves annoying time loss.

The Cajon ultra high-vacuum coupling is another type of seal that can be applied for many GCMS fabrications (Figure 4.18b). The Cajon coupling has the advantage that it can be reassembled as often as is necessary without damage. No translational motion is necessary for component removal, i.e., the tubing does not have to be inserted or pulled back (as is necessary to join or uncouple a Swagelok fitting). Furthermore, if the seal fails, excess torque will usually tighten the leak but if not, the gasket can be easily replaced.

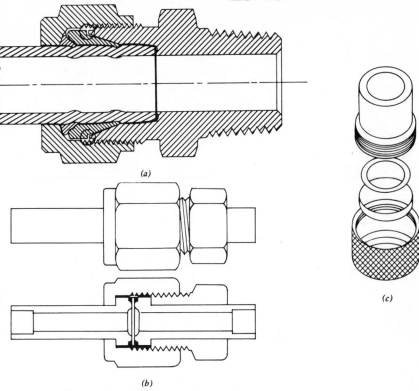

(a)

(b)

(c)

FIGURE 4.18. Vacuum couplings commonly used in GCMS sys-
tems. (a) Swagelok or Gyrolok type (Swagelok Company or Hoke
Manufacturing Co.). (b) Cajon ultra high-vacuum coupling (Cajon
Corporation). (c) Cajon ultra-Torr coupling.

The one major disadvantage of the Cajon fitting is that the body parts
must be welded or hard soldered onto the system tubing. In spite of very
superior performance, this drawback has restricted the use of the Cajon couplings, and the Swagelok type have been highly favored. Nevertheless, when
excessive reassembly of components is expected, consideration should be
given to adapting to the Cajon fitting.

If all-metal joints are not essential, the Cajon 0-ring vacuum coupling
can be used. The only significant advantage of this fitting over the all-metal
one is that a vacuum seal can be obtained with much lower torque. Because
the Viton 0-ring is restricted to applications below 200°C, the metal seal is
more commonly used.

There are also many forms of quick-fit seals that can be applied for
easy GCMS modifications. Most of these fittings use an elastomer gasket, so

that they are restricted to lower-temperature applications. Viton or Teflon can be heated to 200°C and 250°C, respectively, but the compressed gaskets flow badly and cause frequent failure. The Cajon ultra-Torr (Figure 4.18c) is a typical easy-to-use vacuum fitting that can be applied with metal, glass, or plastic tubing. Good vacuum seal is obtained with finger pressure torque.

A variety of all-Teflon couplings, stopcocks, and miscellaneous vacuum fittings are available that have a limited function in the GCMS interface. Occasionally, these parts may be preferred for inertness in the higher-temperature ranges up to 230°C. They are not usually reliable vacuum components over a long term and should be used cautiously on the high-vacuum side of the interface. Their main application is for coupling inert glass systems.

5

The GCMS
Interface

A. INTRODUCTION

Whenever it is necessary to combine two operating systems that function well as separate units, the interface of these two systems becomes the most critical stage of the combined operation. This self-evident truth is probably an extension of "Murphy's Law," which states that "if anything can go wrong, it will go wrong." Murphy only forgot to say where. Some well-known examples of interface problems are accountant/computer, engineering/construction, manufacturing/sales, and administration/faculty. Interface problems of the GCMS system are no exception.

The simple diagram shown in Figure 5.1 is the only easy way to solve the problem, namely, draw a block, label it interface, and assume that the other experts can construct the black box that will perform the necessary miracle. It is too easy to find specialists who will assume full responsibility for their discipline but use the interface as a scapegoat for all that can go wrong. To be an "interface specialist," whether in science, sociology, or business, it is necessary to know the operational problems on both sides of the system and to understand which functions are compatible and which are not.

157

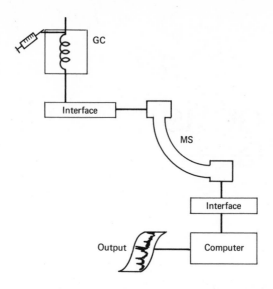

FIGURE 5.1. Schematic diagram of interfaces in the GCMS-Computer system.

B. THE GCMS INTERFACE WITHOUT A MOLECULAR SEPARATOR

1. Standard Vacuum System

The simplest way to introduce gas chromatographic effluent into a mass spectrometer is to split the carrier gas stream at the exit of the column and introduce the small amount of gas that the mass spectrometer can accept directly into the ion chamber. The major portion of carrier gas continues to the gas chromatographic detector. This simple technique was used for the first few years of GCMS analysis[1, 12–14, 181, 182] and, within certain limits of dynamic range and sample size, it is still the most convenient and often the most successful setup for many applications.

Figure 5.2 shows variations that were commonly used in the early GCMS period around 1960. The interface is a simple fine-metering needle valve, similar to those illustrated in Figure 4.15a and d. In Figure 5.2a, the effluent split occurs at a "T" and the quantity flowing to the mass spectrometer is metered by the needle valve. Since the flow through the valve is low (say, 0.2 cm^3 atm/min into a single-pumped mass spectrometer), the time

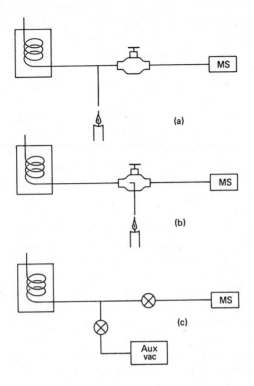

FIGURE 5.2. Schematic diagram of the simplest GCMS interface. The effluent flow is split into two streams and the allowed quantity is directed to the mass spectrometer.

taken to sweep out a 0.06-cm^3 line volume between the "T" and the needle point is about 20 sec. If smooth laminar flow occurs down the tube, peak broadening is not significant, but the approximate 20-sec delay between sample arrival at the chromatographic detector and arrival in the mass spectrometer is sometimes annoying. If eddy currents occur due to irregular pockets in the line, the peak broadening would be excessive. In Figure 5.2b, the split is obtained by drilling a hole through the body to the needle point and welding a piece of tubing onto the valve. The delay volume is thereby completely eliminated.

In both (a) and (b), the valve stem, which might have a dead volume of 0.1–1.0 cc, must be on the mass spectrometer side where the pressure is reduced by several orders of magnitude. The volume flow on the vacuum side is greatly increased and, since $P_A C_A = P_B C_B$, any dead volumes or eddy cavities are rapidly swept out. Even a bellows-type valve with a large dead volume is usually suitable when used in this fashion.

In the third variation (Figure 5.2c), the column outlet pressure is reduced to less than one Torr by applying an auxiliary vacuum to the bypass line.[183-184] The pressure drop occurs primarily across the chromatographic column. To maintain the desired average linear flow through the column, the inlet pressure must also be reduced by one atmosphere. This poses no problem if the standard operating pressure drop across the column is greater than one atmosphere. If it is less than this value, the chromatographic injector and associated fittings must operate at a negative pressure relative to atmospheric pressure, and an additional throttle is placed in the carrier gas line between the supply reducing valve and the chromatograph. These complications usually rule out this technique for use with low pressure-drop columns.

Several advantages are attained when the reduced pressure interface is properly used. (1) Because of the reduced pressure, the volume flow is high, and dead volumes and line volumes are rapidly swept out. Consequently, the interface is easily designed and constructed from standard vacuum fittings and valves. (2) The reduced pressure facilitates control of the flow into the mass spectrometer. Easy metering precision can be attained with simple regulating valves such as Figure 4.15b and c. (3) The total quantity of gas flowing through the chromatographic column is reduced, and in effect, this provides an enrichment of the sample relative to the carrier gas.

From the Poiseillue equation 4.6, the quantity of carrier gas flowing through the column is proportional to the difference of the squares of the inlet and outlet pressures. The ratio of the quantity of carrier gas used under a vacuum condition Q_{vac} to that used under a standard condition Q_{st} is therefore given by

$$\frac{Q_{st}}{Q_{vac}} = \frac{(P_i^2 - P_o^2)_{st}}{(P_i^2 - P_o^2)_{vac}} \tag{5.1}$$

where the subscripts "st" and "vac" refer to standard or vacuum operation. P_i and P_o are the inlet and outlet pressures. If a column is normally operated with 1.3 atmosphere gauge inlet pressure (total of 2.3 atmospheres) and one atmosphere outlet pressure, the corresponding values in the vacuum interface will be 0.3 atm gauge and zero (1 Torr or less). The ratio Q_{st}/Q_{vac} would be $(2.3^2 - 1^2)/(1.3^2 - 0) = 2.5$. This means that for the same linear flow through the column, the vacuum technique would need only 0.4 times as much carrier gas.

The main disadvantage of this interface style is that operation of the column exit under vacuum necessitates readjustment of the chromatograph flow conditions, and the front retention time of an air peak must be matched for the two conditions. The main advantage is freedom from eddy pockets in the interface and hence simplicity of construction. The enrichment factor is particularly useful for low-flow capillary columns since it often permits utilization of up to 50% of the sample without an enrichment device.

Single-stage pumping. The quantity of effluent that can enter the mass spectrometer depends upon the size and mode of the pumping system. A single-pumped mass spectrometer can seldom take more than $0.1-0.3$ cm^3 atm/min unless an exceptionally large, well-designed pumping system is used. Thus, for 0.025-cm ID capillary columns, the simple splitter interface can deliver 10–30% of the total effluent which is sufficient for many mass spectrometric identifications. When the vacuum mode interface is used, the enrichment gives an additional factor of two or more, so that around 50% of the sample can be utilized.

If a 0.32-cm OD packed column is used at a flow of 20 cm^3 atm/min, the mass spectrometer can take only about 1% of the total material. For trace analysis, such a considerable loss of efficiency cannot be tolerated, and an enrichment device is essential. Similarly, if the unknown sample is in a very dilute solution, low efficiency at the GCMS interface must be avoided. However, for many GCMS analyses, a low conversion factor is quite acceptable and because of the simplicity of the splitting valve, some operational advantages are achieved. All the objectives of the analysis should be considered before ruling out the simple method because it is inefficient.

Consider the chromatogram shown in Figure 5.3. The three peaks labeled A, B, and C represent 10%, 1%, and 0.1% of the total mixture. If a simple GCMS interface with only 1% conversion is used, 10^{-3} gm injected into the packed column will deliver 10^{-5} gm of total mixture to the mass spectrometer. A 10% peak eluted with a half-width time of 30 seconds enters the mass spectrometer at a rate of 3×10^{-8} gm/sec at the top of the peak. This amount gives a very strong mass spectrum. The 1% peak corresponds to 3×10^{-9} gm/sec and the resulting spectrum is sufficiently intense for most purposes. The 0.1% peak, corresponding to 3×10^{-10} gm/sec, gives a weak spectrum but the interpretation may be obscured. Frequently, parent ion peaks or other important indicators are not recorded at this level, and identifications made with sample quantities in the ranges 10^{-10} to 10^{-12} gm/sec depend fortuitously upon the nature of the spectrum. (See page 76.)

Because of the high basic sensitivity of the mass spectrometer, a con-

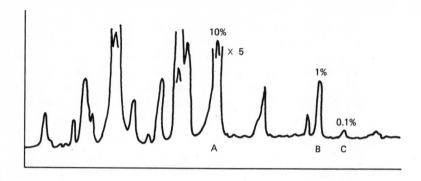

FIGURE 5.3. Simulated chromatogram illustrating the relative importance of small peaks.

siderable range of GCMS analyses can be performed with a simple inefficient splitter interface. The advantages obtained are: (1) simplicity of construction, (2) low cost, (3) easy operation, and (4) minimum exposure of sample to hot active surfaces. Enrichment devices have an important function when properly used to extend the dynamic range of GCMS analyses. However, it is also unfortunately true that many GCMS analyses that didn't need an enrichment device have failed because of thermal decomposition of sample in a complex separator, or due to improper control of the separator vacuum system.

Differential pumping. The most effective method of increasing effluent utilization is with a differentially pumped mass spectrometer (Figure 4.1). By dividing the effluent flow so that only a small fraction (1–10%) enters the analyzer tube, a typical pumping system easily maintains the necessary vacuum level, and up to $1-2$ cm^3 atm/min of the effluent can be utilized (see Tables 3.1 and 4.1). For most experimental conditions, this will mean a gain of at least a factor of 10 over the volume that can enter the single-pumped mass spectrometer.

Any of the simple valve interfaces (Figure 5.2) can be used advantageously with a differential-pumped mass spectrometer. The increased effluent utilization makes it possible to obtain a mass spectral pattern down to the level of 200–300 ppm from a 0.32-cm OD packed column. There are only a few analytical situations that need more efficient use of sample. When these occur, additional gain can be achieved by using a high-speed pumping system or a molecular separator.

2. High-Speed Vacuum System

The recent trend in GCMS interface techniques utilizes a large, high-speed vacuum system on the ion source housing so that as much as 10^{-1} liter Torr/sec (6–8 cm^3 atm/min) of the effluent can pass through the ion chamber. Differential pumping is used on the analyzer section and with a split of 100 to 1, a vacuum of 2 x 10^{-5} Torr is attained with a 50 liter/sec analyzer pumping system. The ion source housing pressure is maintained below 5 x 10^{-4} Torr with a pumping speed to 200 liters/sec (see Table 4.1).

Superficially, a conductance rating of 200 liters/sec appears to be easily attained, but it is important to consider the physical dimensions of the pumps and pumping line that give this performance. One possible configuration is shown in Figure 5.4. The high-vacuum pump is a 10-cm diffusion pump with a speed rating of 750 liters/sec. The other components are chosen to match well with the pump, and the pumping line has been kept at a minimum length consistent with a side port pumping flange. The conductance of a 9-cm orifice is close to 1900 liters/sec for helium, but for a tube length of 45 cm, the tube conductance is only about 580 liters/sec. The net system conductance as calculated from equation 4.3 would be 285 liters/sec for helium.

This discussion shows how much the performance of a pumping system is reduced even when large, high-conductance components are used. If the pumping flange in Figure 5.4 is at the bottom of the ion source housing, the lead to the cold trap could be 18 cm or less and the conductance for helium would then be close to 400 liters/sec. Details of a commercially available large pumping system are shown in Figure 5.5.

The cost for the components shown in Figure 5.4 is around $2000–$2400 exclusive of construction and assembly. The same equipment assembled from 5-cm diameter components would cost about $1000. Thus, the additional $1000 in initial investment buys a factor of 3–4 in sample utilization. Unfortunately, modification of an existing system can seldom be performed for less than $4000. At the present time, several mass spectrometer manufacturers offer a high-speed vacuum assembly as optional equipment for GCMS studies. (Dupont Instrument Products, Bendix Scientific Instruments, Finnigan Corporation, and Nuclide Analysis Corporation.)

A high-conductance differentially pumped system enables utilization of virtually all the effluent sample from a 0.025-cm and 0.05-cm capillary column and close to 75–80% of the sample from a 0.075-cm capillary column. With a 0.32-cm OD packed column operating at a flow rate of 20 cm^3 atm/min, the sample utilization would be 35–40% and further gain will seldom be of any value. Whenever a simple interface is desired to minimize possible thermal decomposition of delicate samples, the large pumping line

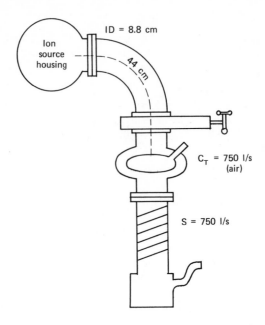

FIGURE 5.4. Illustration of the size of typical components in a large GCMS pumping system.

is quite advantageous. Further increase in sample utilization by this technique is limited by the pressure allowed in the ion source (Table 4.1).

C. THE GCMS INTERFACE WITH MOLECULAR SEPARATORS

1. General Considerations

From the previous discussion it is clear that many GCMS analyses can be performed satisfactorily without the use of enrichment devices. The success of a simple needle-valve interface will depend however, upon the objectives of the analysis and upon the type of vacuum equipment in use. Many situations arise where, due to deficiencies in the pumping arrangement or because of a very wide dynamic range for analysis, the simple interface is not sufficient. Improved performance must then be obtained by using a fractionation technique that increases the ratio of sample to carrier gas in the stream that goes to the mass spectrometer.

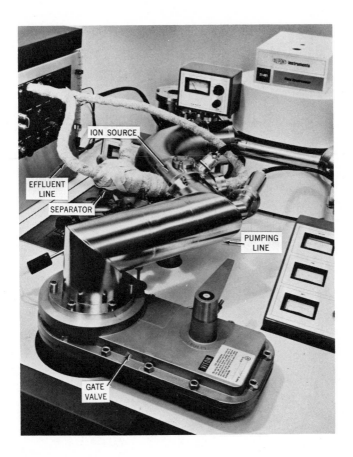

FIGURE 5.5. Photograph of a high-speed pumping system for GCMS analysis. (DuPont Instrument Products, 21–491 mass spectrometer.)

The several types of enrichment devices that have been used are called molecular separators. The processes of enrichment may be classified into three categories: (1) enrichment by effusion through fine pores or a narrow slit; (2) preferential diffusion of carrier gas or of sample through a semipermeable membrane; and (3) fractionation of gases in an expanding jet stream. These various classifications are tabulated for most of the separators in common use in Table 5.1. In addition to those listed, most separators can be used in pure or hybrid two-stage combinations.

TABLE 5.1. Classification of commonly used molecular separators.

		Reference
Effusion separators		
	Sintered glass frit	Watson-Biemann; Markey;	101, 199, 185
		Copet-Evans	
	Stainless steel sinter	Krueger-McCloskey	49
	Porous silver frit	Cree	186
	Porous silver frit	Blumer,	187
	Variable conductance slit	Brunnee-Bulteman-Kappus	188
	Porous silver microseparator	Grayson-Wolf	200
Semipermeable membrane			
(a) carrier removal	Teflon separator	Lipsky-McMurray-Horvath	189
through membrane	Hydrogen-palladium separator	Simmonds-Shomake-Lovelock	190
	Hydrogen-palladium separator	Lucero-Haley	191
(b) sample passage	Silicone polymer	Llewellyn-Littlejohn	192
through membrane	Silicone polymer	Black-Flath-Teranishi	193
Jet orifice			
	Stainless steel	Ryhage-Stenhagen	194
	Glass	Story	195

Some types of separators offer advantages for particular operating conditions or specific samples, and flexibility should be built into the initial setup to permit equipment changes. Unfortunately, many GCMS instruments are operated exclusively with one separator (generally as supplied by the manufacturer), and changes or modifications are difficult. This flexibility is of little importance when the equipment is used essentially on one type of sample. If a wide range of samples is anticipated, provision should be made to facilitate substitution of different separators or a needle-valve interface when appropriate.

The operational parameters used to evaluate the performance of separators are the separation factor and the yield or efficiency.[196] The separation factor N, also often called the enrichment factor, is defined as the ratio of sample concentration in the carrier gas entering the mass spectrometer to the sample concentration in the carrier gas coming out of the chromatographic column. Thus,

$$N = \frac{c_{MS}}{c_{GC}} \quad \text{or} \quad N = \frac{(P_S/P_{He})_{MS}}{(P_S/P_{He})_{GC}} \tag{5.2}$$

where c_{MS} and c_{GC} are the respective sample concentrations or P_S and P_{He} are the partial pressures.

Because the enrichment factor is a ratio, all concentration units give the same result. When units of mole fraction are used, equation 5.2 is the same as the "simplified separation factor" defined by Ryhage[197] namely, $S = N_e/N_0$ where N is the mole fraction. For GCMS studies, gm of sample/ cm^3 atm of carrier gas are frequently the more convenient units.

Other definitions of enrichment factor are sometimes used, and care must be taken to avoid comparing data that have the same name with different meanings. In one of the earlier papers on separators,[198] ten Noever de Braw and Brunnée defined an enrichment factor A as

$$A = \frac{N_e}{N_0} \sqrt{\frac{M_S}{M_c}} \tag{5.3}$$

where M_S and M_c are the molecular weights of the sample and carrier gas and N_e and N_0 are the mole fractions. Later, Brunnée et al. conformed to the definition of enrichment given by equation 5.2[188] As defined by equation 5.3, the enrichment factor may be 4–10 times the value given by equation 5.2. Furthermore, as defined by equation 5.3, the enrichment factor contains the term $\sqrt{M_S/M_c}$, so that a plot of A vs. the molecular-weight ratio M_S/M_C will be linear. This is misleading and the plot of enrichment vs. the molecular-weight ratio is more realistically a square root function. (See equation 5.14 and Figure 5.9.)

Another definition of enrichment factor is "the ratio of detectability... with the helium separator versus the detectability...through the dismantled separator."[186] Because the term "detectability" was not clearly defined, and because the operation conditions were not specified, the reported enrichment factors cannot be compared with those from other studies.

The separation or enrichment factor varies a great deal depending on the type of separator, the gas chromatographic flow, the vacuum system efficiency, and the molecular weight of the sample. For single-stage effusive-type separators and jet separators, N may be as low as 1.1 and as high as 20. For separators that use a semipermeable membrane, N may be as low as 3–4 or as high as 10^3. An enrichment factor should be specified as due to a single stage of enrichment, and the flow parameters for that stage must be clearly stated. The enrichment of a two-stage separator will be the product of the enrichment factor for each stage.

The separation factor is an important parameter to evaluate the fractionation efficiency, but it is so dependent upon the flow and vacuum conditions that it is not always a significant indication of the success of the experiment. A large separation factor does not mean that a large percent of total sample gets into the mass spectrometer. Consequently, the efficiency or yield Y, is more important for evaluation of the overall system. Efficiency is very simply defined as the percent of the total sample that enters the mass spectrometer. Thus,

$$Y = \frac{Q_{MS}}{Q_{GC}} \times 100\% \tag{5.4}$$

where Q_{GC} and Q_{MS} are the quantity of sample that leaves the chromatograph and the quantity that enters the mass spectrometer.

Although the definition of yield is independent of any separation process, it is algebraically related to the separation factor by the equation,

$$N = \frac{Y}{100} \times \frac{V_{GC}}{V_{MS}} \tag{5.5}$$

where V_{GC} and V_{MS} are the respective carrier gas volumes measured at 760 Torr. From equation 5.5 it is clear that the maximum theoretical value for N occurs with a 100% yield and is inversely proportional to the fraction of total carrier gas (V_{MS}/V_{GC}) that enters the mass spectrometer. A 100% separator yield can be a reality only for the hydrogen/silver-palladium separator in which case N may be very high (page 206). For any other separators in current use, some sample is always lost in the separation process.

Occasionally, N values are quoted in the literature without specifying the significant volume ratios and as such, the data cannot be compared with any other system. In addition, some reports quote N values that exceed the theoretical maximum given by the volume ratio. Such a discrepancy may be due to differences in definition of terms, and greater care must be taken to assure definition of operational parameters in the conventional manner. Errors might also arise in measuring the small quantity of gas entering the mass spectrometer or in comparing the mass spectral signal obtained with and without the separator. Some of the techniques for determination of Y can have a discriminating effect on the results, and it is important to appraise the evaluation methods with respect to specific operating conditions (see page 206).

2. The Effusion Separator

The porous glass separator. One of the first separator systems to be widely used was the glass frit separator introduced by Watson and Biemann in 1964.[101] This apparatus, shown in Figure 5.6, consists of an ultrafine porosity sintered glass tube enclosed in a vacuum envelope with glass capillaries at the entrance and exit to provide flow restriction. The effluent from the gas chromatograph enters through the entrance restrictor, causing a drop in pressure in the sintered tube to about one Torr. The effluent gas is split into two streams. One of these streams effuses through the porous tube; the other enters the mass spectrometer.

If the pressure of the gas in the sintered glass is sufficiently reduced to permit molecular flow through the pores, the rate at which a gas effuses to the exhaust vacuum will be inversely proportional to the square root of the molecular weight and directly proportional to the partial pressure of each component. Thus, the quantity of any gas Q going through the porous glass is given by

$$Q = kp\sqrt{1/M} \qquad (5.6)$$

where p is the partial pressure of the component, M is the molecular weight, and k is a constant that depends on the conductance of the porous tube. (The pressure on the outside of the porous tube is assumed to be sufficiently reduced so that back effusion is negligible.) The ratio of the quantity of

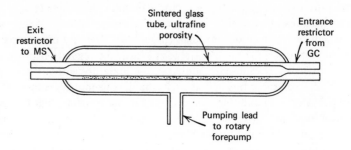

FIGURE 5.6. Watson-Biemann effusion separator.[101]

sample Q_S to the quantity of helium carrier gas V_{He} that goes through the frit is

$$\left(\frac{Q_S}{V_{He}}\right)_{frit} = \frac{P_S}{P_{He}} \sqrt{\frac{M_{He}}{M_S}} \qquad (5.7)$$

where V_{He} is the quantity of carrier gas conveniently expressed in units of cm^3 atm/min. Consequently, a fractionation of sample to carrier gas is obtained which depends upon the inverse ratio of the square root of the molecular weights.

For molecular flow to occur through the porous tube, the mean free path of the gas must be large in comparison with the diameter of the pores (see page 125). The average pore size in ultrafine sintered glass is about 10^{-4} cm, so that from equation 4.2, the allowed pressure in the fritted tube could be as high as 50 Torr, corresponding to $L = 10^{-4}$ cm. In practice, 1–10 Torr is preferred to avoid adverse effects from the pore size distribution and to avoid some flow in the transition range approaching the viscous flow region.

If this fractionation process is to be of any value in sample enrichment, it is essential that the gas flow through the exit restrictor to the mass spectrometer is not similarly fractionated, i.e., viscous flow must occur through the exit capillary. To obtain viscous flow, the mean free path of the gas must be small compared with the diameter of the exit restrictor. The rate at which gas flows down the tube under this condition will be given by equation 4.6.

For a gas mixture undergoing viscous flow, the flow is independent of the molecular weight, and all components are transported with equal velocity. The quantity of a particular component will be given by the product of the total gas flow and the component mole fraction. Thus,

$$(Q_S)_{MS} = (Q_{Total})_{MS} \left(\frac{p_S}{p_{He} + p_S} \right) \quad (5.8)$$

Since the carrier is always in considerable excess over the sample (i.e., $p_{He} >> p_S$), $Q_{Total} \cong V_{He}$, and $(p_{He} + p_S) \cong p_{He}$. Therefore,

$$\left(\frac{Q_S}{V_{He}} \right)_{MS} = \frac{p_S}{p_{He}} \quad (5.9)$$

and it is seen that the ratio of the quantity of sample to the quantity of helium that enters the mass spectrometer depends only on the partial pressure of each component in the fritted tube.

By dividing equation 5.7 by equation 5.9, the ratio of sample lost through the porous tube to sample entering the mass spectrometer is shown to be

$$\frac{Q_{frit}}{Q_{MS}} = \frac{V_{frit}}{V_{MS}} \sqrt{\frac{M_{He}}{M_S}} . \quad (5.10)$$

Since

$$Q_{Total} = Q_{MS} + Q_{frit} \quad (5.11)$$

the yield is defined by equation 5.5 as

$$Y = \frac{Q_{MS}}{Q_{MS} + Q_{frit}} \times 100\% . \quad (5.12)$$

Substituting in the value of Q_{frit} from equation 5.10 gives the yield percent as

$$Y = \frac{1}{1 + \frac{V_{frit}}{V_{MS}} \sqrt{\frac{M_{He}}{M_S}}} \times 100\% . \quad (5.13)$$

Similarly, since $N = (Y/100)(V_{GC}/V_{MS})$ (equation 5.5), appropriate substitution of equation 5.13 gives

$$N = \frac{V_{GC}}{V_{MS} + V_{frit}\sqrt{M_{He}/M_S}} \tag{5.14}$$

Equation 5.14 shows that for an effusive separator, the maximum separation factor occurs as V_{MS} approaches zero (Figure 5.7), at which point

$$(N_{max})_{effuse} = \sqrt{M_S/M_{He}} \quad . \tag{5.15}$$

This expression for $(N_{max})_{effuse}$ is independent of volume flow although a zero carrier gas flow to the mass spectrometer is implied. From equation 5.13, it is seen that this condition also implies zero yield. As defined in equation 5.15, $(N_{max})_{effuse}$ should not be compared with the theoretical $(N_{max})_{100\%}$ suggested on page 169 for a separator with 100% efficiency. The latter is indeed dependent on volume flow rates.

Interface performance of an effusive separator can be evaluated by consideration of the variation of Y and N as a function of the flow rate and molecular weight. The influence of different gas flow rates to the separator and to the mass spectrometer is shown in Figures 5.7 and 5.8. In Figure 5.7, the separation factor is larger for larger effluent flow and for lower flow into the mass spectrometer. However, this is not the parameter that must be optimized. In Figure 5.8, the yield is shown to be significantly greater for lower effluent flow rates and for larger amounts into the mass spectrometer. Unfortunately, the flow to the mass spectrometer cannot be whimsically varied, so that the maximum allowed by the vacuum system should always be used. The chromatographic flow can be varied to optimize yield but only over modest limits. Figures 5.7 and 5.8 are valuable for comparison with experimental performance.

Figures 5.9 and 5.10 give the effusive separation factor N and yield Y as a function of molecular weight at various flow ratios. It is noted that the separation factor decreases only a small percent with decrease in the flow ratio V_{GC}/V_{MS}, but the percent yield is greatly increased with decreased flow ratio. This apparent improved efficiency is not due to the separator but rather due to the improved performance of the vacuum system which permits increased V_{MS}. In the higher molecular range, the yield-vs.-MW curve increases only slightly with increasing molecular weight.

The porous glass separator has been widely distributed as an accessory

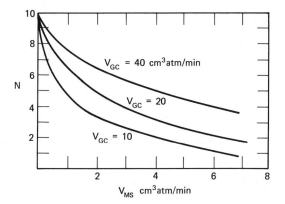

FIGURE 5.7. Variation of effusive separation factor N as a function of the volume of carrier gas entering the mass spectrometer. Molecular weight of sample = 400. Carrier gas = helium.

by several mass spectrometer manufacturers and is probably the most commonly used enrichment device. It is fairly cheap to purchase or make, and the glass surface provides reasonable inertness to prevent decomposition of delicate compounds. However, the conventional design can be optimized for only one operating condition, and the fragile glass construction is subject to occasional breakage. One interesting variation has been proposed by Markey.[199]

Porous metal and ceramic separators. Several types of metal and ceramic porous separators have been proposed[185–187, 200, 201] that offer ruggedness, convenient interchange or replacement of components, compactness, and an inert surface reasonably free from decomposition of labile compounds. The sinters are made of silver, stainless steel, or ceramic and can be silanized when adsorption or decomposition effects indicate the necessity. Three variations are shown in Figure 5.11a, b, and c.

There is no reason to believe a priori that different separation factors would be obtained for different porous materials, provided that the flow conditions are standardized. However, metal sinters are available in a wide range of pore sizes and thicknesses, and separators have been fabricated using pore diameters ranging from 5×10^{-4} cm to 10^{-5} cm, thus providing a variation of the conductance of the effusion path. More uniform pore size distribution is available in metal sinters which reduces the possibility that a significant quantity of sample will be pumped out through larger pores where

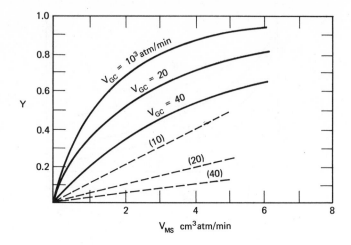

FIGURE 5.8. Variation of effusive separator yield Y as a function of the volume of carrier gas entering the mass spectrometer. Molecular weight of sample = 400. Carrier gas = helium. Dotted lines show yield if no separator is used.

viscous flow might prevail. These same advantages are claimed for the porous ceramic separator (Figure 5.11a), but little data is available.

The compact design of these metal separators greatly facilitates mounting in an interface oven or transfer line. Frequently, the reduced dead volume is claimed to be an operational advantage. Actually, this could not be significant provided the separator pressure is in the molecular flow range of 0.1–10 Torr. The effluent flow through the frit will be approximately 1.3 x 10^{-1} liter Torr/sec (10 cm^3 atm/min), so that from equation 2.2, $Q = PC$, the frit conductance must be in the range 0.1–1 liter/sec.[200] The frit volume is seldom greater than 2–3 cm^3. Hence, the characteristic pump-out constant V/C is about 10^{-2} sec and provides a very adequate margin to avoid peak broadening. The smaller frit dead volume is not necessary to avoid loss of chromatographic resolution, but it can help in some cases by reducing the active area.

The modular construction of these separators permits easy replacement of components, particularly in the Grayson-Levy microseparator (Figure 5.11b) and the Krueger-McCloskey stainless steel fritted tube separator (Figure 5.11c). The advantages of this convenience will be discussed in the next section.

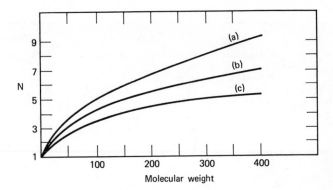

FIGURE 5.9. Variation of effusive separator factor N as a function of molecular weight. Carrier gas = helium. Flow ratio V_{GC}/V_{MS}: (a) = 100, (b) = 20, (c) = 10.

Operational problems of porous diaphram separators. One of the disadvantages of the nonvariable all-glass separator and similar fixed-enrichment devices, is that once the dimensions of the porous frit and restrictors have

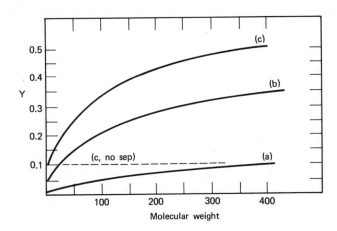

FIGURE 5.10. Variation of effusion separator yield Y as a function of molecular weight. Carrier gas = helium. Flow ratio V_{GC}/V_{MS}: (a) = 100, (b) = 20, (c) = 10.

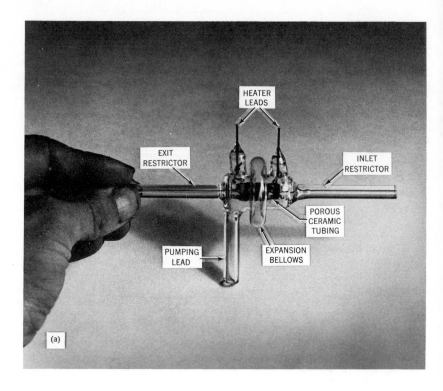

FIGURE 5.11. Variations of Watson-Biemann effusive separator. (a) Ceramic separator (Theta Corporation, Media, Pa., distributors).

been chosen, the operation can be optimized only at one set of flow conditions. For high efficiency of sample utilization, the maximum flow into the mass spectrometer should be established and must be maintained. This flow maximum determines the maximum percent of total effluent that can go into the separator for specified separator conductances. If more effluent goes to the separator, the allowed pressure in the mass spectrometer will be exceeded. If less effluent goes into the separator, maximum sample utilization is not attained.

For example, consider a separator that has been designed for optimum performance with 10 cm^3 atm/min (1.3×10^{-1} liter Torr/sec) of total effluent and delivers the allowed 0.2 cm^3 atm/min to a single-pumped mass

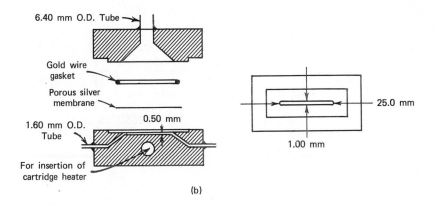

6.40 mm O.D. Tube

Gold wire gasket

Porous silver membrane

1.60 mm O.D. Tube

For insertion of cartridge heater

0.50 mm

25.0 mm

1.00 mm

(b)

FIGURE 5.11. (cont.) Variations of Watson-Biemann effusive separator. (b) Porous silver membrane microseparator.[200] A, from gas chromatograph; B, 1/16 in. standard stainless steel tubing, silver soldered; C, 0.02 in. flat silver gasket; D, heli-arc weld; E, to mass spectrometer; F, to vacuum; PS, pressure seal.

spectrometer. The pressure in the porous diaphram region would be close to 1 Torr. If this interface is used with a gas flow of 20 cm^3 atm/min, the pressure in the separator would be 2 Torr. The increased pressure would probably not destroy the condition of molecular flow through the pores, but the mass spectrometer pressure would be four times higher. (Viscous flow through the exit restrictor requires a p^2 term for the total quantity of gas as determined by equation 4.6.) Optimum mass spectrometer performance would not be possible and the mass spectral signal would probably be decreased rather than increased. Thus, it would be necessary to split the effluent before it enters the separator and an automatic factor of 0.5 reduction is suffered for the overall system efficiency.

If the interface is used with the same amount of sample and a gas flow of 5 cm^3 atm/min, the pressure in the separator would then be 0.5 Torr. The decreased pressure would have no effect on the molecular flow through the pores, so that the conductance of sample effusing through the frits is still independent of the helium pressure (equation 5.6). However, assuming that the necessary condition of viscous flow occurs through the exit restrictor, the quantity of sample to the mass spectrometer can be determined by substituting for $(Q_{total})_{MS}$ from equation 4.6 into equation 5.8, whereby

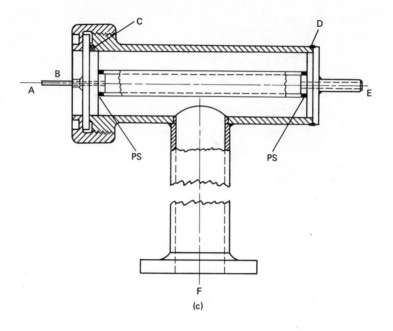

(c)

FIGURE 5.11. (cont.) Variations of Watson-Biemann effusive separator. (c) Porous stainless steel tubing separator.[49] A, from gas chromatograph; B, stainless steel tubing; C, silver gasket; D, heli-arc weld; E, to mass spectrometer; F, to vacuum; PS, pressure seal.

$$(Q_S)_{MS} = \frac{0.16}{\eta} \frac{d^4}{l} (P_2{}^2 - P_1{}^2) \left(\frac{P_S}{P_{He} + P_S} \right) \qquad (5.16)$$

Assuming $P_2{}^2 \gg P_1{}^2$, $P_{He} \gg P_S$, and $P_{He} = P_2$, equation 5.16 can be rewritten

$$(Q_S)_{MS} = kP_{He}P_S \qquad (5.17)$$

where $k = (0.16/\eta)(d^4/l)$. The total sample to the mass spectrometer is seen to be a linear function of the carrier gas pressure. Thus, the quantity of helium through the exit restrictor is 1/4 of the optimum. However, the pressure of sample in the frit is the same and because the pressure of helium is only one-half of the previous example the mole fraction is doubled. Thus, the sample quantity to the mass spectrometer is reduced by a factor of 2.

This represents a loss of separator efficiency and should be avoided by using an exit restrictor or effusion path of variable conductance.[49, 187, 188] A different separator can also be used but this involves annoying hardware changes.

In some inflexible situations it is profitable to add make-up gas to the effluent flow so that the total quantity of helium going to the separator is still at the desired maximum. Intuitively, this trick is not sound because, of course, the objective of the separator is to remove helium. However, if the separator is operating below the optimized maximum pressure, then the quantity of sample entering the mass spectrometer can be increased with a make-up gas (equation 5.17). The conductance of sample through the frit by molecular flow is still independent of helium pressure.

The effectiveness of this ploy can be best appreciated by calculating the separator yield from equation 5.13 with and without make-up gas, and comparing the results with the yield obtained using a variable-exit restrictor. It will be assumed that a 0.05-cm capillary column is used with a chromatographic flow of 5 cm^3 atm/min and that the optimum flow to the mass spectrometer is 0.2 cm^3 atm/min. Four cases will be considered.

Case 1. Optimum flow to MS without use of separator .

V_{GC} = 5 cm^3 atm/min
V_{MS} = 0.2 cm^3 atm/min
MW 200, Y = 4%
MW 400, Y = 4%

Case 2. Inefficient use of separator so that only 0.05 cm^3 atm/min goes to MS (4x less than maximum allowed).

V_{GC} = 5 cm^3 atm/min
V_{MS} = 0.05 cm^3 atm/min
MW 200, Y = 6.5%
MW 400, Y = 9.1%

Case 3. 5 cm^3 atm/min of effluent plus 5 cm^3 atm/min of makeup gas so that flow to MS is optimum at 0.2 cm^3 atm/min.

MW 200, Y = 12.5%
MW 400, Y = 17%

Case 4. Efficient use of separator so that 0.2 cm^3 atm/min goes to MS
without any make-up gas.

V_{GC} = 5 cm^3 atm/min
V_{MS} = 0.2 cm^3 atm/min
MW 200, Y = 22%
MW 400, Y = 30%

It is seen that when the effusive separator is used inefficiently as in
Case 2, the yield is low, particularly for molecular weights below 200. In the
example, make-up gas gives about two times better yield (Case 3), but this is
still about another factor of two less than that obtained by optimum operation as in Case 4.

Careful consideration is always given to assure molecular flow through
the pores of an effusion separator, but curiously, the exit restrictor seldom
receives the careful consideration it deserves. To obtain viscous flow through
this tube, the diameter must be 5–10 times larger than the mean free path
as calculated by equation 4.2. If the diameter is too small, molecular flow
or intermediate flow will take place.

If the vacuum in the separator is 0.5 Torr, then from equation 4.2,
$L = 10^{-2}$ cm and the diameter of the exit tube should be 5 x 10^{-2} cm or
larger. The allowed quantity of helium to the mass spectrometer is about
1.3 x 10^{-2} liter Torr/sec (1 cm^3 atm/min) for a differential-pumped mass
spectrometer and 1.3 x 10^{-3} liter Torr/sec for a single-pumped system.
The length of the exit tube can be calculated using these parameters in equation 4.6. It is assumed (quite safely) that $P_2^2 \gg P_1^2$. The viscosity of
helium at 200°C is 267 μpoise. Thus, for d = 0.05 cm and Q = 1.3 x 10^{-2}
liter Torr/sec, $l \cong 0.1$ cm; for Q = 1.3 x 10^{-3} liter Torr/sec, $l \cong 1.0$ cm. If
the diameter of the tube were increased to be 10^{-1} cm, the respective tube
lengths would be 1.6 cm and 16 cm, respectively.

Calculations on viscous flow parameters are not always reliable because
of the fourth power dependence on diameter, the square power dependence
on pressure, and the variation of viscosity with changes in temperature. Furthermore, equation 4.6 is not valid for short tubes. The significance of
these calculations is the order of magnitude. For example, it was just shown
that a short restrictor (~ 0.1 cm) could be used with a differential-pumped
mass spectrometer receiving 1 cm^3 atm/min of effluent. A typical needle
valve would have a flow path in this order of magnitude and could therefore be used as an exit restrictor metering the effluent to a differential-
pumped mass spectrometer. However, for a flow of only 0.1 cm^3 atm/min,
the opening of the needle valve might be less than the necessary minimum

of 5 x 10^{-2} cm and intermediate or molecular flow would occur through the needle valve.

It is apparent that the dimensions of the two critical parts of the separator, namely the porous tube and the exit restrictor, must be carefully considered in the light of each application. To have molecular flow through the sinter, the pressure must be low enough so that $L > d_{sinter}$. To have viscous flow through the exit, the dimensions of the exit restrictor must be large enough so that $L \ll d_{exit}$. This further emphasizes the need for optimization of the conditions in the effusive separator at all times.

The variable-conductance separator. To avoid the inefficient operation encountered when a fixed-conductance porous diaphram separator is used with different flow rates, Brunnée et al. devised an effusion-type separator with a variable-conductance effusion path.[188, 202] This apparatus (Figure 5.12) operates on the same principle as the sintered frit separators, but the effusion occurs through a small variable slit formed between a flat coverplate and the edges of two annular rings about 2 cm in diameter. The opening is varied from 0 to 5 x 10^{-3} cm by a micrometer spindle on the flat cover plate. The plate can be completely closed so that the mass spectrometer can receive up to 100% of the sample from a low-volume flow column. When columns with higher flow rates are used, the opening is adjusted to give the maximum desired flow to the mass spectrometer. In this way, the pressure in the separator is always maintained at the optimum operating level. When the cover plate is fully opened, the separator acts as a bypass valve, and excess solvent or abundant components can thus be diverted.

The molecular conductance of the effusive path is proportional to the area of the tiny slit opening. With a plate-to-edge separation of 10^{-4} cm the conductance is about 0.3 liter/sec for helium at 200°C. Thus, at a flow of 20 cm^3 atm/min, the pressure is 0.8 Torr which provides a comfortable operating range in which effusive flow is maintained. This margin of safety is easily attained with a slit effusion path but it is essential since the probability of molecular collision is considerable in the direction of the slit length.

The viscous conductance of the exit restrictor is constant for all effluent flow rates as long as the pressure in the separator is adjusted to the optimum. The exit conductance should be designed for about 3 x 10^{-3} liters/ sec for a single-pumped mass spectrometer and about 3 x 10^{-2} liters/sec for a differentially pumped mass spectrometer. The conductance of the restriction for a single-pumped or differentially pumped mass spectrometer would thus differ by about one order of magnitude. The exact value is not important since minor adjustments can be made in the effusive flow.

The Brunnée separator has been operated efficiently over a flow range

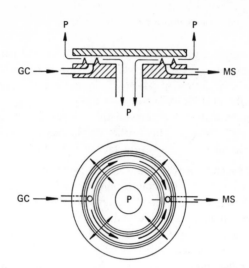

FIGURE 5.12. Brunnée separator with variable effusion path.[188]

from 1 cm³ atm/sec to 50 cm³ atm/sec. This wide operating range is quite advantageous. In addition, the separator can be conveniently constructed in a variety of shapes and sizes. The parts can be disassembled for cleaning, polishing, or gold plating. Because of the small interior surface, decomposition and tailing are minimized.

The purpose of a variable effusion path is, as explained above, to attain maximum efficiency by maintaining the maximum allowed flow into the mass spectrometer. It is also possible to achieve this objective with a variable-exit restrictor. This method was used by Blumer[187] and by Krueger and McCloskey[49] in their metal sinter separators using a fine-metering valve. In effect, their equipment also constitutes a variable-conductance separator.

The limitation of using a fine-metering valve as the exit restrictor was previously discussed. Great care should be taken to be certain that the valve opens enough to permit viscous flow. If the quantity of effluent to the mass spectrometer is small (around 0.1 cm³ atm/min), most fine-metering valves will not be cracked open very wide and molecular flow is highly probable.

A variable viscous flow valve can be made from a slotted rod that slides down a tube (Figure 5.13a). If the tube and rod are metal, galling is a considerable danger but is avoided if the rod has 2–3° of taper. The taper also gives a wider operating range. If a Teflon rod can be used, a valve of this type would operate with minimal trouble and assure viscous flow

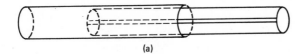

(a)

FIGURE 5.13. Variable viscous flow exit restrictors. (a) Slotted rod.

to the mass spectrometer. One method of arranging this type of flow restrictor is shown in Figure 5.13b.

Dead volumes at this stage of the interface are seldom troublesome because of the relatively large conductances. A dead volume on the separator side will be swept out by the sinter conductance which is generally around 100 cm^3/sec. A dead volume on the mass spectrometer side would be swept out by the ion chamber conductance (with consideration for the conductance of the connecting line) which is generally more than 500 cm^3/sec. Thus, the condition $V/C < 0.1$ sec is easily met in this type of valve.

3. The Jet Separator

The different rate of diffusion of different gases in an expanding supersonic jet stream can be used to accomplish fractionation of gaseous mixtures.[203-206] Ryhage has designed and tested a jet separator based on this principle to produce efficient enrichment of sample to helium in a gas chromatographic effluent.[194, 197, 207] The commonly used two-stage model of the Ryhage separator is shown schematically in Figure 5.14a. Construction details of a commercial single-stage separator are shown in Figure 5.14b.

Effluent gas from the chromatograph passes through a restricted orifice d_1 and expands rapidly in the evacuated area between d_1 and a second orifice, d_2. The freely expanding jet establishes a pressure gradient in its interior. Diffusion flow in this region for each component is a function of the molecular weights and is proportional to the diffusion coefficient D. Thus, a separating effect occurs provided that viscous flow conditions are maintained. The core of the jet stream is enriched in the heavier component, and concentrated effluent enters the skimmer or collecting orifice d_2. The peripheral portion is enriched in the lighter component and pumped away. Similar enrichment occurs in the second-stage jet region between d_3 and d_4.

The theoretical separation factor for an expanding jet is difficult to establish, and in practice, it is simpler to determine the yield empirically. Ryhage studied the variation in efficiency for several combinations of different-

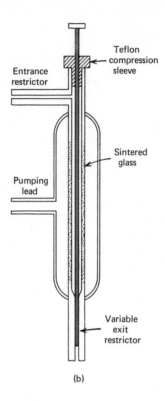

Teflon
compression
sleeve

Entrance
restrictor

Sintered
glass

Pumping
lead

Variable
exit
restrictor

(b)

FIGURE 5.13.   Variable viscous flow exit restrictors. (b) Adjustable stainless steel rod in exit restrictor.

size jet diameters.[197] It was shown that with this remarkable separator, 80% of a methyl stearate peak (MW 298) passed through the first jet stage and furthermore, 50% of that amount passed through orifice d_4 in the second stage. The total system yield of 40% was essentially the same for a gas chromatographic flow of 27 cm^3 atm/min and 50 cm^3 atm/min.

For a chromatographic flow of 30 cm^3 atm/min, an analyzer pressure of 1.2 x 10^{-6} Torr was recorded by Ryhage. It was stated that this corresponded to a flow of 0.25 cm^3 atm/min (0.32 x 10^{-2} liter Torr/sec) to the mass spectrometer. However, that would require a pumping speed of 2700 liters/sec (C = Q/P) and since the quoted speed of the mass spectrometer diffusion pump was 550 liters/sec, it is doubtful that the system conductance was significantly more than 150–200 liters/sec. Nevertheless, this inconsis-

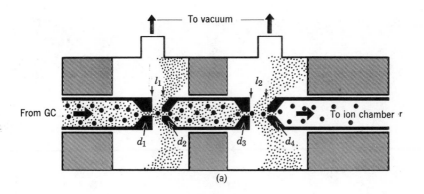

FIGURE 5.14. (a) Schematic diagram of Ryhage jet separator.[194]

tency does not detract from the excellent yield values which were measured
independently of the mass spectrometer pressure.

The orifice dimensions recommended by Ryhage were $d_1 = 10^{-2}$ cm,
$d_2 = 2.5 \times 10^{-2}$ cm, $d_3 = 10^{-2}$ cm, and $d_4 = 3 \times 10^{-2}$ cm. The distance
separations were $l_1 = 1.5 \times 10^{-2}$ cm and $l_2 = 5 \times 10^{-2}$ cm. The first expansion chamber was evacuated with a mechanical pump rated at 2.6 liters/sec.
With 30 cm^3 atm/min of effluent, the measured pressure in the pumping arm
was 0.1–0.2 Torr. The second expansion volume was pumped with a 150-
liters/sec diffusion pump and the measured pressure was about 10^{-3} Torr.
From these data and estimated pumping tube dimensions, it can be calculated
that approximately 26–28 cm^3 atm/min of effluent was separated out by the
first expansion jet and 2–4 cm^3 atm/min by the second expansion jet. As
previously stated, the quantity to the mass spectrometer was about 0.25 cm^3
atm/min. The split ratios at each stage will remain approximately the same
for other effluent flow volumes.

The high measured yield demonstrates excellent enrichment at high
effluent flow rates (30 cm^3 atm/min) and low mass spectrometer flow rates
(0.2 cm^3 atm/min). It is also likely that, with minor dimensional changes,
one stage of jet expansion would remove sufficient carrier gas for use with
a differential mass spectrometer accepting 1–2 cm^3 atm/min. A one-stage
separator can also be used for enrichment of the low effluent flows from a
capillary column,[207, 208] but marginal operation may be encountered.

The average pressure in the expanding jet must be high enough to
ensure a viscous flow condition in the jet core. Becker has shown that when

FIGURE 5.14. (b) Construction details of a stainless steel single-stage jet separator (DuPont Instrument Products Division).

the condition of viscous flow is lost, the separation factor drops rapidly to a value of one.[203]

In Ryhage's paper,[197] pressure measurements were made in the pumping arm. These measurements are useful to calculate the quantity of gas pumped away from the peripheral jet surface in each compartment as shown above, but they cannot be related in any simple fashion to the pressure in the jet stream. For example, in the first compartment the measured pressure in the pumping arm was 10^{-1} Torr. Since $L = 5 \times 10^{-2}$ cm and $d_1 = 10^{-2}$ cm, the condition that $L \ll d_1$ in the jet would not be met. Although the available data are insufficient to calculate the true feed pressure from the chromatograph to the jet, reasonable estimates (within an order of magnitude) are $P_1 = 30$ Torr and $P_3 = 0.3$ Torr. Thus, in the first jet, $L \approx 1 \times 10^{-4}$ cm, or two orders of magnitude less than d_1. In the second jet, $L \approx 10^{-2}$, which is essentially equal to the jet diameter d_3. Reduced efficiency would be anticipated for this stage.

Ryhage's data for methyl stearate indicated an 80% yield in stage 1, a 50% yield in stage 2, hence an overall yield of 40%. Assuming that the effluent is divided 28 cm^3 atm/min to the first pump, 2 cm^3 atm/min to the second pump, and 0.25 cm^3 atm/min to the mass spectrometer, the corresponding separation factors calculated from equation 5.5 are $N_1 = 12$, $N_2 = 4$, and $N_T = 48$. The estimated efficiency is seen to be three times less in the second stage. The reduced efficiency may be related to the drop in efficiency observed by Becker when operating a jet separator at feed pressures of about 0.5 Torr.[203]

It is apparent from the foregoing that the jet separator will function most efficiently at high flow rates (above 10 cm^3 atm/min). When the quantity of gas is less than 3–4 cm^3 atm/min ($\sim 4 \times 10^{-2}$ liter Torr/sec), the jet pressure may be too low to maintain supersonic flow conditions, and the separation factor will approach a value of one, i.e., no effective enrichment. Make-up gas is often used to give improved performance. For low flow conditions, the effusive separator has comparable efficiency, but can be operated with less danger of completely losing the separating effect.

Most jet separators are constructed of stainless steel, but one type of glass jet separator has been proposed for work with thermally labile compounds (Figure 5.15). This single-stage separator has been tested by Bonelli et al.[195] using mixtures of inert gases. The enrichment data are plotted as a function of the square root of the molecular weight in Figure 5.16. A reasonable straight-line agreement suggests that for these inert gases the effect is partly diffusion controlled. However, a straight-line dependence on the square root of the molecular weight as shown cannot continue indefinitely since the maximum enrichment as defined by equation 5.5 is V_{GC}/V_{MS}. For the data of Bonelli et al., $N_{max} = 19$.

These data cannot be related to the performance of any other jet separator since differences in jet alignment or orifice diameters would considerably change the measured enrichment. It is unfortunate that there are no data that compare the efficiency of different jet separators. It also is lamentable that no definitive data have been published that define the efficiency of jet separators as a function of orifice dimensions.

The metal jet separator provides an interface in which effluent surface contacts are made only with stainless steel tubing. Decomposition effects should be no more than those caused by standard chromatograph exit plumbing. The system can be silylated with conventional techniques, but gold plating of the chamber surfaces is impractical in that only the tubing

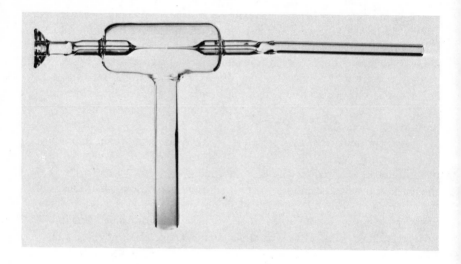

FIGURE 5.15.   All-glass single-stage jet separator (Finnigan Corporation, Sunnyvale, California).[195]

and not the bulk surfaces touch the sample. The all-glass jet separator is particularly attractive for work with delicate samples because it avoids the active surfaces that might occur on a glass sinter or stainless steel parts.

The nature of the supersonic expanding jet, the pumping speed involved, and the jet core sampling technique all contribute to fast throughput, and hold-up by dead volume is absent. There have been no reports of chromatographic peak broadening by a jet separator. Occasionally, the jet orifices become clogged. This may happen due to thermal decomposition of excessive solvent, sample, or column bleed. Particulate matter must be avoided at all times.

4. The Semi-Permeable Membrane Separator

The Teflon separator. Lipsky, Horvath, and McMurray have described an unusual form of separator in which helium is preferentially removed through a thin Teflon film thus enriching the organic vapors inside the Teflon tubing.[189, 196] At temperatures below 200°C, helium permeability into Teflon

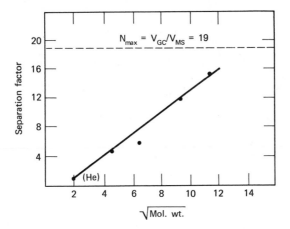

FIGURE 5.16. Separation factor as a function of the square root of the molecular weight for an all-glass jet separator. $V_{GC} = 15$ cm^3 atm/min., $V_{MS} = 0.8$ cm^3 atm/min.[195]

is essentially zero, but above 250°C, sufficient helium diffuses through the Teflon to give significant sample fractionation.

The apparatus of Lipsky et al. is shown schematically in Figure 5.17. The basic separating surface is provided by a 200-cm length of thin-walled (0.01 cm) Teflon capillary tubing of 0.025 cm internal diameter. The tubing is coupled to short pieces of stainless steel capillary to facilitate the vacuum connections at each end of the separator. The coil of Teflon tubing is housed in a glass jacket which is pumped by a moderate-size (500 liters/min) rotary pump.

The precise mechanism of this separator is not clearly understood. One concept considers that at higher temperatures, tiny pores develop in the Teflon that permit an effusive type of gas flow through the membrane. This mechanism may occur in part, particularly above 300°C, but the necessary gaseous conductance of the porous Teflon seems to be very high relative to the impervious nature of the Teflon below 200°C. Furthermore, as used by Lipsky et al. (i.e., with 120 cm of stainless steel 0.025-cm capillary at either end of the Teflon tubing, and with a throttle valve between the separator and mass spectrometer), it is questionable whether or not the average pressure in the Teflon tube would be low enough to give effusive fractionation. High pressure is also implied in the work of Grayson and Wolf, who found that control of the mass spectrometer pressure required a considerable restriction after the separator.

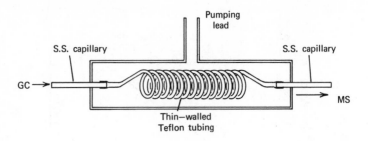

FIGURE 5.17. Schematic diagram of Teflon separator.[189]

It is also probable that the heated Teflon film behaves as a semipermeable membrane with considerable discrimination against organic material. The permeation of helium through a membrane is known to increase exponentially with temperature. Over the limited range of 250–325°C, Lipsky et al. found a linear decrease in the amount of helium that went through to the mass spectrometer, but since the quantity going through the Teflon film was not specified, a diffusion function can not be established.

At the same time, the yield for repeated injections of the same-size sample was shown to increase 30% over this temperature range even though the total quantity of effluent to the mass spectrometer was reduced by four. Assuming viscous flow through the exit restrictor, the data indicate the unexpected result that five times less organic sample diffuses through the Teflon membrane at the higher temperature.

With the paucity of data on the Teflon separator, the mechanism of fractionation cannot be quantitated, and it is difficult to suggest an anticipated theoretical yield. Lipsky and coworkers report experimental yields of 40–70% and enrichment factors up to 200,[209] but the flow rate was not clearly specified, so the significance of the data is lost. Effluent flow of 2–30 cm^3 atm/min was mentioned but no specific value was related to the quoted yields.

Grayson and Wolf[196] reported yields from a Teflon separator in the range of 80–90% for effluent flow rates of 15 cm^3 atm/min and 5 cm^3 atm/min. However, these high yields were obtained with flow of 2.0–2.5 cm^3 atm/min ($\sim 3 \times 10^{-2}$ liter Torr/sec) into the mass spectrometer, and the resulting enrichment factors were in the range of only 1.4–5.7, depending on the temperature, the flow rate, and the compound. The enrichment factor was only 25% larger with isononane (MW 128) than with ethane

(MW 30). A definite increasing trend was observed from 280°C to 320°C, but the change was not dramatic and amounted to only a 5–10% larger value of N.

Although the Teflon separator is easily constructed, it has not been widely used. The high yields reported by Lipsky et al. and by Grayson and Wolf are not necessarily practical for general applications. It is questionable that the system could be used successfully where the flow to the mass spectrometer, V_{MS}, is low. In the work of Grayson and Wolf, a high flow (2 cm^3 atm/min) was used for the purpose of evaluating the separator, but this flow could not be applied for standard GCMS analysis in a single-pumped mass spectrometer. A few of the laboratories that have tried a Teflon separator have reported relatively easy fracture of the thin Teflon at 350°C or irreversible occurrence of pinholes. Peak distortion and time delays are also reported.[187, 196]

The silicone membrane separator. The silicone membrane separator attains sample enrichment as a consequence of preferential passage of organic vapors through a thin polymer barrier. The principle was first suggested for GCMS enrichment by Llewellyn and Littlejohn who demonstrated that very high separation factors could be obtained while retaining satisfactory yield.[192] The merits of this separator have been investigated by several laboratories and used successfully in a variety of applications.[193, 210–212] Various modifications are shown in Figure 5.18. The membrane separator is easily constructed or if preferred, several manufacturers supply either one or two stage models (Varian Associates, Palo Alto, California; Cangal Inc., Lafayette, California; Environmental Devices Corp., Saratoga, California).

In operation, exit gas from the chromatograph passes through a small chamber that has a thin silicone rubber membrane supported on a fine metal mesh or glass sinter. The inorganic carrier gas is quite insoluble in the silicone polymer and most of it passes through the exit of the chamber. On the other hand, organic material has a considerable solubility in the silicone polymer, and diffuses through the film to the other side where it flows to the mass spectrometer. In a single-stage model (Figure 5.18b and c), the outlet is at one atmosphere, and no auxiliary pumping is needed. In a two-stage membrane separator (Figure 5.18a), the exit of the first stage is at one atmosphere and a rotary pump is used at the second stage.

The rate of conductance of a gas through a polymer membrane is a function of the specific solubility, the diffusion constant of the gas, and of the area and thickness of the membrane. The rate at which the partial pressure of a component is depleted is thus given by

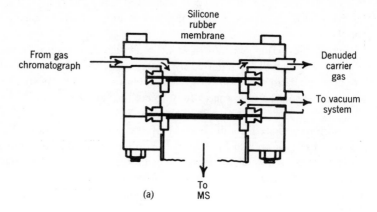

FIGURE 5.18. Variations of the silicone membrane separator. (a) Two-stage separator (Llewellyn and Littlejohn).[192]

$$ - \frac{dp}{dt} = \frac{A}{V_c} \frac{P}{l} (SD) $$ (5.18)

where

S = specific solubility of the gas in the membrane
D = diffusion constant of the gas in the membrane
A = area of membrane exposed to gas
l = thickness of membrane
V_c = volume of the cavity in the separator.

It is assumed here that the concentration gradient dc/dx can be approximated by P/l and that the back pressure is negligible.

For a sample component

$$ \left(\frac{p}{p_0} \right)_S = e^{-\frac{A}{V_c} \frac{(SD)_S}{l} t} $$ (5.19)

and for the carrier gas (helium)

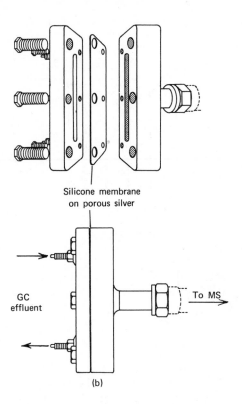

Silicone membrane
on porous silver

GC
effluent

To MS

(b)

FIGURE 5.18. Variations of the silicone membrane separator.
(b) Single-stage separator (Black et al.)[193]

$$\left(\frac{p}{p_0}\right)_{He} = e^{-\frac{A}{V_c}\frac{(SD)_{He}}{l}t} \qquad (5.20)$$

The expression $(p/p_0)_S$ represents the fraction of sample still in the carrier
stream after a time t. The fraction that has gone to the mass spectrometer
is therefore $1 - (p/p_0)_S$, so that the yield fraction is given by

$$Y = 1 - e^{-\frac{A}{V_c}\frac{(SD)_S}{l}t} \qquad (5.21)$$

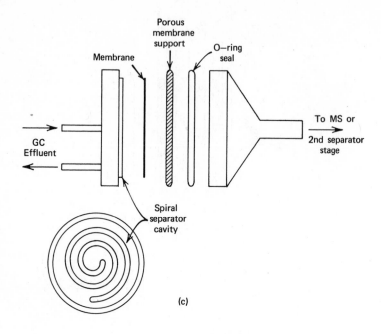

FIGURE 5.18. Variations of the silicone membrane separator. (c) Single-stage spiral cavity separator (Grayson and Wolf).[210]

For the spiral cavity separator proposed by Wolff and Grayson (Figure 5.18c), smooth laminar flow occurs, and the exposure time and cavity volume are therefore related to flow rates V_{GC} by the equation

$$V_{GC} = \frac{V_c}{t} \tag{5.22}$$

and hence,

$$Y = 1 - e^{-\frac{A}{l}\frac{(SD)_S}{V_{GC}}} \tag{5.23}$$

From equation 5.5, the enrichment can be written

$$N = \frac{V_{GC}}{V_{MS}} \left(1 - e^{-\frac{A}{l}\frac{(SD)_S}{V_{GC}}}\right). \tag{5.24}$$

For most applications, the permeability of helium, $(SD)_{He}$, will be small, so that the change in helium pressure across the membrane cavity is insignificant. For this condition, the quantity of helium entering the mass spectrometer per unit time is the product of the conductance (i.e., permeability) and pressure as given by equation 2.2. Thus,

$$Q_{He} = \left(\frac{A}{l}\,(SD)_{He}\right)p_{He}. \tag{5.25}$$

Since p_{He} is approximately one atmosphere for a single-stage membrane separator, the quantity of helium entering the mass spectrometer depends only on the permeability parameters, $A/l\,(SD)_{He}$ and is approximately independent of flow rate.

The validity of the approximation has been well established by Grayson and Wolf[210] and by Hawes et al.[211] However, Hawes et al. did report increased helium flow to the mass spectrometer at a chromatographic flow rate of 150 cm^3 atm/min. The observation implies an increase in the helium pressure and is consistent with equation 5.25.

The helium permeation data observed by Grayson and Wolf are shown in Figure 5.19. For their separator (membrane area = 15 cm^2, thickness = 0.0025 cm), the helium transport at 100°C is essentially 2.4 cm^3 atm/min over the chromatographic flow range of 10–60 cm^3 atm/min. The helium permeation is dependent on temperature and, as is shown in Figure 5.19b, can vary from 1–4 cm^3 atm/min in the range 50–200°C.

These data are extremely valuable in that they permit an a priori estimate of the desired separator area for specific applications. For example, if a single-pumped mass spectrometer capable of handling 0.2 cm^3 atm/min of carrier gas is to be interfaced with a single-stage membrane operating at 200°C, the maximum active membrane area should be 0.75 cm^2. As will be shown below, this area would lead to a low yield, so that a two-stage interface system would be preferred.

To calculate the yield of a membrane separator using equation 5.23, it is necessary to know the solubility and diffusivity at the operating temperature. Unfortunately, very little applicable data is available. Llewellyn and Littlejohn[192] gave conductance data for a few inorganic gases and octane at 30°C. A yield calculation for octane using their data in equation 5.23 gives

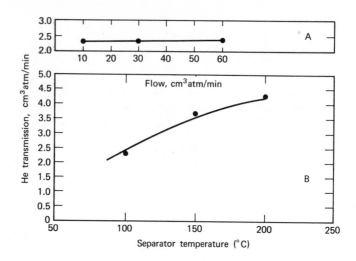

FIGURE 5.19. Helium permeation through a silicone membrane. Area = 15 cm², thickness = 0.0025 cm. (a) Relatively invariant permeation as a function of V_{GC}. Temperature = 100°C (b) Variation of permeation as a function of temperature. Flow = 30 cm³ atm/min.[210]

$Y = 95\%$ at $V_{GC} = 60$ cm³ atm/min. In good agreement, Llewellyn and Littlejohn reported 90% experimental yield in the first stage.

Grayson and Wolf[210] reported yields for a number of compounds through a one-stage membrane separator at 100°C and at flow rates of 10, 30, and 60 cm³ atm/min. Membrane conductances can thus be calculated using equation 5.23 in the form

$$\frac{A}{l} (SD)_S = - V_{GC} \ln(1 - Y). \tag{5.26}$$

The values calculated for A/l $(SD)_S$ are reasonably constant provided Y is less than 0.9. At higher yields, equation 5.26 is very sensitive to small measurement errors, and the calculations are not reliable.

A few of the conductance values averaged from the three sets of flow data of Grayson and Wolf are given in Table 5.2. The numbers are specific to that particular dimethyl silicone polymer at 100°C, but the data give some indication of conductances that might be expected for other conditions or other compounds. Unfortunately, this sketchy information fails to define the effects of boiling point, operating temperature, and compound functionality.

TABLE 5.2. Conductance at 100°C of methyl silicone membrane for organic compounds.[1]

Compound	Average Calculated Membrane Conductance		Experimental Yield % for VGC		
	Boiling Point, °C	$\frac{A}{l}$ (SD)S cm³ atm/min	10 cm³ atm/min	30 cm³ atm/min	60 cm³ atm/min
Hexane	69	12	57	35	17
Carbon tetrachloride	77	15	70	60	22
Trichloroethylene	87	29	85	64	38
1-Pentanol	138	32	89	69	42
1-Decene	171	84	98	94	75
Methyl heptanoate	172	55	97	86	56

[1]Active area = 15 cm², thickness = 0.0025 cm (Grayson and Wolf).[210]

Hawes et al.[211] presented raw data for dodecane through a dimethyl silicone membrane as a function of flow rate over the range 15 cm^3 atm/min to 150 cm^3 atm/min. Although the temperature was not specified and data were not given as a yield percent, by extrapolation to zero flow, the raw signal corresponding to 100% yield can be estimated and the raw data can thus be expressed as a yield. When transformed in this manner, the log plot of $1 - Y$ vs. $1/V_{GC}$ gives a straight line as predicted from equation 5.26 (Figure 5.20).

Equation 5.23 can be of considerable value in designing a membrane separator to meet specific operating conditions. In Figure 5.21, the yield Y is plotted as a function of the exponent factor $(A/l)[(SD)_S/V_{GC}]$. It is noted that as long as the exponent factor is greater than two, the yield is 90% or better, and reasonable variations in V_{GC} will not seriously change the performance. However, if the exponent factor is only 0.3 to 1.0, the yield is low and very susceptible to changes in chromatographic flow. Thus, using as a guide the conductances given in Table 5.2 (for 15 cm^2 area and 0.025 cm thickness), it is possible to estimate the expected yield for given values of V_{GC} and A. Of course, as demonstrated by Figure 5.19, the upper value of A is limited by the quantity of helium that can be accepted by the mass spectrometer.

Black et al.[193] prepared a membrane separator by coating a thin porous silver support (hole size 2 x 10^{-4} cm, thickness 4 x 10^{-3} cm) with a solution of phenyl methyl siloxy copolymer. This polymer was chosen to reduce the quantity of helium that permeated through to the mass spectrometer, but no comparative data are given. The thickness of the film, as measured microscopically, was 5 x 10^{-4} cm, but because the solution might penetrate the pores of the support during the curing process, the effective diffusion thickness could be close to 1–4 x 10^{-3} cm. The membrane thickness in other separators is in the range of 3 x 10^{-3} to 10^{-2} cm so that the additional thickness in the porous silver supported membrane could account in part for reduced helium pressure. Three separators of this type were made with active areas of 1.6 cm^2, 3.2 cm^2, and 6.4 cm^2, but no comparative data were given.

The most glaring disadvantage of the membrane separator is the severe temperature dependence. To have high solubility S, the temperature should be as low as possible; to have high diffusivity D, the temperature should be as high as possible. This means that an optimum temperature exists for each compound that will depend upon the boiling point, the molecular weight, and the compound functionality. Even if data were available to assess this optimum, it is unlikely that the data could be used to set up a meaningful temperature program for a complex unknown.

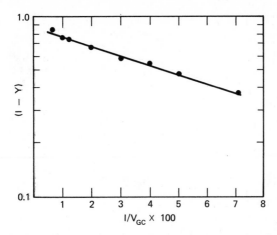

FIGURE 5.20. Plot of $(1 - Y)$ vs. $1/VGC$ for membrane separator. Points calculated from data of Hawes et al.[211] Estimated temperature, 125°C. (Not specified in original paper.)

For organic materials, solubility is the more important parameter, and a higher yield is obtained by operating at the lowest possible temperature. Care must be taken to avoid slow diffusion through the membrane which results in tailing of the chromatographic peak and serious loss of component separation. Grayson and Wolf suggested that the temperature of the separator should be kept 50–70°C below the boiling point. When this condition was met, their separator (15 cm^2 area) gave yields greater than 90% and only 10–20% broadening of the chromatographic peak was reported. The necessity for establishing the best operating temperature was dramatically shown by Black et al. who ran an alkane series C_6–C_{14} with the separator at 84°C, 152°C, and temperature programmed from 30–200°C. Their data (Figure 5.22) show considerable peak broadening when the component boiling point is more than 70°C above the membrane temperature. The isothermal run at 150°C shows poor sample yield due to the lower solubility. Best results were obtained by the temperature program. A convenient compromise was suggested by Hawes et al. who placed their separator in the chromatographic oven so that it was temperature programmed at the same rate as the column. This technique will be quite useful in many applications, but the peak broadening in their chromatograms (Figure 5.23) suggests that the chromatographic column temperature may be a bit lower than that desired in the separator.

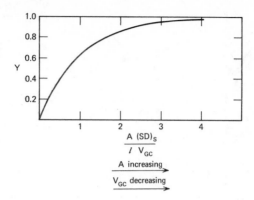

FIGURE 5.21. Yield of membrane separator as a function of
exponent factor $(A/l)[(SD)_S/V_{GC}]$ (Grayson-Wolf design).[210]

The rate controlling step for transfer of sample through the membrane
is the migration of dissolved molecules through the bulk of the polymeric
material, and this parameter must be controlled if sharp chromatographic
separations are to be obtained. When the organic material dissolves in the
membrane, transport through the polymer may be considered as a random
walk movement. For a one-dimensional system, the mean-square displace-
ment of a molecule after a time t is given by

$$\overline{x}^2 = 2Dt \qquad (5.27)$$

so that the average time τ for transport of a molecule across the membrane
thickness l is

$$\tau = \frac{l^2}{2D} . \qquad (5.28)$$

For most membrane separators, the thickness of the membrane l, is much
smaller than the lateral dimensions, so that the one-dimensional expression
5.28 gives a reasonably accurate description.

A time of approximately 6τ assures that most of the molecules have
passed through the membrane. For a typical membrane thickness of $2.5 \times 10^-$
cm, it is necessary that $D > 10^{-6}$ cm^2/sec. Since most organic substances have
a diffusion constant of that order of magnitude, the danger of serious peak
broadening is always imminent. A higher temperature promotes faster dif-

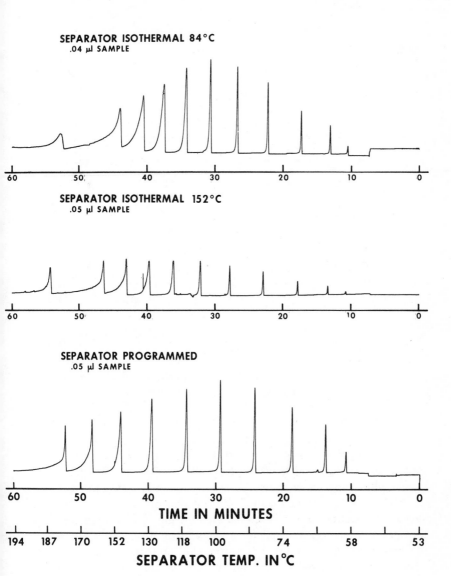

FIGURE 5.22. Total ion chromatograms of n-Alkanes through a membrane separator.[193]

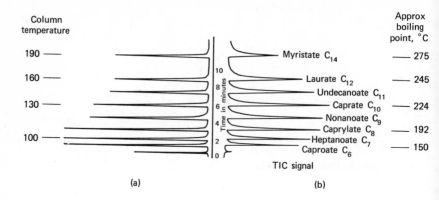

FIGURE 5.23. Comparison of chromatograms of methyl esters. (a) Flame ionization detector. (b) Total ion monitor. Membrane separator placed in oven and temperature programmed 80–180°C at 8°/min.[211]

fusion and maintains sharp chromatographic peaks as indicated in Figures 5.22 and 5.23, but the solubility, and hence the yield, is decreased. The maximum temperature that will attain the chromatographic separations must be established for effective use of a membrane separator.

Most membrane separators have a cavity volume V_C in the range 0.01–0.4 cc. The time constant for exponential sweep out of this volume will be V_C/V_{GC} where V_{GC} is the effluent flow rate. If a smooth laminar flow is maintained through the cavity, only one characteristic time period is needed to clean the sample out of the cavity; if eddy currents occur, the sweep out is exponential and needs 6–8 time periods. When a separator with $V_C = 0.4$ cm^3 is used with an effluent flow of 60 cm^3 atm/min, V_C/V_{GC} would be about 0.4 sec and even if eddy currents occurred, peak broadening would be negligible (2–3 sec maximum). However, if a low flow column is used, for example, 5 cm^3 atm/min from a 0.05-cm ID capillary, the time constant through a cavity volume of 0.4 cm^3 would be close to 5 sec. Even for smooth laminar flow, this could not be accepted, but if turbulence occurs, the peak would be broadened by 20–30 sec. Fortunately, this facet of peak broadening in a membrane separator can be minimized by proper cavity design. The spiral cavity used by Grayson and Wolf is particularly favorable and should be considered if low flow rates will be encountered.

Black et al. performed an excellent analysis of the loss of theoretical plates suffered when a substance passes through a membrane separator. Using capillary columns 0.075 cm ID by 150 meters long, theoretical plate values based on peaks measured directly by a flame ionization detector were in the range 120,000–180,000. When the peaks were measured by the total ion monitor after passing through the separator, the values were about 2–3 times lower. This loss may be acceptable with the excellent separations attained when using high-resolution columns, but for many applications, the peak broadening of a membrane separator would be prohibitive.

The membrane separator is capable of giving a very high yield over a large range of effluent flow rates. The outlet of the first stage is effectively at atmospheric pressure, so that the residual sample in the carrier gas can be directed to any conventional gas chromatographic detector for auxiliary detection.

The disadvantages of the membrane separator are: (1) temperature must be carefully programmed to optimize yield and reduce tailing. (2) Peak broadening is excessive for some applications. (3) The maximum safe operating temperature is 230°C. (4) The danger of membrane rupture is always imminent. The membrane separator is gaining increased popularity because of potentially high yields, but great care must be taken to avoid the inherent difficulties mentioned above. In addition, in many designs, the physical dimensions have not been optimized to give the best performance at the anticipated flow rates and consequently, yields are no better than can be obtained with other separators.

The hydrogen/silver-palladium separator. Palladium-silver alloys possess the unique property of being highly permeable to hydrogen and virtually impervious to all other gases and vapors. This characteristic of palladium suggested its use as a semipermeable membrane separator.[190, 191, 213] Such a system can be designed with a high conductance for removal of hydrogen with a resultant large enrichment factor N. Because the membrane is impervious to all other organic and inorganic vapors, the yield is always 100%.

The hydrogen/palladium separator system was first proposed for use in extraterrestrial analytical systems.[214] The allowed payload in such an application imposes severe restrictions on the material, size, weight, and power of the components, and virtually eliminates the use of a liquid diffusion pump. If a small low-capacity, low-speed ion pump is to be used as the acceptable vacuum source, the separation factor for the GCMS separator must be extremely high.

Lucero and Haley discussed the theoretical aspects of a hydrogen/palladium separator and developed equations using design parameters to express the separation efficiency.[191] Their treatment was overly simplified by the assumption that the concentration of hydrogen in the membrane does not change down the length of the palladium tube. Although this is admittedly a loose approximation, the results permit an estimation of the physical dimensions that will give very high enrichments. Some caution must be used in applying the theoretical equations since the expressions permit attainment of an infinite enrichment whereas in practice, the approach to complete separation must be asymptotic.

Simmonds et al. tested the efficiency of a small hydrogen/palladium interface for general GCMS applications.[190] The separator was fabricated from a 75% palladium/25% silver tube that was 23 cm long and 0.025 cm ID with a wall thickness of .0062 cm. The volume of this separator is 1.2×10^{-2} cm^3 and the active surface area 1.8 cm^2. The palladium alloy was coiled into a helix and enclosed in a glass vessel with inlet and outlet tubes to admit external flushing gases.

Since the palladium separator does not permit organic vapors to pass through, and capture adsorption effects are negligible, the yield must be 100% for all intents and purposes. The enrichment factor is therefore the theoretical maximum. Thus,

$$N_{Pd} = N_{max} = V_{GC}/V_{MS} . \qquad (5.29)$$

The work of Simmonds et al. was directed primarily to space flight applications and it was convenient for them to define a specific palladium separator efficiency given by equation 5.30.

$$Pd \ sep. \ effic. = \frac{(H_2 \ flow \ in) - (H_2 \ flow \ out)}{(H_2 \ flow \ in)} . \qquad (5.30)$$

By appropriate substitution of V_{GC} and V_{MS}, it can be shown that the palladium separation efficiency is related to the conventional enrichment factor by

$$Pd \ sep. \ effic. = \frac{N-1}{N} . \qquad (5.31)$$

Data plotted on the basis of palladium separator efficiency give a convenient method to evaluate the parametric conditions leading to 100% efficiency or $N = \infty$. However, for mundane terrestrial applications, data expressed as the conventional enrichment factor permit a more facile comparison with other separators and also, a more realistic evaluation in terms of the needs of a standard laboratory. In Figure 5.24, interpolated data have been recalculated from Figure 2 of the paper by Simmonds et al.[190] The plot of separation factor N versus effluent flow rate V_{GC} shows that efficiency does indeed increase very rapidly at the lower flow rates. The condition for $N \to \infty$ is that the hydrogen conductance of the palladium-silver membrane must be at least $8-10$ times faster than the effluent flow V_{GC}. At higher temperatures, the conductance increases, thereby extending the range of the separator with respect to chromatographic flow.

Figure 5.24 would indicate that the useful range of the specified separator was limited to $2-4$ cm^3 atm/min in the temperature range 200–250°C. However, this restriction is a function of the area of the palladium surface and the permeability through the membrane can be increased by using a larger-diameter, longer-length tube or thinner membrane. In another application (high sensitivity gas chromatographic detection using gas transmodulation), Lovelock et al. designed a palladium-silver separating system (transmodulator) with an internal area of 8.5 cm^2 and length of 60 cm.[213] The hydrogen flux through this separator was in the range of $60-120$ cm^3 atm/min at temperatures of 200–250°C. It is thus apparent that the hydrogen/palladium separator can be designed for use with most GCMS flows. The yield is always effectively 100%, and the separation factor can be selected to maintain the mass spectrometer operating pressure well below the operating maximum.

One serious objection to general use of the hydrogen/palladium separator is the probability of catalytic reduction of the unknown during passage through the membrane. The possibility of this danger was recognized by the first proponents of the method, and reasonable studies were performed by Lovelock et al.[213] to determine the extent of chemical change. Fortunately, even at 250°C, sensitive compounds such as alkenes, aldehydes, and nitriles are not reduced unless the functionality occurs in conjugation with another unsaturated center. Those compounds that reduce do so specifically and essentially quantitatively. Consequently, firm identification of the reduction product by GCMS can be used to deduce the previous identity of the unknown. Confirmation by retention time is particularly valuable. For example, the identification of ethyl benzene could be due to the reduction of styrene; identification of propanenitrile could be due to the reduction of acrylonitrile. Such uncertainties are resolved by reference to retention indices. On the other hand, molecules such as 2-methyl-2-hexene, cyclohexanol, furan, 2,4-pentadione, benzonitrile, and furfural were quite stable at 250°C.

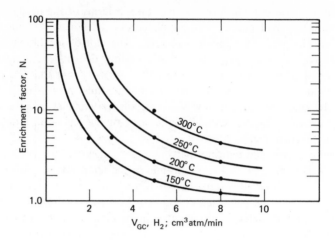

FIGURE 5.24. Enrichment factor as a function of flow rate V_{GC} for a small hydrogen/palladium separator. Surface area, 1.8 cm^2. Length, 23 cm. Points are calculated from Figure 2, reference 190.

Many sulfur compounds and iodine compounds will poison the palladium surface and severely reduce the hydrogen permeation. The effect is temporary for small exposures but when it occurs, V_{MS} is increased and pressure-activated safety devices may shut off the mass spectrometer. Such a trip generally terminates the GCMS run. At high concentrations of sulfur or iodine compounds, the palladium may be permanently inactivated and a reconditioning process is necessary.

When hydrogen carrier gas is completely removed at temperatures above 300°C, dehydrogenation reactions may take place on the palladium surface.

5. Measurement of Separator Efficiencies

Determination of separator efficiency is made by measurement of four parameters:

(1) chromatographic carrier gas flow rate to the separator V_{GC};

(2) the carrier gas flow rate to the mass spectrometer V_{MS};

(3) the quantity of sample entering the separator Q_{GC};

(4) the quantity of sample entering the mass spectrometer Q_{MS}.

Thus, in principle, evaluation of separator performance should be relatively simple and straightforward. In practice, most separators are operated with blind faith in the performance characteristics determined by other laboratories under different operating conditions. Even many of the papers describing separator applications fail to specify all four of the necessary evaluation parameters.

If the objectives of the investigation do not need the maximum possible sample utilization, the measurement of separator yield and enrichment factor is not important. If ultimate performance is necessary, or if the system performance does not seem to be up to expectations, simple measurement of the four efficiency parameters permits diagnosis of the problem.

Chromatographic effluent flow, V_{GC}. Measurement of carrier gas flow from the column to the separator is simple and straightforward, particularly if the column outlet is at ambient pressure as in conventional gas chromatography. Any flow meter suitable for the determination of effluent flow in normal chromatographic procedures serves the function in GCMS. Typical GCMS experimental setups in which the column exit may be at atmospheric pressure are shown in Figure 5.25.

For systems such as (a) or (b), the total flow is first determined at the entrance to the chromatographic detector with the metering valve or switching valve closed to the mass spectrometer. A second measurement is made with the valves open and the normal operating flow going to the mass spectrometer and separator. The difference of these two flows establishes the quantity of effluent entering the separator V_{GC}.

Incidentally, inherent in this measurement is the realization that the restrictor or metering valve leading to the separator must not have a conductance greater than the total flow through the chromatograph. Occasionally, this poses a problem when columns of different-size diameter are interchanged. Whenever unexpected leaks appear, the possibility of air back flow from the chromatographic detector should be suspected.

For a membrane separator (c), the flow rate to the separator is usually the normal chromatographic flow rate. Measurement is conventional and can be made by disconnecting the separator from the column. If this process is repeated frequently, a switching valve can be provided at this point.

If the chromatographic column is connected directly into the interface without splitting a fraction to another detector, the pressure at the end of the column may be different from one atmosphere, and gas flow through the column will be changed accordingly. Such an arrangement has been commonly used with jet separators, effusion separators, and the Teflon separator. The pressure at the column exit is usually less than one atmosphere and conse-

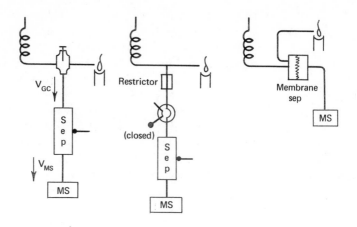

FIGURE 5.25. GCMS interface systems with the column exit at one atmosphere. (a) Variable flow to separator. (b) Fixed flow to separator. (c) Total flow to membrane separator.

quently, unless a pressure measurement is made at this point,[49] the quantity of carrier gas per minute exiting the column cannot be accurately established.

To have the same average linear flow rate through the column, and hence the same retention time, the column input pressure must be reduced by the same absolute amount as the exit pressure. It was shown on page 160 that this can result in an "enrichment" of a factor of three or more. However, if the input pressure is unchanged when the column exit is at a reduced pressure, the quantity of carrier gas through the column will be increased and a lower sample yield to the mass spectrometer may occur. Many of the papers describing separator performance do not specify the column exit pressure or whether the input pressure was changed for the GCMS run. As a result, it is difficult to assess the significance of the reported yield and enrichment data, or to make meaningful intercomparisons.

GCMS operation with the column under reduced pressure is basically sound and results in a lower helium flow with comparable chromatographic results.[183, 184] There is no indication of a theoretical plate loss that is significant in the interpretation of GCMS data. However, for separator evaluation, the system should be operated in a mode similar to those shown in Figure 5.25 and corrections are thus avoided.

When it is necessary to know the quantity flow of effluent in a reduced pressure operation, the value can be calculated using equation 5.1 in the same manner as was demonstrated on page 160. In the previous case, the outlet pressure under vacuum, $(P_0)_{vac}$, was assumed to be zero. In some systems it may be a few hundred Torr above zero, and unfortunately, it is often inconvenient to measure this pressure. For such cases, it may be necessary to measure the volume of the gas coming out of the separator and mass spectrometer forepumps by techniques described below.

Flow to the mass spectrometer V_{MS}. The gas flow into the mass spectrometer can be measured directly by means of a flow meter attached to the exit port of the mass spectrometer forepump. Simple bubble-type flow meters are commonly used. The procedure is generally successful, but because the gas is pumped out in spurts, the accuracy is not high. This adverse effect can be damped by using a large displacement-type flow meter with a relatively long sampling period. The success of the technique varies from laboratory to laboratory and also depends on the nature of the forepump.

Calibration of the ionization gauge with metered quantities of helium entering the mass spectrometer is another practical method for determination of V_{MS}. A very fine metering valve can be used to draw small quantities of helium from a reservoir at a measured flow rate. Figure 5.26 suggests two methods for this measurement. In (a), the reservoir is flushed with helium and closed off at one atmosphere pressure. The metering valve is opened to admit the desired flow as indicated on the ion gauge, and the bubble meter is activated to obtain the V_{MS} measurement. Several flow rates can be measured to obtain the gauge calibration curve for helium. Method (b) utilizes a difference method.[215] A standard flow is set with the metering valve closed. The desired helium flow is then metered to the mass spectrometer and the flow through the bubble is remeasured. The difference of the two flow measurements gives the value of V_{MS}.

The above techniques are used successfully to obtain V_{MS}, but the easiest and most practical way is by calculation using equation 2.2, $Q = PC$ (where $Q = V_{MS}$). Curiously, the straightforward calculation is seldom used. The vacuum conductance C of any GCMS system must be known if the overall performance is to be understood and evaluated. (See Chapter 4.) The ion gauge reading (multiplied by a factor of 7 to give helium pressure) is observed during each operation. The product PC thus gives the value for V_{MS} within a factor of 10–20%, and higher accuracy is not needed. In spite of the simplicity of this calculation, many researchers have gone through considerable effort to obtain direct measurement of V_{MS}. Many readers have suffered unnecessary frustration because simple vacuum system parameters are not specified, and pressure readings are not corrected or even stated as uncorrected.

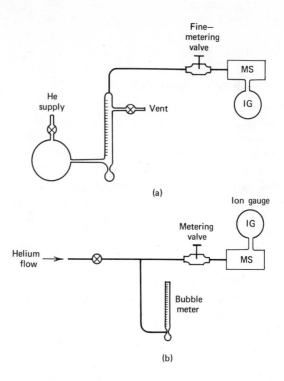

FIGURE 5.26. Apparatus for calibration of ion gauges as a function of helium flow rate. (a) Direct measurement. (b) Difference method.[215]

Quantity of sample to separator, Q_{GC}. The amount of sample entering the separator is given from the amount injected into the chromatograph multiplied by an appropriate factor to account for any stream split to an auxiliary detector. The accuracy of determination is limited by the precision of sample injection and measurement of stream split ratio. If V_{GC} is established as described above, Q_{GC} is easily decided.

Quantity of sample to mass spectrometer, Q_{MS}. The best way to establish the amount of sample from a chromatographic peak that enters the mass spectrometer is to compare the integrated peak area as measured by the mass spectrometer with the known mass spectrometer sensitivity. The chromatographic signal can be measured on the total ion monitor or by setting the mass spectrometer on a selected mass peak (e.g., mass 74 for a methyl ester). The mass spectrometer sensitivity can be determined as described in Chapter 2.

For example, suppose the electron multiplier is adjusted so that 10^{-9} gm/sec of methyl laurate from the batch inlet gives a 10-cm chart deflection due to mass 74 ions. A solution containing 10^{-7} gm of methyl laurate is then injected onto the chromatographic column and the total flow introduced to the separator. If the chromatographic peak due to the mass 74 signal is observed to be 7.5 cm high and 50 sec wide at the base, the signal at the top of the peak corresponds to a sample rate of 0.75×10^{-9} gm/sec. For a peak that is approximately Gaussian, the area can be estimated from the height x 1/2 base width. Thus, the total amount of methyl laurate entering the mass spectrometer is 0.75×10^{-9} gm/sec x 25 sec = 1.9×10^{-8} gm total. The yield would be $1.9 \times 10^{-8}/10^{-7} \times 100\% = 19\%$. (A more sophisticated determination of area can be made if the shape of the chromatographic peak is not close to Gaussian.)

The above technique is the easiest way to obtain the separator yield. The only pitfall, assuming stable mass spectrometer operation, may occur when helium is added simultaneously with the sample from the batch inlet for determination of the mass spectrometer sensitivity. In some experimental systems, the helium flows through the same line that connects the inlet leak. If the helium pressure at this point is high enough to permit viscous flow of helium back through the leak, the rate of sample effusion from the inlet is reduced (i.e., sample will be pushed back). Observation of a background peak will indicate whether diminished sample signal is due to a viscous back flow or a modified mass spectrometer condition. If it is the former, a correction can be applied. If it is the latter, the mass spectrometer focusing parameters should be reoptimized or the helium pressure should be reduced. (See page 228.)

Another popular method for measuring the mass spectrometer sensitivity uses the total integrated signal obtained from a known weight of sample introduced on a direct probe.[49, 187, 189] This procedure is fundamentally sound but can be in error if the probe is not completely butted into the port of the ionization chamber (see page 13). If sample effuses away from the ion chamber, the calculated yield will be high. Many separator yields reported from data using this method may be too high, but it is not possible to check the results a posteriori.

The separator yield can also be determined by comparing the integrated mass spectral signal from a separator sample with that obtained when the sample is introduced via a simple split (Figure 5.27). In (a), the effluent flow to the mass spectrometer V_{MS} is adjusted to the desired level with the splitter valve open. An aliquot sample is injected and the integrated signal S_1 is recorded. In (b), the same effluent flow V_{MS} is obtained by closing or throttling

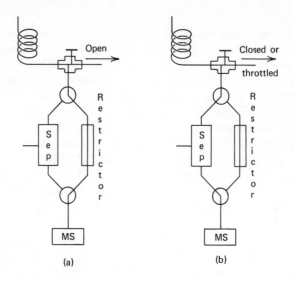

FIGURE 5.27. Convenient experimental set-up for determination of separator efficiency. (a) Simple split without using separator. (b) Sample enriched in separator.

the splitter valve. The same sample is injected but the signal S_2 is greater than S_1 due to the enrichment process. Since V_{MS} is adjusted to be the same in both cases, the ratio of the two signals gives the enrichment factor. Thus,

$$N = S_2/S_1 \tag{5.32}$$

The beauty of this technique is that, since V_{MS} is the same in both cases, the mass spectral signal is a direct measurement of the concentration of sample entering the mass spectrometer. Hence, when expressed as the ratio S_2/S_1, the measured enrichment factor is independent of V_{GC}, the split ratio, and the absolute value of V_{MS}. Of course, V_{GC} and V_{MS} must be known to obtain the yield using equation 5.5. The technique is very convenient when several separators are to be tested. It is easily adapted for test measurements using gas mixtures such as helium and xenon, or helium and freon. Because of the additional valves and lines, the technique is inconvenient for testing of only one separator.

Ryhage used an excellent collection method to obtain the yield of sample that goes through the first and second stage of his jet separator.[197] The interface was disconnected from the mass spectrometer. Sample was collected in a glass U-tube 0.25 cm x 60 cm, cooled in liquid nitrogen, and evacuated with a small diffusion pump. The condensed material (methyl stearate) was dissolved in solvent an an aliquot reinjected into the chromatograph for quantitative assay. The yield through a two-stage jet separator was in the range of 40–50% for an effluent flow rate of 27 cm^3 atm/min. A 7% correction for trapping efficiency was determined from a run without the separator.

Krueger and McCloskey[49] collected a standard sample in the tip of the introduction line at room temperature with the separator vacuum turned off. After condensation of the sample, the vacuum was turned on and the tip introduced to the ion chamber where the sample was rapidly heated and vaporized into the source in about one minute. The integrated ion current was used to establish the mass spectrometer response. The data obtained in this manner were reportedly reproducible but no mention was made of trapping efficiency. If significant losses occurred in the condensation step, the calculated separator yield would be too high.

With a Llewellyn-type membrane separator, the sample concentration can be measured with two nondestructive chromatographic detectors placed before and after the separation process. The depleted signal of the second detector gives a measure of the sample that has permeated the membrane, provided the quantity of helium is essentially constant. Although this method is probably the best technique for obtaining Q_{GC} and Q_{MS}, it is limited to the one specific separator.

6. Comparison of Different Separators

No one expects a single gas chromatograph to perform all chromatographic applications. No one expects a single mass spectrometer to perform all mass spectral applications. Yet it is commonly felt that one separator should perform over the entire range of GCMS analysis. The effectiveness of every separator depends on the chromatograph (flow conditions), the chemical sample (temperature), and the mass spectrometer (pumping speed). Considering the extreme variations that are encountered for these three systems, it is difficult to imagine any single-stage separator that would be proficient for all GCMS operations.

The choice of a separator can be made easier by considering the application in terms of the effluent flow rates into the separator V_{GC} and into the mass spectrometer V_{MS}. Since these operating conditions must be met, other demands of temperature, inertness, and time lag should be satisfied as secondary needs.

A high effluent flow rate may be defined as any value of V_{GC} above 15 cm^3 atm/min; low effluent flow would be from 1 cm^3 atm/min up to 15 cm^3 atm/min. High mass spectrometer flow would be 0.7–8.0 cm^3 atm/min; low mass spectrometer flow would be 0.07–0.7 cm^3 atm/min. These arbitrary divisions are crude but permit convenient classification of separators.

Separators for use with high V_{GC}, high V_{MS}. Large separation factors are not of prime importance when the mass spectrometer can accept a high V_{MS} quantity. The ratio of V_{GC}/V_{MS} will be in the range 15–60, and separation factors of 10–20 will thus give yields of approximately 25–60% which is more than sufficient for most samples.

A separator operating with high effluent flow must conduct gas out of the active separator cavity fast enough to avoid loss of efficiency due to pressure build up. In principle, a design with a larger conductance is always possible, but the restraints of size convenience have to be recognized. For example, a typical glass frit separator of 15–20 cm length might have a conductance of 3 x 10^{-2} liter/sec so that at 15 cm^3 atm/min flow (2.0 x 10^{-1} liter Torr/sec), the pressure would be close to 7 Torr. Any significant increase in flow rate would lead to a pressure increase which might cause viscous flow through the frit. If a flow of 50 cm^3 atm/min is desired, the frit of this separator would have to be 60–70 cm long. With metal frits, thinner walls and higher porosity eliminate this inconvenient restriction.

If an effusion separator is used for high V_{GC} and V_{MS} conditions, the expected yield can be calculated from equation 5.13. For a V_{GC}/V_{MS} ratio of 15, the yield will be 35–40% in the molecular weight range 200–400, but for $V_{GC}/V_{MS} = 30$, the yield would be reduced to 20–25%. A fixed effusive separator suffers even greater loss of yield if operated at a mass spectrometer flow rate below the optimum (see page 179). The variable separator of Brunnée[188] is greatly preferred when different effluent flow rates are encountered.

Higher efficiency can be obtained from effusive separators by operating two in series as a two-stage system.[198] The overall yield Y_T of a separator combination such as is shown in Figure 5.28 is given by the product of the yield for each stage, Y_1 and Y_2. For optimum advantage, the ratio of the gas split (i.e., $(V_{frit}/V_{MS})_1$ and $(V_{frit}/V_{MS})_2$) should be approximately equal, but the function is relatively flat over large variations of the flow ratios, so that the two-stage effusive separator is easily applied to advantage. For any separation device used in two stages,

$$Y_T = Y_1 Y_2 .$$

(5.33)

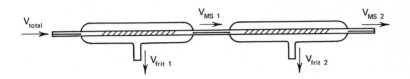

FIGURE 5.28. Two-stage effusion separator.[198]

Substituting in equation 5.13 for effusive separators gives

$$Y_T = \frac{1}{1 + 2\,\dfrac{V_{frit}}{V_{MS}}\sqrt{\dfrac{M_{He}}{M_S}} + \left(\dfrac{V_{frit}}{V_{MS}}\right)^2 \dfrac{M_{He}}{M_S}} \qquad (5.34)$$

where $(V_{frit}/V_{MS})_1 = (V_{frit}/V_{MS})_2$.

If the flow parameters in Figure 5.23 are

$$
\begin{aligned}
V_{GC} &= 30 \text{ cm}^3 \text{ atm/min} \\
V_{frit\,1} &= 24.5 \text{ cm}^3 \text{ atm/min} \\
V_{MS\,1} &= 5.5 \text{ cm}^3 \text{ atm/min} \\
V_{frit\,2} &= 4.5 \text{ cm}^3 \text{ atm/min} \\
V_{MS\,2} &= 1.0 \text{ cm}^3 \text{ atm/min}
\end{aligned}
$$

then $Y_1 = Y_2 = 1/1 + 4.5\sqrt{He/MW}$. For MW = 200, $Y_1 = Y_2 = 62\%$ and $Y_T = 38\%$. For MW = 400, $Y_T = 48\%$. The yield for the two-stage separator is thus seen to be about twice as large as that for a one-stage effusive separator used with the same overall flow rates of V_{GC} and V_{MS}.

The fixed two-stage effusive separator shown in Figure 5.28 is useful for a wide range of chromatographic conditions. However, it is optimized only at the V_{GC} value that gives the maximum V_{MS} allowed. When sample needs demand optimization over a wide range of chromatographic conditions (e.g., V_{GC} from 10–100 cm^3 atm/min), the second stage should be variable. Either the Brunnée system[188] or the variable viscous exit (see page 182) is suitable.

A one-stage jet separator is an excellent system for operation at high V_{GC} and high V_{MS}. Ryhage showed that at a flow of about 30 cm^3 atm/min, yields of 80–85% could be obtained through the first stage.[197] The estimated values of the enrichment factor and of V_{MS} were 12–20 and 2–4 cm^3 atm/min (page 185). A 20% reduction of the diameter of the jet receiving hole would reduce V_{MS} a factor of 2–3 (assuming viscous flow) and give very little loss of yield. Thus, an optimized jet separator could be designed from empirical data for specified flow conditions. Loss of efficiency caused by operation at nonoptimized conditions will not be as important due to the high yield.

The one-stage membrane separator is also ideal for use at high column and mass spectrometer flow rates. Grayson and Wolf[210] obtained yields of 85–95% at effluent flow rates of 30–60 cm^3 atm/min. The enrichment factors ranged from 10–30 when the temperature was properly adjusted. This particular separator had an active area of 15 cm^3 and passed helium at a rate of 2–4 cm^3 atm/min. For a first approximation, helium permeation is proportional to the area, whereas sample permeation is given by equation 5.23. Thus, if the active area is reduced a factor of 4, V_{MS} would be 0.5 cm^3 atm/min but a yield of 90% would be reduced only to 45%. (See Figure 5.21.)

The Teflon separator was originally proposed for high-temperature work (300°C) with high V_{GC} and V_{MS} values.[189] The reported yields are satisfactory but leaks frequently develop during operation. Because of satisfactory performance of other separators, the Teflon system has not gained popularity.

The hydrogen/palladium separator is not intended primarily for high V_{MS} operation, but this would not exclude its application here. However, it would generally be avoided because of the ever present danger of unknown catalytic effects.

Separators for use with high V_{GC}, low V_{MS}. Quite obviously, this is the most difficult GCMS flow condition that can be encountered. The situation arises when it is necessary to couple a packed column at a high flow rate with a single-pumped mass spectrometer. The ratio of V_{GC}/V_{MS} may be 100–500. Depending upon the nature of the sample, the quantity available, and the desired information, a highly efficient separator may be necessary.

The only single-stage separator capable of giving a high yield with low carrier gas to the mass spectrometer appears to be the hydrogen/palladium separator. The permeation rate of hydrogen through a thin palladium foil is around 60 cm^3 atm/min/cm^2 at 300°C. Lovelock et al. have discussed design parameters that will give the desired permeation rate.[213] Thus, in principle, the system gives 100% yield of organic material with an extremely low flow

V_{MS} to the mass spectrometer. The disadvantage of catalytic activity has been discussed and although always present, it is not always serious. The hydrogen/palladium separator should be considered whenever an extreme V_{GC}/V_{MS} ratio is encountered and ultimate sample utilization is essential. The proposed extraterrestrial applications are excellent examples. It is questionable that our worldly needs will exceed the ability of other alternative separator systems.

The available data on the other separators indicate that a single-stage design is not practical for high enrichment. For example, suppose the chromatographic effluent flow rate is 50 cm^3 atm/min and the maximum flow to the mass spectrometer is 0.1 cm^3 atm/min (1.3 x 10^{-3} liter Torr/sec). A simple effluent split without separator gives only a 0.2% yield (V_{GC}/V_{MS} = 500). A single-stage effusive separator operating with these flow requirements gives 2% yield, N = 10 (MW 400). The diaphram separator described by Grayson and Wolf would have the active area reduced a factor of 20 to deliver 0.1 cm^3 atm/min, and the yield calculated from equation 5.23 using the Grayson and Wolf data as a standard would be 12%, N = 60. A single-stage jet separator with the dimensions suggested by Ryhage would pass excess gas to the mass spectrometer, but this quantity can be reduced a factor of 20 by reducing the diameter of the receiving jet port (by approximately 1/2 assuming viscous flow conditions prevail). The first-stage yield reported by Ryhage was above 80%, so that, depending on the concentration gradient in the jet core, a yield of 6–12% might be anticipated from the smaller single-stage jet. Unfortunately, the reduced jet will be very susceptible to being plugged.

All of the above yields can be increased by using the separator in two stages. Ryhage showed that his two-stage jet separator could give a yield of 40–50% with V_{GC} = 27 cm^3 atm/min and V_{MS} = 0.25 cm^3 atm/min.[197] He also showed that the yield could be increased by increasing the flow rate. This could be interpreted to indicate that the pressure in the second jet stage was too low to give efficient enrichment. For that particular separator, optimum total flow was in the range V_{GC} = 30–35 cm^3 atm/min.

Llewellyn and Littlejohn first introduced the two-stage membrane separator for use with an ion pump vacuum system.[192] Yields were reported in the range 30–50%. With proper adjustment of the pumping speed in the second stage, the helium pressure is kept to about 2 Torr, and consequently, the amount of helium dissolving into the second membrane is about 400 times less than in the first. The value of V_{MS} is around 0.01 cm^3 atm/min ($\sim 10^{-4}$ liter Torr/sec), so that enrichment factors as high as 600–1000 are easily attained.

The two-stage membrane separator is highly efficient, but if high chromatographic efficiency is important, alternative methods should be investigated. The second stage of membrane separation introduces an additional time lag. Black et al. found that the chromatographic theoretical plates of a 150-meter x 0.075-cm ID capillary column were reduced a factor of 3–5 during the enrichment process through a one-stage membrane separator.

Considerable improvement in the performance of effusive separators is attained by two-stage operation. This was shown to have practical application for high V_{GC}, high V_{MS} conditions. The improvement is less striking for low V_{MS}. The calculated yield of an optimized two-stage Watson-Biemann separator is 6.2% for V_{GC} = 50 cm^3 atm/min and V_{MS} = 0.1 cm^3 atm/min. For the same flow rates, a three-stage effusive separator would have a yield of 12%, but such a contraption is considered impractical and gives rise to unnecessary experimental nuisance.

Table 5.3 summarizes the performance that can be expected from various GCMS connections when used in the difficult condition of high V_{GC}, low V_{MS}.

Separators for use with low V_{GC}, low V_{MS}. The common GCMS set-up in which a capillary column is connected to a single-pumped mass spectrometer is classed in the category low V_{GC}, low V_{MS}. The ratio V_{GC}/V_{MS} is in the range 10–100 and many research problems can be solved without using a separator. For example, suppose a flow of 5 cm^3 atm/min is used in a SCOT column and 0.1 cm^3 atm/min goes to the mass spectrometer. Even though only 1/50th of the sample is being utilized, reasonably good mass spectra could be obtained from peaks that are only 0.1% of the total sample (see page 161). If sample utilization must be higher, then a separator is necessary.

A single-stage effusion separator will give the additional yield desired for most applications in this flow range. Even for an unfavorable V_{GC}/V_{MS} ratio of 100, the yield in the molecular weight range of 200–400 is 7–10%. For V_{GC}/V_{MS} = 10, the yield will be greater than 40%. The effusion separator is preferred for simplicity of operation and economy. The pressure conditions necessary for molecular flow in the effusion path and viscous flow in the exit restrictor are easily attained. A double-stage effusion separator will occasionally be useful when V_{GC}/V_{MS} is around 100 (Y will be doubled), but is of little additional value when V_{GC}/V_{MS} is around 10.

A single-stage jet separator is often used when effluent flow is low,[207, 208] but caution must be taken to assure that the supersonic flow condition of the jet is maintained. If the internal pressure of the jet gets too low, the enrichment effect drops to zero[203] and the separator is, in effect, only an expensive splitter. Make-up gas is often added to improve this marginal performance.

TABLE 5.3. Yield and enrichment from various separators with total effluent flow of 50 cm^3 atm/min and flow to mass spectrometer of 0.1 cm^3 atm/min.

Separator	Yield, %	Separation Factor
No separator, sample split 500/1	0.2	1
1-stage effusion	2^1	10
1-stage jet	$6-12^2$	30–60
1-stage silicone membrane	12	60
1-stage H_2/palladium	100	high[3]
2-stage effusion	6	30
2-stage jet	$\sim 40^4$	48^4
2-stage silicone membrane	$30-50^5$	$300-500^6$

[1] Molecular weight, 400.
[2] Estimated, data not available.
[3] Depends on dimensions and operational mode of separator. May exceed 1000.
[4] Data for V_{MS} = 0.25 cm^3 atm/min.[197]
[5] Exact data not available.
[6] Flow to mass spectrometer $\sim$ 0.05 cm^3 atm/min.

Single-stage membrane separators have proven to be effective for low-flow applications provided the active membrane area is reduced so that helium permeation is low.[193, 211, 212] Excellent yields (20–40%) can be obtained even with V_{GC}/V_{MS} ratios close to 100. Of course, modest peak broadening and temperature limitations must be accepted.

The hydrogen/palladium separator is excellent for low-flow conditions if very high yields are essential.[190] The danger of catalytic activity must be accepted and the application will generally be highly specialized.

Separators for use with low V_{GC}, high V_{MS}. This is the idealized GCMS combination in which chromatographic flow has been kept low and the mass spectrometer vacuum system has been designed with high conductance. The V_{GC}/V_{MS} ratio may be anywhere from 1–10 and additional enrichment will seldom be necessary. Separators should be avoided unless an unusual analytical situation indicates the need for the additional yield factor. Even so, operation of the column under reduced pressure[183] will usually gain a factor of 2–4 in sample concentration (page 160).

When a separator is necessary, the enrichment factor does not have to be high. A simple effusion separator gives yields above 40%. The jet separator might be less effective if the pressure in the jet is too low. Membrane separators can be used, but their inherent operational difficulties are no longer offset by the need for a high separation factor.

Hybrid two-stage separators. From the foregoing discussion, it is apparent that different separators have specific advantages when applied in the proper manner. It was shown that pure two-stage separators give improved enrichment, but the characteristics that make a specific separator applicable in the first stage are not necessarily preferred for the second stage. Membrane separators and the jet separator can be designed for high efficiency at high chromatographic flow rates. The effusion separator is usually more effective at low flow rates.

The frequent need for increased separator yield suggests that the best performance may often be attained by using a hybrid two-stage system. Grayson and Wolf[210] designed a very practical hybrid separator that couples a high-flow silicone membrane in the first stage (Figure 5.18c) to the low-flow Blumer[187] effusive separator in the second stage. Efficient operation was attained at flow rates up to 60 cm^3 atm/min. When temperature is optimized, the first-stage yield is 90% or greater with 2–4 cm^3 atm/min of helium permeation to the second stage. Because of the low flow to the frit, the enrichment factor N in the second separation was less than 2, but yields were 50–60%. The overall yield Y_T was 30–40% with enrichment levels N_T of 20–30. Minor time lag effects were reported, but these are minimized by having only one stage of membrane enrichment.

Another practical hybrid combination is a jet-effusion separator. The jet system operates effectively at high flow rates with excellent yield. At lower flow rates, maximum efficiency is not attained, but more important, the separation effect can be lost completely if the gas is too dilute.[203] By combining the jet with an effusive separator, the second stage will operate more efficiently at reduced pressure.

It is questionable whether other hybrid combinations would be advantageous. The hydrogen/palladium separator can be designed and operated to remove virtually all the hydrogen gas so that a second stage is of no value. The Teflon separator has not been extensively investigated, but in view of reported operational problems (cracking of Teflon at high temperature), a combination seems to offer no preference over the two hybrid systems suggested above. A variable effusive separator would give easier operation in either of the suggested associations. The combination of a variable effusion separator with a fixed effusion separator is practical (page 181), but yields would be lower than for the jet-effusion or membrane-effusion unification.

6

Operational Techniques of the GCMS System

A. DEVELOPING THE GAS CHROMATOGRAPHIC METHOD

To attain satisfactory performance of a GCMS combination, operation of the gas chromatograph, the interface, and the mass spectrometer must all be optimized. Neglecting the demands of any one of these components will give poor results or even complete experimental failure. As a point of habit, the interface tends to be considered first; mass spectrometric needs are considered second; and finally, if at all, the possible modification of the chromatographic operation is casually examined. This nonchalant regard for chromatographic optimization seems to arise from familiarity rather than uncertainty. More is the pity. Fully one-third of GCMS failures result from this neglect.

Before GCMS analysis of a sample is obtained, a conventional gas chromatographic run should be made. The best column for maximum separations should be selected, and the temperature and flow conditions optimized.

The objectives of the GCMS research problem should now be examined. How much sample is available? What peaks are already identified? Is mass analysis needed for all peaks of interest? What percent of the total sample is solvent? Are there any involatile residues that reduce the real quantity of volatiles? The answers to these questions are nothing more than standard knowledge of the sample and of the research problem, but unless they are known, neither the chromatography nor the GCMS run can be performed for maximum achievement.

222

When the answers are known, the most favorable interface technique can be determined and the probable success of the GCMS run can be evaluated. The ratio V_{GC}/V_{MS} gives the factor by which the sample amount is reduced when split to the mass spectrometer. If the smallest peak of interest is not sufficient to be in the useful range (probably between 10^{-11} and 10^{-10} gm/sec), a separator is necessary.

What type of separator? The answer to that question depends upon the factor by which the smallest peak to be mass analyzed fails to meet the minimum sample level. If it is a factor of 10 times too small, the separator enrichment factors should be 10 or more; if it is a factor 100 times too small, the separator enrichment factor should be 100 or more. It can be seen from Table 5.3 that even with an extreme V_{GC}/V_{MS} ratio of 500, an enrichment factor of 100 is not easily attained. A two-stage jet or membrane separator would be most effective, although a hydrogen/palladium separator could be designed if that separator system is applicable.

If the yield of the smallest peak is still not sufficient, consideration must be given to the quantity of effluent flow. A gas chromatographic method using 50–60 cm^3 atm/min may be modified for efficient separation at 15–20 cm^3 atm/min by using higher temperatures. Modifying the chromatography to meet the demands of the GCMS does not give extreme gains, but for a chemical system needing maximum sample utilization, every factor must be considered.

Needless to say, no separator or interface technique can give more than 100% yield. If calculation of the total quantity of sample in the small peaks indicates that the amount is insufficient, enrichment must be performed outside of the GCMS combination. Section trapping is often very effective. The collection efficiency for small pure peaks may be very low, but by collecting a poorly separated section containing 15–20 components, the small peaks are retained in good yield and noticeably enriched in the process. In some cases, it is advantageous to do the section trapping with one stationary phase and the GCMS analysis with another.

Prior separation by chemical classes is an extremely valuable process when GCMS analyses are to be performed on complex mixtures.[216-217] Acid/base/neutral separations or polarity separations can lead to type enrichment of two to three orders of magnitude. Figure 6.1 compares the chromatogram of a total hop oil and that of the oxygenated fraction separated by column chromatography.[217] The total oil is dominated by terpene hydrocarbons, particularly humulene, myrcene, and farnesene. The total oxygenated material is in the range of 1–5%, so that minor oxygenated peaks cannot be analyzed by GCMS. Furthermore, the abundant hydrocar-

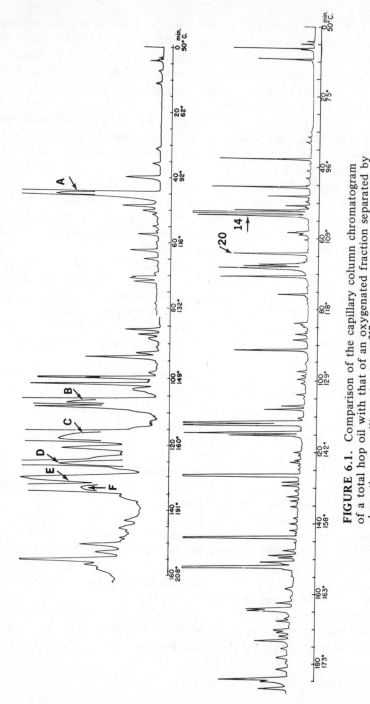

FIGURE 6.1. Comparison of the capillary column chromatogram of a total hop oil with that of an oxygenated fraction separated by absorption on a silica gel column.[217]

bons would make the spectral interpretation of the minor oxygenated components quite difficult. Column separation of the polar fraction permits GCMS analysis of components that are two orders of magnitude less abundant.

The sharpness of a chromatographic peak is an important factor in determining the utilization of the sample in GCMS. A component peak delivering 10^{-8} gm total to the mass spectrometer will yield 10^{-9} gm/sec for a peak of half-width of 10 sec, but only 10^{-10} gm/sec for a peak of half-width 100 sec. Some of the factors that influence peak sharpness are (1) % load of a stationary phase, (2) carrier gas flow rate, (3) temperature, and (4) type of column or column diameter. Although it is not possible to vary these parameters indiscriminately, one should examine the chromatographic method to be sure that obvious points have not been overlooked.

For most GCMS work, stationary phase load should not exceed 4–5%, and 1–2% will often give acceptable separations with sharper peaks.[112] Occasionally, there could be a need for a large sample charge that will dictate use of 10–20% stationary phase loads. However, the increased sample charge may be self-defeating, and the increased amount of low abundance components will be of minimal value if the chromatographic peaks are broadened.

To keep the chromatographic peaks sharp for GCMS analysis, the column temperatures should be as high as can be allowed, consistent with the necessary degree of separation. Unfortunately, at high temperatures, column bleed often interferes with mass spectral interpretation and thus restricts the choice of stationary phase (see page 90). In addition, thermal instability of the sample may preclude use of higher temperatures (see page 235). Temperature programming[117] is commonly used because of the wide range of volatility encountered in GCMS samples, and a fast temperature program is often advantageous to obtain sharp chromatographic peaks.

Flow programming is used in conventional gas chromatography as a means of increasing the rate of component elution without imposing excess temperatures.[111] The danger of excess column bleed and thermal decomposition is thus conveniently reduced. In developing a GCMS method, every instinct is directed towards reducing the volume of gas, and there is a natural aversion to increasing the flow. However, it has been shown repeatedly that many GCMS analyses do not need maximum sample utilization, and there may often be an order of magnitude of sample excess. Flow programming should be considered as a means of trading off quantity of sample to reduce bleed and decomposition. However, even a factor of four or five increase in the flow rate may not cause significant waste of sample. If thermal instabilities in the column or sample have dictated a low temperature limit, the later chromatographic components will be eluted slowly with reduced maximum peak intensity. By flow programming, the peaks can be eluted faster and the peak intensity will be increased accordingly provided it is consistent with good chromatographic separation. This is illustrated in Figure 6.2. In (a), the peak is eluted with a low temperature and low flow rate and, consequently, spreads out with a half-width of about 150 sec and a shallow peak height. In (b), the temperature is the same but the flow is increased a factor of three so that the peak has a half-width of 50 sec and three times higher peak intensity.

If the sharper peak is now split through a simple nonenriching interface, the concentration of sample to the mass spectrometer will be effectively the same as that for the shallow peak. However, if a separator interface is used, the higher-intensity peak will give increased concentration over a shorter time period.

The problems of varying chromatographic flow rate in most fixed separators have been thoroughly discussed (page 175), and often a fixed separator will be incompatible with a chromatographic flow program. A nonvariable effusive separator can be optimized only at the high flow range. If optimized for the lower effluent flow range, the quantity of helium going to the mass spectrometer at higher flow will exceed the allowed amount. A variable effusive separator of the Brunnée type is needed to accomodate a chromatographic flow program. If this separator is used at a flow of 20 cm^3 atm/min through the effusive path and 1 cm^3 atm/min to the mass spectrometer, the yield will be close to 25% (calculated from equation 5.13). If the flow

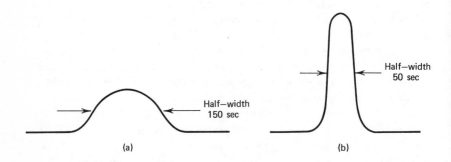

FIGURE 6.2. Use of flow programming in GCMS. (a) Peak taken at 20 cm^3 atm/min. (b) Peak taken at 60 cm^3 atm/min. With effective use of a separator, peak concentration of sample to the mass spectrometer can be higher at the higher flow rate.

through the effusion path is increased to 60 cm^3 atm/min and the flow to the mass spectrometer kept at 1 cm^3 atm/min, the yield will be 11%. However, since the flow program will have increased the peak height by a factor of three, the concentration of sample to the mass spectrometer will be equivalent to a 33% yield as compared with the previous low flow yield of 25% obtained from the flat peak.

A similar problem is expected if a flow program is used with an optimized jet separator. If the separator is already passing the maximum quantity of gas to the mass spectrometer at the lower flow rate, the flow program would not be allowed. However, because of their high efficiency, jet separators are frequently used with a low quantity of carrier gas to the mass spectrometer, and thus the system can accomodate higher effluent input. As a consequence, flow programming can give peak sample concentration of 2–4 times higher.

The membrane separator is excellent for chromatographic flow programming[210] since the quantity of carrier gas to the mass spectrometer does not vary significantly with flow (page 195). On the other hand, the efficiency of transfer of organic material with varying flow depends upon the operating conditions. From Figure 5.16 it is seen that if the yield is 95%, $(A/l)(SD)/V_{GC}$ equal to 3, increasing the flow by a factor of 3 decreases the yield to 62%, $(A/l)(SD)/V_{GC}$ equal to 1. Since the peak intensity will be three times greater (Figure 6.2), the sample rate to the mass spectrometer at the top of the peak is approximately two times greater than that obtained at the lower flow. However, if the membrane separator is being used at low efficiency, say 50% yield, a threefold flow increase will decrease the yield to 20% and the partial pressure of sample received at the mass spectrometer is then essentially the same in spite of the increased concentration of the peak.

Close correspondence between the conventional chromatographic run and the GCMS run should be maintained to assure a reliable intercomparison of pattern and retention indices. For convenience, the outlet of the column should be at atmospheric pressure during the GCMS run (page 158). If the column outlet pressure must be reduced, the inlet pressure should be adjusted to give the same retention times (i.e., the same average linear velocity) as are obtained in normal operation. As an approximate rule, the absolute pressure drop across the column should be constant. For example, if the pressure at the front of the column is 1.3 atmospheres (gauge reading) for normal operation, the gauge pressure would be reduced to 0.3 atmospheres for operating the column exit at a vacuum level. This adjustment results in a fairly close match of retention times[183, 184] and minor variations can be corrected by more closely matching the retention time of a light solvent peak or air peak.

B. ESTABLISHING THE MASS SPECTROMETRIC OPERATING PRESSURE

To obtain maximum sample utilization in GCMS analysis, the mass spectrometer should be operated at the maximum allowable pressure. The only exception to this rule occurs when using a highly efficient separator that removes almost all the carrier gas (e.g., hydrogen/palladium or two-stage silicone membrane).

The various factors that necessitate high vacuum in mass spectrometry were discussed in Chapter 4. Rules were formulated that permitted an estimation of the maximum pressure limit in the three distinct sections of the mass spectrometer (ion chamber, ion source housing, and analyzer)(Table 4.1). However, it was clearly pointed out that because of the uncertainties in the physical significance of mean free path in an ion/molecule optical system, the formulae are only approximate. The allowed quantity of gas may vary a factor of two or three from the calculated values and hence, the desired operating pressure must be confirmed empirically.

The maximum operating pressure in a mass spectrometer can be established by observing the output signal from a known sample as the helium pressure is increased. The signal will be constant up to the maximum allowed pressure. The procedure is simple, but there are a few pitfalls. Sample can be added from a standard inlet system or from a probe, but both of these sources may be subject to error, and a steady background peak is often more reliable.

The signal from a probe standard sample may fluctuate due to uneven volatilization, and the operator is then uncertain as to the exact point at which detrimental pressure effects are observed. The batch inlet system should be ideal for such a function, but sometimes the sample signal is observed to decrease with increasing pressure before the background signals decrease, thus indicating that less sample is going to the mass spectrometer. As was pointed out on page 211, this occurs when the molecular leak is in the line that conducts the effluent gas. At higher flow rates, the pressure of carrier gas may be sufficient to cause some viscous flow of helium back into the batch inlet with a consequent reduction of sample effusion. The occurrence of this back diffusion effect can be checked by close observance of background signals. However, it is more convenient simply to use a background signal as the monitor. For the maximum pressure test, a mass scan is not important, and it is convenient to focus on one peak and observe the signal on a strip chart recorder. The multiplier gain can be set quite high to give a reasonable-size signal for background peaks. Best results are obtained by focusing at higher mass ranges, above 100 if feasible.

The test peak used to monitor pressure effects should have constant intensity with increasing helium pressure up to the point at which the mean free path of the ion is too short. In practice, electrical field aberrations are caused by the high charge density of carrier gas ions, and these aberrations change the focusing properties of the ion source. Electrical adjustment is often necessary to maintain a constant peak height. Sometimes the peak is displaced on the mass scale, necessitating magnet adjustment. With care, these effects should not interfere with the determination of maximum pressure, but it is important to recognize the existence of these problems. Since many mass spectrometers have different focusing properties when operated with a pressure of 10^{-3} Torr or more in the ion chamber, it is essential that the focusing adjustments are made under the conditions that will be used in the GCMS run.

When the instrument is thus optimized, the height of a background monitor peak will remain constant as the pressure is slowly increased until the ion mean free path is too short for good transmission. The ion signal decrease and the resolution of the mass spectrometer will be noticeably lower. The pressure at this point represents the maximum operating level of that particular mass spectrometer. For safe and efficient operation, the GCMS runs should be performed at approximately one-half this pressure level.

The limiting pressure determined experimentally should be within a factor of two of that calculated from equation 4.2 ($P = 5 \times 10^{-3}/L$). These calculations were summarized for different mass spectrometer configurations in Table 4.1. When the pressure is established, the conductance of the pumping system can be used to calculate the total quantity of gas that may enter the mass spectrometer ($Q = PC$). (Remember, if the ion gauge is calibrated for air, the reading will be 7–8 times lower than the true pressure). When the exact value of V_{MS} is thus known, the efficiency of the GCMS interface can be estimated and the demands on a separator are easily appraised.

C. TEMPERATURE CONTROL IN THE GCMS INTERFACE

1. General Considerations

There is no mystery in specifying the temperature desired in the GCMS interface. As a first standard operating rule, all interface parts should be maintained at the temperature of the column. If the column is temperature programmed, the interface temperature can also be programmed or held at the maximum column temperature, whichever is more convenient.

Actually, the column temperature is higher than that needed in the interface. Any organic material has a lower vapor pressure in equilibrium with the chromatographic liquid phase than in equilibrium with a clean metal or glass surface. As a result, the interface temperature can be lower than the maximum column temperature. The exact difference allowed is not well established and depends upon the nature of the substrate, the effluent compound, and the history of the interface surfaces. An empirical rule that works fairly well is that the interface temperature can be lower than the column temperature by 15–20°C for each 100°C of column temperature (i.e., column 100°C, interface 80°C; column 300°C, interface 250°C). From first appearances, there would seem to be no problem. In real life, extremes of interface temperature, either too high or too low, are one of the most common causes of GCMS abortion.

The GCMS interface can be heated in a small oven or separately wrapped with heating tapes. In either case, all parts of the lines leading to and from the interface components must be at the proper temperature. A typical interface system is shown in Figure 6.3 with six separate temperature regions.

Regions 2, 4, and 5 represent the connecting lines between the column, the interface, and the ion chamber. Conventionally, these regions are brought to temperature with heating tapes or a coil of heating wire. Thermostatic control is unnecessary, but a thermocouple or other measuring device is extremely important for each section. Lamentably, this simple measurement is often ignored or performed inadequately. Provision should be made to monitor the temperature of a heated line at three points: (1) on the line, (2) right at the point of exit from or entrance to an oven or mass spectrometer, and (3) especially at any bends or elbows. It is not uncommon to find an elbow that is 25–30°C lower than the rest of the line.

Region (5) is inside the mass spectrometer envelope and if a thermocouple is not provided by the manufacturer, it is usually difficult to add one. If the temperature of this line is unknown, the spectroscopist must increase (or decrease) the heater input until the chromatographic peaks are observed to stop (or start) tailing. This seemingly inadequate procedure is widely used with satisfactory results.

The separator and valves may be heated by tapes as described above or housed in a separate oven. In both cases the temperature should be carefully monitored. Because of the physical bulk of the interface device, heat may be carried out more rapidly than by the small line pieces, and significantly lower temperatures will result. Even a convection-circulated air oven may have wide temperature variations, and themocouple measurements should be made at the top and bottom of the oven, on the separator, and on all valves. As an extreme example, a tall narrow separator oven without forced circulation can have 50–100°C differential between the top and bottom. Chromatogra-

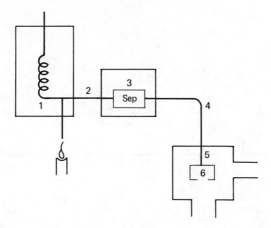

FIGURE 6.3. Schematic diagram of the GCMS system showing six common temperature regions.

phic peaks will tail badly unless additional circulation or localized heat input is provided.

The occurrence of cold spots can be reduced by careful design of the effluent flow system. Figure 6.4 shows the method used to heat two commercial interface systems. In Figure 6.4a, most of the parts are enclosed in a metal heating block. Region 2, between the column and separator, is enclosed in the block, and a cold spot is unlikely. Region 4 is so small that it is effectively nonexistent. The temperature of region 5 inside the mass spectrometer is controlled by internal heaters. This well-designed commercial system will be free from cold spot problems. However, it offers no advantage over the more open interface shown in Figure 6.4b, provided the temperature is carefully monitored and controlled. The delay time introduced by 30–40 cm of vacuum line is negligible.

The ionization chamber, region 6, must be considered as part of the system for proper temperature control. If the column is operated up to 300°C, the ionization chamber must not be lower than 250°C or sample condensation will occur. To minimize the possibility of variations in the mass spectral pattern over several months, the operating temperature should always be the same (see page 54). This suggests that the maximum temperature that will ever be needed should be established as early as possible in the history of the mass spectrometry laboratory, and this temperature should be used whenever possible. If special sample requisites demand a higher or lower

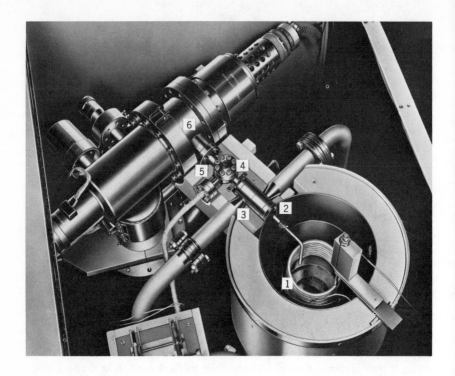

FIGURE 6.4. Photographs of two commercial GCMS interface systems. (a) Interface parts heated in oven or metal blocks (LKB Instruments, Inc.).

temperature, the change in the spectral pattern is not too serious and must be accepted in deference to the needs of the sample.

One way to avoid cold spots in the interface lines is to use an excessively high temperature. Unfortunately, this may lead to the other temperature problem, namely, thermal decomposition in the interface. Many organic materials are cracked by contact with a hot metal surface. The extent of the breakdown for a given compound depends on the type of metal and its previous history. Often a new clean interface surface will give no discernible disintegration. When used for a few weeks, minor sample damage will cause a buildup of thermal products (tars, carbon, etc.) which further catalyze the decomposition to a harmful level. Because of the unpredictable nature of thermal cracking, interface and line temperatures should be kept as low as possible consistent with proper flow of effluent sample.

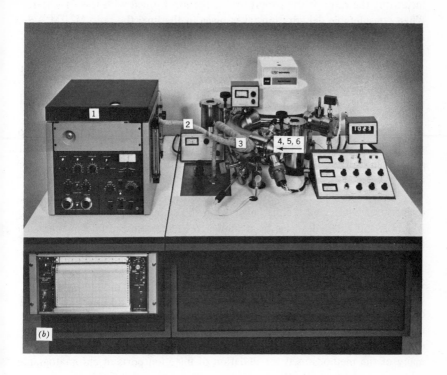

FIGURE 6.4. Photographs of two commercial GCMS interface systems. (b) Interface parts wrapped with heating tapes (Du Pont Products Division).

Sample decomposition can be reduced by use of inert interface materials such as stainless steel, glass, sapphire, or gold.[188] In addition, since much of the sample injury occurs on active catalytic sites, silylation of the interface surfaces will greatly reduce this damage.[218]

For many chemical systems, it has been shown that a large slug of sample (10–20 μgm) will act as its own conditioning agent, and only a small fraction of the total is consumed on the active interface sites. Of course, this means that a small sample (less than 1 μgm) will be totally lost, a severe consequence often suffered in microchemical manipulations. Brunnée et al. demonstrated the effect for different surface conditions by passing a specified flow of cholesterol through capillaries of different material and measuring the ratio of the parent ion to the ion formed with loss of water.[188] When considerable dehydration takes place in the line, the ratio $M^+/(M - 18)^+$ is very low. When the dehydration sites are used up, the relative intensity of the M^+ ion increases due to reduced decomposition (Figure 6.5).

If a very small quantity (subnanograms) of a sensitive biochemical compound must be measured, it is possible to protect the sample by adding an excess of a similar homolog or a deuterated derivative. The additive material should elute shortly in front of, or simultaneously with, the unknown. The compound of interest must have a mass peak that is not present in the added carrier material so that this peak can be used as a chromatographic monitor for good quantitative detection of the sample.[219-221] (See also selective ion monitoring, page 252, and Watson, page 396.)

Similar effects have been observed for other materials. The minimum quantity of unaffected sample depends on the stability of the compound. Alcohols and aldehydes have been shown to be more susceptible to thermal breakdown than ketones, esters, olefins, or ethers.[218]

2. Interface Silylation

Since thermal reactions are partly due to catalytic activity of a small number of sites, silylation of chromatographic components has become a common method for decreasing substrate reactivity.[49, 187, 188, 218, 222-225] The techniques are easily applied to interface parts and, with minor modifications, many of the methods used for silylation of injection ports or chromatographic substrates can be applied directly to the components of a GCMS interface. Dimethyldichlorosilane, $(CH_3)_2SiCl_2$, has been used with satisfactory results although some reports indicate that all reagents of this type $((CH_3)_3SiCl)$ are not completely acceptable.[49] Periodic conditioning with additional reagent (e.g., Silyl-8, Pierce Chemical Company, Rockford, Ill.) assures a longer deactivation lifetime.

McLeod and Nagy[218] found that direct in situ silylation of interface parts with N,O-bis-(trimethylsilyl)-acetamide, $CH_3C[OSi(CH_3)_3] = NSi(CH_3)_3$, gave improved performance. The silylating reagent was slowly added through a modified injector located at the entrance of a Watson-Biemann separator. The interface was heated to 150°C, and the mass spectrometer vacuum system was used to remove excess solvent and reagent. With an untreated interface, reactive compounds such as linalöol, menthol, and geranial could not be analyzed for less than 10^{-6} gm of sample. The treated surface permitted analysis at the 10^{-9} gm level. One possible advantage of McLeod's method is that all interface parts, including the ion chamber, receive a similar silylation treatment.

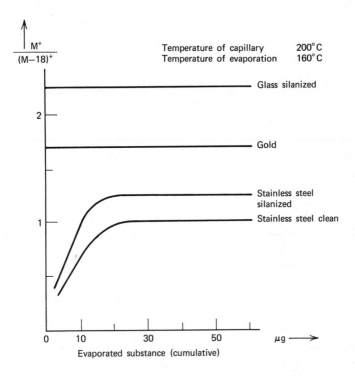

FIGURE 6.5. Influence of type of material and surface treatment of the connecting lines on the decomposition of cholesterol.[188]

The success of a silylation procedure depends upon the compounds involved in the analysis and on the past history of the interface surfaces. Frequently, a special technique works well in one laboratory but fails in another, and predictions on the effectiveness of a procedure are difficult to make. Often two or three treatments may be necessary before an inert surface is attained. In general, greater success is attained with the more active reagents such as N,O-bis-(trimethylsilyl)-acetamide or N-methyl-N-(trimethylsilyl)-trifluoroacetamide[222] but other reagents and methods should be considered when standard processes fail.

3. Derivative Formation

Synthesis of volatile derivatives is an important way to decrease the overall temperature that must be used in a GCMS run. The volatility of polar compounds can be increased by reaction with reagents that combine with groups such as hydroxyl, carboxyl, amine, and amide. Esterification, methoxylation, acetylation, and reduction have all been applied in one form or another. Recently, exotic combinations of reagents have been employed for special applications (Table 6.1).[4-6, 40, 43, 44, 112, 118, 119, 226-235]

Utility of derivatization depends upon prior knowledge of the sample. In the analysis of a complex essential oil, it is uncommon for a chemist to use any conversions more extensive than simple methyl esterification. On the other hand, for analysis of steroids, amino acids, carbohydrates, bile acids, etc., volatile derivatives are essential. Due to preliminary class separations, chromatograms with more than 12–15 peaks are not always encountered, so that data obtained from GCMS runs on derivatives can be sorted and interpreted.

A partial list of derivative reactions and reagents is given in Table 6.1. The selection of examples is intended to be generally informative, and many useful processes with important specific applications are absent. A more comprehensive survey can be obtained from recent review articles.[40, 43, 44, 233]

Derivatization is an extremely valuable technique, but it is imperative that the chemist also applies his knowledge of the sample in an intelligent fashion. Each chemical system will be different, and general rules cannot be applied at random. In the final determination, the decision to derivatize and the choice of reagent is up to the chemist. Knowledge of GCMS methods and techniques has little bearing, although interpretation of the spectra may be aided by judicious choice of derivative.

Table 6.1 shows that derivatization is applied to many classes of low-volatility compounds, particularly those encountered in biological studies. However, the number of different types of reagents in common use is relatively low, and silylation methods are highly preferred.

The utility of a derivative in GCMS depends on the nature of the mass spectrum obtained, and parent ion enhancement is important if it can be achieved. Trimethyl silylation does not usually result in a large increase in parent ion intensity, but loss of CH_3^+ is quite prominent and serves equally well to establish the molecular weight. If the unsilylated compound is of the type that fragments extensively (e.g. sugars), the TMS derivative also fragments extensively and M^+ or $(M - 15)^+$ ions may not be observed in the mass spectrum.

Derivative methods can also be used to obtain additional structural information when the mass spectrum of an unknown cannot be fully interpreted.

TABLE 6.1. A partial list of derivatization reagents used in GCMS analysis.

Reactant group	Reagent	Reaction product	Reference
R$-$OH	hexamethyl-disilazane, tri-methyl-chloro-silane, etc.	R$-$O$-$Si(CH$_3$)	225
R$-$CO	methyl hydroxylamine	O-methyloximes	226
	(CF$_3$C anhydride structure)	enol trifluoro-acetates	
RCOOH	CH$_2$N$_2$	RCOOMe	4, 232
	N,O-bis (trimethyl-silyl) acetamide	RCOOSi(CH$_3$)$_3$	
	other silylation reagents		
	H(CF$_2$)$_n$CH$_2$OH	RCOOCH$_2$(CF$_2$)$_n$H	
RNH$_2$, RR$'$NH	(CF$_3$C anhydride structure)	$\underset{R}{\overset{R'}{>}}$NCOCF$_3$	234
Thiamine metabolites (OH, COOH, NH$_2$)	silylation reagents	$-$O$-$Si(CH$_3$)$_3$ $-$COOSi(CH$_3$)$_3$ $-$NHSi(CH$_3$)$_3$	
Hydroxy keto steroids	methyl hydroxylamine and silylation reagent	CH$_2$OSi(CH$_3$)$_3$ | C=NOCH$_3$	228

TABLE 6.1. cont.

Reactant group	Reagent	Reaction product	Reference
17,21-dihydroxy 20-ketosteroids	dimethyl methoxychloro-silane	(cyclopentane steroid fragment) $O=C$, $Si(CH_3)_2$ bridged via two O	231
Corticosteroids	methyl boronic acid	(cyclopentane steroid fragment) $=O$, $O-BCH_3-O$ bridge	230
Phosphoryl-ethylamine	silylation reagent	$TMSO$, $TMSO$ — P $=O$ — $O-(CH_2)_2NTMS$	230
Disaccharides	borohydride or borodeuteride plus trimethyl silane	reduced to alditols and silylated	229
Sugars in lipopoly-saccharides	hexamethyl-disilazane and trimethyl-chlorosilane	persilylated sugars	235
Amino acids	butanol or propanol, trifluoroacetic anhydride	N–TFA ester	222, 223
	silylation		224

In some cases, the location of an oxygen function (keto, hydroperoxy, epoxy, carboxylic acid) can be established from the spectrum of the reduced compound. Further structural features are obtained by the use of deuterated reducing agents. For example, differentiation between a $(1 \rightarrow 3)$ or $(1 \rightarrow 4)$ glycosidic linkage can be obtained by reduction of disaccharides with borodeuteride.[229] Specific site deuteration on double bonds, enolized ketones, or amino acids has also been of value. Double-bond locations have been determined by a variety of different derivatives.[227]

D. SCAN TECHNIQUES

1. Scan Rate as a Function of Resolution and Frequency Response

The quality of the mass spectral records obtained during a GCMS analysis is closely related to the rate of scanning of the mass spectra. If the spectra are scanned slowly, a representative ion signal is measured for each mass peak, and adequate electrical filtering can be used to remove spurious background spikes and noise signals. However, the mass spectral pattern will be severely distorted due to changing sample concentration during the elution of the chromatographic peak. Figure 6.6 shows the distortions that occur in the mass spectral pattern of a chromatographic peak when it is scanned too slowly. In the example, the base of the peak is assumed to be six scan periods wide. One-half a scan period is needed to return to the start for the next scan.

These aberrations are serious, particularly when dealing with a series of similar compounds (e.g., terpenes), so that different structural features of the homologues are sought. The aberrations are not too serious if type identification only is desired. Even so, identification could not be positive until confirmed by other data. Clearly, six scan periods across the base is too few to assure representative spectra. As a rule, the scan should be fast enough so that there are at least ten periods across the base of the chromatographic peak. If the time constants of the apparatus deny this condition, the recorded spectra must be corrected for the concentration change.[236]

If the spectra are scanned very quickly, the change in sample concentration during the scan is insignificant. However, a fast response time is necessary for each component of the detector/recorder system, and only minimal electrical filtering is possible. If the response time of the slowest recorder or amplifier component is not fast enough, the records suffer decreased resolution and decreased sensitivity. Figure 6.7 illustrates this effect.[131] The mass spectrum of the mercury isotopes and other background peaks was scanned at

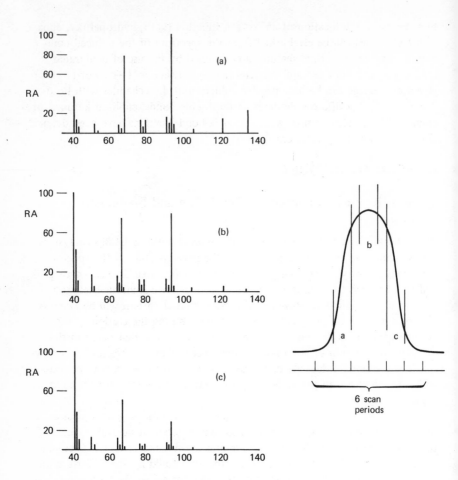

FIGURE 6.6. Distortion of the mass spectrum of myrcene due to the variation of concentration that occurs while scanning a chromatographic peak. (a) Scanning up the peak. (b) Scanning across the top. (c) Scanning down the peak.

a resolution of 400. The galvanometer natural frequencies of 1600 cps and 1000 cps were the limiting time constants in the detection system. When scanned at 0.7 sec/decade (i.e., 0.7 sec to scan from mass 20–200, 50–500, etc.), a satisfactory mass spectrum was recorded with no significant signal diminuation. At 0.35 sec/decade, the resolution was downgraded from 400

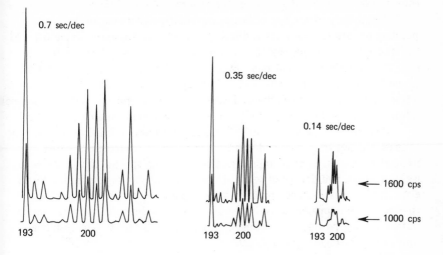

FIGURE 6.7. Fast scan of mercury isotope peaks. (a) 0.7 sec/decade. (b) 0.35 sec/decade. (c) 0.14 sec/decade. Top trace from 1600-cps galvanometer. Bottom trace from 1000-cps galvanometer.

to about 200, and the intensity was only 60% of the static signal. At 0.14 sec/decade, the record is completely unsatisfactory.

The preferred rate for the mass spectral scan is thus seen to be limited on either side. Fortunately, the frequency response of most amplifier/recorder systems can be set so that the scan is fast enough for most chromatographic conditions. As a rule, the total scan period should be 1/5 to 1/10 of the chromatographic peak half-width, i.e., in the range 2–6 seconds. For oscillographic recording of mass spectra at a resolution of less than 1000, scan periods of 5–6 seconds are preferred, corresponding to a rate of 3–5 sec/decade. At higher resolutions (8,000–10,000), compromise is necessary, and even scan rates as slow as 8–15 sec/decade impose severe demands on the amplifier and recorder system.

The necessary frequency characteristics of the amplifier/recorder apparatus depend on the rate at which a mass peak is traversed during the scan. For a mass resolution of 500 and scan rate of 3 sec/decade, the time taken to cross a mass peak will be about 2 msec. A frequency response of 1500 cps will assure a good record. To scan at a mass resolution of 10,000 at the same scan rate of 3 sec/decade, the frequency response must be close to 30,000 cps. This response rate is higher than that of most practical mass spectral recorders, thus necessitating slower scan rates for high-resolution GCMS.

A summary of the approximate amplifier/recorder frequencies corresponding to various scan rates and resolution settings is given in Table 6.2. At the scan rate of 2 sec/decade, any resolution greater than a few hundred imposes severe frequency demands. Even at 5 sec/decade, a resolution of 1000 would be the practical maximum for 1500 cps galvanometers. Higher resolutions must have correspondingly faster recording equipment, and conventional galvanometer oscillographic recorders are not considered suitable or practical.

Banner[237, 238] and associates[239] presented a more sophisticated treatment of mass spectral frequency and intensity response characteristics. Using a triangular peak shape as an approximation of a mass spectral signal, the time lag, distortion, and decreased sensitivity were calculated corresponding to the electrical band-pass of the amplifier/recording apparatus. The dynamic resolving power and apparent sensitivity were expressed through the parameters of static resolving power, system time constant, and scan time. The results are summarized in Figures 6.8 and 6.9.

These two figures are extremely useful for establishing the scan limitations of a specific mass spectrometer system. For example, suppose a mass spectrometer is set for resolution 1000. It is seen in Figure 6.8 that the curve for 1000 resolution starts to decrease when t_{10}/τ is about 3×10^4. If the frequency response of the amplifier galvanometer is 2000 cps, $\tau = 1/2000$ sec and $t_{10} = 15$ sec. This establishes the fastest scan period/decade for that galvanometer at that resolution without noticeable peak distortion and sensitivity loss. If the chromatographic peaks are narrow and a scan rate of 5 sec/decade is used, then $t_{10}/\tau = 10^4$. The dynamic resolution is reduced to 700 and, from Figure 6.9, the peak intensity will be only 70% of the maximum static value. For optimum performance, a faster galvanometer is necessary.

This method of analyzing dynamic response permits the operating characteristics of any GCMS system to be evaluated provided the static resolution and frequency response are known. Since the efficiency of the output apparatus is so dependent on the ratio t_{10}/τ and on the resolution, it is obvious that one should never use a faster scan or a higher resolution than is deemed necessary. The consequences of excess ambition in these categories are summarized in Table 6.3 for selected conditions. Data are taken from Figures 6.8 and 6.9. The losses in sensitivity and resolution caused by a fast, incompatible scan rate are quite apparent.

TABLE 6.2. Amplifier/recorder frequency response characteristics for various mass spectral scans and resolution settings.

Scan Rate	Resolution	Frequency Response, cps
2 sec/decade	200	800
	500	2,000
	2,000	8,000
	10,000	40,000
5 sec/decade	200	300
	500	800
	2,000	3,000
	10,000	16,000
10 sec/decade	200	160
	500	400
	2,000	1,500
	10,000	8,000

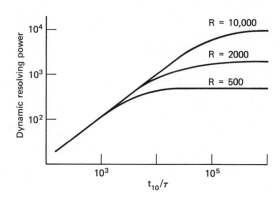

FIGURE 6.8. Resolving power as a function of the scan rate/frequency response ratio, t_{10}/τ. Maximum resolution corresponds to static or slow scan operation.[236, 237]

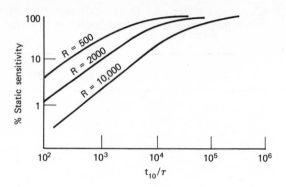

FIGURE 6.9. Sensitivity as a function of the scan rate/frequency response ratio, t_{10}/τ. Maximum sensitivity corresponds to static or slow scan operation.[236, 237]

2. Scanning the Mass Spectrum

The scan function, m = f(t). Since the three types of mass spectrometers used in GCMS studies (quadrupole, time-of-flight, magnetic sector) operate on different physical principles, the relationship of mass with time, i.e. the scan function, is different for each instrument. Electronic changes can be made in the control circuit to modify the scan function, but the ion peak width also becomes a function of time. Except for a few specialized applications, the "natural" scan function is preferred, i.e. the scan function that gives the same time width for all mass peaks.

The commonly used natural scan functions of the three instruments are:

Quadrupole, linear scan:

$$m = m_0 + k_1 t \tag{6.1}$$

Time of flight, hyperbolic scan:

$$m = m_0 + k_2 t^2 \tag{6.2}$$

Magnetic sector, exponential scan :

$$\text{high to low mass, } m = m_0 e^{-k_3 t} \tag{6.3}$$

$$\text{low to high mass, } m = m_0 e^{k_4 t} \tag{6.4}$$

TABLE 6.3. Loss of resolving power and sensitivity caused by excess scan rate.

	Static Resolving Power $(RP)_s$	Dynamic Resolving Power $(RP)_d$	Apparent Sensitivity (% of static response)
$t_{10}/\tau = 5000$	250	225	90
e.g., $t_{10} = 3.3$ sec	500	350	70
$\tau = 1/1500$ sec	1,000	450	55
	2,000	600	37
	10,000	700	10
$t_{10}/\tau = 10,000$	250	250	99
e.g., $t_{10} = 5$ sec	500	450	85
$\tau = 1/2000$ sec	1,000	700	70
	2,000	900	55
	10,000	1,500	20

In the above equations, m is the mass at time t, m_0 is the mass at the start of the scan, and k_n is the scan constant. The graphical representation of these functions[240] is given in Figure 6.10.

In the mass spectrometer, the spread of the ion beam or ion bundle is independent of mass to a first approximation. The so-called natural scan function is thus preferred because it records a peak time width approximately independent of mass. The consequences of these relationships are shown in Figure 6.11. For a magnetic up-scan (a), the time difference between two adjacent mass peaks is much shorter at the higher masses, and the peaks are crowded. If the scan is electronically linearized as in (b), mass assignment may be facilitated, but the peak width is changed. This is an inconvenient condition for GCMS where the optimum scan rate is determined by the time width of a peak. However, with the linear scan function of a quadrupole (c), each peak is the same width over the entire mass range. This, incidentally, gives rise to a different resolution over the entire scan range, so that the mass range must always be specified with the resolution of a quadrupole spectrometer.

The time t_{10} required to scan a mass decade is almost universally used as a standard of scan rate. However, from equations 6.1 to 6.4, it is apparent

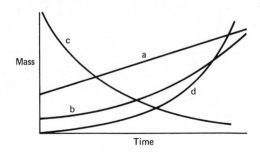

Mass

Time

FIGURE 6.10. The natural scan functions, m = f(t), for the three common GCMS mass spectrometers. (a) Quadrupole. (b) Time-of-flight. (c) Magnetic sector, down scan. (d) Magnetic sector, up scan.[240]

that the expression for t_{10} is different for each type of mass spectrometer. A natural significance for t_{10} exists only for the exponential scan in which t_{10} is independent of the starting mass m_0. The equations for t_{10} can be derived from equations 6.1–6.4 by substituting the relation $m = 10\ m_0$.

Quadrupole:

$$t_{10} = 9\ m_0/k_1 \tag{6.5}$$

Time-of-flight:

$$t_{10} = 3\,(m_0/k_2)^{1/2} \tag{6.6}$$

Magnetic sector:

$$t_{10} = 2.3/k_3. \tag{6.7}$$

If the three types of mass spectrometer were set up to scan the mass decade 20–200 in 4 sec, then the decade 40–400 would be scanned in 8 sec with a quadrupole, 5.7 sec with a time-of-flight, and 4 sec with the magnetic exponential scan.

Manual scan. Manual actuation of the scan circuits is still widely used for conventional GCMS runs even of long duration. Manual operation has one advantage in that the scan of shoulder peaks or peak tops can often be started at the most propitious time. When manual scan is used with a good oscilloscope display (page 264), the best starting time can be selected easily, and sharp peaks

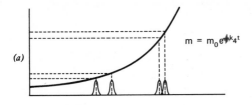

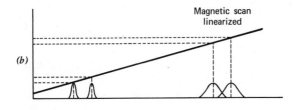

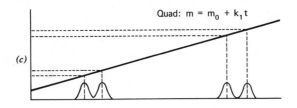

FIGURE 6.11. Relative width of mass spectral peaks obtained from different scan functions. (a) Magnetic exponential scan. (b) Magnetic scan, electronically linearized. (c) Linear quadrupole scan.[240]

($\sim$ 10 sec width) that might be missed or poorly scanned are properly recorded. Manual scan is convenient when only a small number of scans are needed.

Cyclic scan. Recent trends toward automation of GCMS analyses and data processing have popularized the use of a cyclic scan mode.[241, 242] A repetitive scan system continuously flashes the mass spectral ion current across the collector of the electron multiplier, and with a fast continuous recording device, maximum representative data are obtained. If the GCMS data are to be computer processed, the cyclic scan mode offers several advantages, and for many magnetic mass spectrometers, it is often a necessity.

The advantages of cyclic scan/continuous recording are:

(1) It permits automatic data acquisition. Typical GCMS runs span 1/2–2 hours. Automatic scan methods release the operator for other duties.

(2) A data record is obtained on each cycle, not just those selected by the operator. To obtain a spectrum for each scan, however, the recording apparatus must have a large capacity. Magnetic tape or direct computer system storage may be used,[96–99, 241] but an oscillographic recorder is unsuitable. Of course, a cyclic scan can be used without recording each cycle if the operator wishes to start the recorder (oscillograph, tape recorder, or computer) at the start of each chromatographic peak. However, this technique loses the advantage of automatic operation.

(3) The cyclic scan mode conditions or stabilizes the magnet and gives a more reproducible mass vs. time scan function.[241] This is a necessity for computer reduction of data from many magnetic sector instruments. For a time-of-flight or quadrupole mass spectrometer, the stability and fast recovery of the electrical scanning system are unaffected by the scanning operational mode.

(4) Parallel signal output to an oscilloscope permits a convenient display of the mass spectrum. (See page 264.)

(5) If an on-line computer is not available, the data are stored on a magnetic tape in a form compatible with subsequent computer treatment.[96–99]

The disadvantages of cyclic scan/continuous recording are:

(1) It necessitates a large data storage unit (discussed above in (2)).

(2) The scan cannot be started at a precise time. On sharp chromatographic peaks, 2–3 sec may be critical in obtaining the best scan position. Taking the scan with such precision can be accomplished only by using an independent oscilloscope display and manually actuating the scan and recording circuits when desired.[143]

(3) The span of the spectral record must be set at a safe limit above the highest anticipated mass. If most compounds in the run are at molecular weights less than 200, an upper limit of 400 means wasted recording space. Such a waste is critical if the total data exceeds the capacity of the tape unit or computer storage. For overlapping chromatographic peaks, the longer scan period also means wasted time and hence lost data.

If a cyclic magnetic scan is performed at a rate of 2–4 sec/decade and 1–2 sec is used for magnet recovery, the total time per cycle will be 4–6 sec.

Figure 6.8 shows that for resolution of 500–1000, the system response must be around 5 x 10^3 to 10^4 cps. This is too fast for most oscillographic recorders but is a reasonable range for magnetic tape recorders or on-line computer data acquisition. Of course, the mass spectrometer amplifier must be modified for broad band recording (see page 58).

Analog tape recorders for use with GCMS systems can be obtained with a frequency response of 10–20 kilocycles/sec.[96-98] The dynamic range of a single track (about 100 to 1) is insufficient for the broad range of typical mass spectral signals, but this is easily circumvented by using multiple track recording at different gain levels. Additional channels are available (anywhere from 7–15 channels is common) to record a reference timing signal and total ion current signal if desired. The analog tape thus provides a means of recording a large number of GCMS spectra either continuously or by manual selection.

One disadvantage of magnetic tape records is that the operator does not have visual spectra until the tape is processed in some manner. If a tape contains single-scan selected spectra (manual operation), these can be played back at a slow speed into an oscillographic recorder for preliminary study and the tape later processed by an off-line computer. When used in this manner, the magnetic tape offers the advantage of slower playback and hence reduces the frequency response of the hard copy writing device.

If a tape record is obtained by continuous scan, it is unlikely that every spectrum in the complete chromatogram would be needed. The desired spectra cannot be selected easily since there is no file code for reference. Processing of a continuously recorded analog tape is thus quite difficult, and alternative procedures are preferred.

Direct digital recording of cyclic scan mass spectra provides a convenient alternative.[99, 243] The frequency response for analog to digital (A/D) conversion and the necessary digital sampling rate are easily compatible with low-resolution mass spectral scanning (see page 286). Peak time width for most scan rates and low resolution settings will be 1–3 msec, so that digital sampling rates of 3–10 KHz give a representative signal selection across a peak. For high-resolution studies (R = 10,000, scan rate 10 sec/decade), the digital sampling rate should be at least 20 KHz.

Digital spectra can be recorded on magnetic tape with a packing density as low as 200 bits per inch (bpi) for a sampling rate of 3 KHz, but for the preferred digital sample rates of 8–16 KHz, the packing density must be 600–1600 bpi. A trigger circuit is used to start and stop the recording process and to write an "end-of-file" message after each scan. Analog mass spectra can be regenerated by D/A conversion and recorded on a suitable strip chart for quick visual inspection. However, the real function of the continuous cyclic scan system is to obtain data for easy computer processing. The digital record is usually treated in this fashion, and the spectral record is drawn or printed as a normalized mass marked spectra after data reduction.[99]

When an on-line computer system is available, digitized spectra from a cyclic scan go directly to the computer. In most configurations, the data are partially reduced and output to a digital tape or other storage device for subsequent processing. The details of computer treatment of GCMS data are discussed in Chapter 7.

The consecutive "jump" scan. One obvious time waste in fast scanning of mass spectra is that much of the scan time is spent between peaks where there is little or no information of interest. An idealized scan system[95, 244, 245] would make a fast jump from one preset peak position to the next and spend a few hundredths of a second on the top of the peak instead of the millisecond or less that would be spent scanning over it. In this way a factor of 10–100 can be gained in sampling time and hence in the minimum detectable ion current. The increased time is particularly advantageous for very small samples in which the sensitivity is statistically limited by the number of ions collected in the peak. A computer control circuit can be programmed to give an equal time period to each mass, or alternately, the system can be programmed to count a specified number of events, and record the counting time as an inverse intensity factor. The maximum dwell time on small peaks can thereby be increased an additional factor of 3–5.

These various concepts of scanning, i.e., analog, digital, and jump, are illustrated[245, 246] in Figure 6.12. An analog chart gives a record of the data which is considerably smoothed by the relatively slow response of the amplifier and oscillograph galvanometers. The digital scan recorded by a computer system samples the amplifier output via the A/D converter about 10 times across each peak. Because of the faster response in the amplifier and sampling system, no mechanical smoothing is obtained, and the spectrum is a truer representation of the ion signals (see page 288). At conventional scan rates, the time used for each digital sample is about 10^{-4} sec. In the jump scan, the mass spectrometer goes from one consecutive mass to another in a fraction of a millisecond and sits on the top of the peak for 10–30 msec.

The jump scan is much like a digital scan with a programmed scan period and preset sampling positions. The dwell time on each peak is longer and hence increased sensitivity can be attained. However, there are a few disadvantages and restrictions.

(1) Jump scan is not practical with a varying magnetic field. Electrostatic scan can be used on a magnetic sector mass spectrometer, but over a wide range, considerable mass discrimination results (page 38). For magnetic

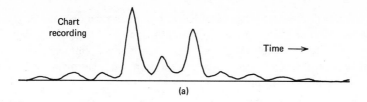

(a)

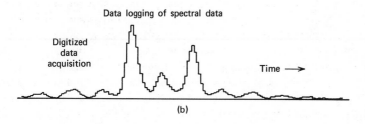

(b)

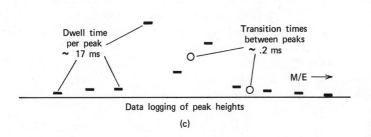

(c)

FIGURE 6.12. Representation of three different scan modes.
(a) Analog scan. (b) Digital scan. (c) Consecutive jump scan.[244]

mass spectrometers, use of jump scan is restricted to a small number of close
mass neighbors. (See next section on multiple ion detection.)

(2) Control by a dedicated computer must be used to effect jump scan.
A calibration run is necessary, but because this is necessary for any computer/
mass spectrometer system, it is not a singular disadvantage. The calibration
must be stable, since there is no way for the computer to establish a peak
center during or after the scan.

(3) Jump scan is restricted to low-resolution mass spectra. The calibration is made on nominal masses, and the mass defect of different hetero ions
must be covered by using a broad sensing window.

(4) Doubly charged ions recorded at half mass values are not registered. These ions do not normally have important diagnostic value, so this is a minor inconvenience. Metastable ions are also excluded, but since conventional time-of-flight and quadrupole spectra do not display metastable ions, the point is not significant.

Multiple ion detection or selective ion detection. One of the techniques used in early GCMS studies with inadequate mass spectrometers (i.e., slow scan, no electron multiplier) was to select a specific mass and run a chromatogram by monitoring that one mass. The process was repeated using as many mass assignments as were deemed necessary[247] and the data were combined to give a partial mass spectrum. For chemical systems with only a few possible functional groups, such as a mixture of aromatic and aliphatic hydrocarbons or ester mixtures, the method proved to be very useful. For polyfunctional mixtures, the relatively small number of different ions that could be feasibly covered in the repeated runs was an unacceptable restriction.

In recent studies, especially work on extracts of biological systems, certain advantages have been attained by monitoring from one to six or eight selected ions.[221, 248-254] This process, popularly known as multiple ion detection (MID), is a special application of jump scan in which only the signals from a few specified masses are recorded. The technique is applied by using prior knowledge of the unknown to select the masses of interest. If the operator knows that only a few mass peaks will be used in the final data interpretation, the mass spectrometer can be programmed or set to scan only those values. The data are recorded in the form of a chromatogram (often called a mass chromatogram or mass fragmentogram) on a multichannel recorder using one channel for each mass selected. Applications occur in the monitoring of suspected drugs, pesticides, or metabolites, or when following the course of chemical or biochemical reactions, particularly of isotopic species.

Figure 6.13 shows the results of analysis of an isotopic mixture of sugars in which the ion curents from the deuterated and undeuterated species were recorded by a selective scan method.[248] The partial separation of the isotopic pair is illustrated by the chromatographic record as an interesting serendipitous observation, but the real intent of the experiment is to obtain good quantitative data for the two components. The experiment was performed with a magnetic deflection instrument modified so that the accelerating voltage could be switched back and forth repetitively from one voltage to another, corresponding to the selected masses. Monitoring the fragment ions at masses 217 and 220 (from pent-0-trimethylsilyl-β-glucose and the D_7 derivative) gave an accurate measurement of the amount of deuterated specie

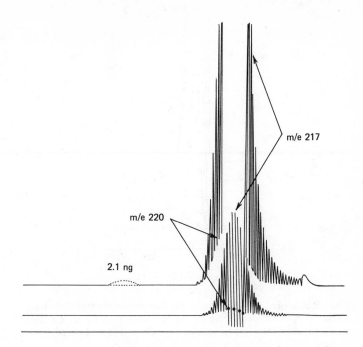

m/e 217

m/e 220

2.1 ng

FIGURE 6.13. Multiple ion detection chromatogram of mass 217 and 220 of TMSi glucose and TMSi glucose, D_7.[248]

in the sample (2.71% measured, 2.67% calculated).

Because of the switching process, the data consist of a series of sweeps that define a chromatographic peak on the sweep envelope. When more than two ions are monitored, the sweep traces often obliterate the mass chromatogram. Various methods are used to dampen or filter this undesirable hash. (See also Watson, page 396.) A more recent example of multiple ion detection[221] is shown in Figure 6.14. Three attenuated traces are recorded for two specified ions. Mass 392 is the parent ion of the heptafluorobutyl derivative of homovanillic acid (a metabolite of dopamine), and mass 395 is the trideutero equivalent. The dotted lines due to mass 392 were obtained from a sample with 0.48 nmole/ml of homovanillic acid. The results suggest a detection limit lower than 5 pmole/ml from a cerebral spinal fluid.

Any GCMS mass spectrometer can be modified for multiple ion detection. However, most magnetic instruments are restricted in the mass range and the number of ions that can be selected. A range of 20% (e.g., 240–300)

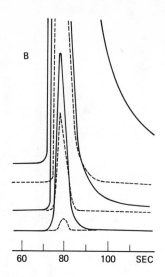

FIGURE 6.14. Multiple ion detection chromatogram of the heptafluorobutyl derivative of methyl homovanillate (MW 392) and trideutero methyl homovanillate (MW 395). Dotted line is mass 392; solid line is mass 395.[221]

and 2–4 selected ions is typical. On the other hand, the quadrupole or time-of-flight mass spectrometers can be adapted for multiple ion detection over any mass range with up to 6–8 selected ions. Figure 6.15 shows a mass chromatogram obtained on a quadrupole mass spectrometer using an automatic peak selector. The mixture contained lidocaine, monoethylglycinexylidide (MEGX, a metabolite of lidocaine), and trimecaine. The gas chromatographic retention time gives a first indication that the material may be present. The ratio of the selected masses at 58, 86, and 120 confirms the identification.

The specific information obtained by multiple ion detection methods is also present in a continuously recorded mass spectrum, but to have such a clear presentation, the operator must perform considerable effort in data reduction. Even more important, selective scan methods are advantageous when the sample size is very small and all parts of the GCMS system must be optimized. Since a fast scan is not used, the time spent on each mass peak can be increased to as much as 0.1–1 sec and the lowest limit of detectable current approaches 10^{-17} to 10^{-18} amp (10–100 ion/sec). This current level provides sufficient ions to satisfy statistical demands. With a multiplier gain of 10^6 to 10^7, the electron current from the multiplier to the electrometer will be around

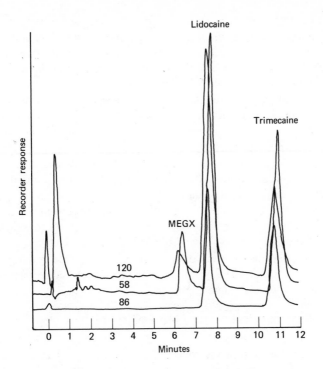

FIGURE 6.15. Multiple ion detection chromatogram of lidocaine, the metabolite monoethylglycinexylidide (MEGX), and trimecaine. The ratio of the signals for masses 58, 86, and 120 confirms the peak identification.[253]

10^{-11} to 10^{-12} amp, and an input resistance of 10^9 to 10^{10} ohm gives an adequate voltage (1–100 mV) to drive a suitable recorder. Sensitivity is determined primarily by the methods of handling the sample and baseline stability.[255]

Photoplate recording. Although analysis of conventional chromatographic samples (10^{-9} gm or more) at conventional mass resolution (~ 500) poses no unsolvable problems, fast GCMS scans at high mass resolution or for ultra trace analysis require compromise of the scan method. The jump scan techniques discussed in the previous sections suggest that a no-scan system utilizing a photoplate would also be advantageous. The details of photoplate ion detection were discussed on page 68. For the most part, a photoplate is unpopular because of the time-consuming photo development process and the expensive comparator equipment needed for reading and making accurate mass measurement at high resolution. As a consequence, very few laboratories use high-resolution GCMS routinely and prefer instead the more sophisticated computer methods used with electrical detection.

Occasionally, a particular sample will demand maximum sensitivity at 10,000 resolution or more, and the integrating properties of the photoplate may gain an order of magnitude in sensitivity. (See Habfast et al., page 408.) In addition, the photoplate is still the only practical method to obtain GCMS spectra in 2–3 sec at 10,000 resolution or higher. GCMS spectra have been obtained at resolution up to 40,000, but these applications are rarely encountered in routine analytical procedures.[102, 103]

It is doubtful that the photoplate would have any advantages in low-resolution GCMS runs. The value of the integrating property for enhancing sample utilization (sensitivity) is virtually lost due to the fact that a low-resolution mass peak is traversed or scanned in a longer time period and hence, more ions are collected. The wet developing process and subsequent plate reading would only be an annoying nuisance.

E. MONITORING THE CHROMATOGRAPHIC RUN

It is a good rule to monitor the course of the GCMS run to assure that the separation process is occurring as expected. There are various ways of obtaining this information, each with modest advantages. Almost none of the methods exclude the others, and sometimes two or three monitor processes are useful (e.g., flame detector, total ion monitor, and oscilloscope display). The monitor trace provides the record relating a given mass spectral scan with a particular chromatographic peak and gives the location of the scan on that peak. With the exception of routine computerized GCMS analyses, the simultaneous chromatogram is considered a must. With computerized systems, it is often sufficient to regenerate a total ion chromatogram from the mass spectral traces. (See page 299.)

1. Conventional Gas Chromatographic Detectors

The discussion on interface techniques gave some advantages for methods that split a fraction of the effluent to an auxiliary detector (Figure 5.2, page 159), which provides a valuable chromatographic record of the GCMS run and facilitates the process of positioning the mass spectral scans on the chromatographic peak. The sensitivity must be compatible with mass spectral sample levels, so that a hydrogen flame detector is most commonly used.

Occasionally, a thermistor-type thermal conductance detector will be suitable. If the sensitivity of the detector is known, knowledge of the split ratio can be applied to give an additional check on the interface efficiency.

The arrival time of the sample peak in the ion chamber generally lags the chromatographic detection by 2–6 sec, depending on the length of the interface connections. An operator soon learns to utilize the delay to select the best time to scan the spectrum. In some GCMS systems, a timelag has been deliberately built into the interface. McCloskey[49] placed a chromatographic split at a point which would be about 20–30 sec before the peak reached the end of the column. This permits a good inspection of the peak shape before the sample enters the ion chamber, and the mass spectra of shoulder or fused peaks can thus be properly recorded.

Even when a total ion monitor chromatogram is obtained, many chemists prefer to have the conventional trace as well. Comparison of the hydrogen flame record with the total ion monitor record reveals possible decomposition or loss of separation efficiency in the interface or in the mass spectrometer.

2. Total Ion Monitors

There are several methods of monitoring the chromatographic effluent in a GCMS run which use some part of the mass spectral ion current as a signal. Since this is completely representative of conditions in the ion chamber, it provides an important part of the overall GCMS record. When inconsistent results are obtained, comparison of this record with the conventional chromatogram (obtained either prior to or during the GCMS run) will help to pinpoint the source of trouble.

Most modern magnetic mass spectrometers are built with a beam monitor to facilitate initial start-up operations and focus adjustment optimization. Usually, the beam monitor is located between the ion source and the magnetic sector (Figure 6.16) and can be set to intercept 0–100% of the total beam. The ion signal goes to a high-gain electrometer amplifier circuit (variable input resistance in the range 10^{10} to 10^{12} ohms). The beam is monitored on a meter and a total ion chromatogram is obtained on a potentiometric recorder.[101]

The beam monitor gives a simple method for observing the progress of the chromatographic run. Since the ionization efficiency for organic compounds is 7–10 times higher than for helium (Table 3.3), the appearance of organic material gives rise to an increased ion monitor signal. However, because the quantity of helium is high, the current due to helium ions is excessive compared with the organic ion signal, and the excess voltage signal

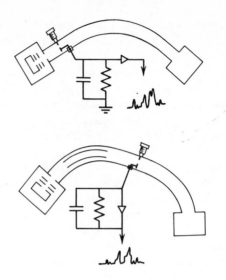

FIGURE 6.16. (a) Position of total ion beam monitor in magnetic mass spectrometers. In a double-focusing mass spectrometer, the monitor may be located before or after the electric sector.

due to helium must be bucked-out with a stable variable voltage source. Even so, the sensitivity is severely limited by small fluctuations in the helium current, and sometimes the ion monitor signal may be 100 times less sensitive than the mass spectral record obtained by scanning. The main source of this fluctuation is uneven pumping, and use of a mercury diffusion pump will often smooth out this noise, thus making the monitor more reliable and more sensitive.[256]

If the electron gun of the mass spectrometer is operated at an energy below the 24 eV ionization potential of helium (nominally at 20 eV), the signal at the total ion monitor will be due only to water, air, and organic compounds. Thus, only a small background signal needs compensation, and the ion monitor can be operated with higher sensitivity to the organic ions. However, in this mode of operation, the mass spectrometer system has been changed to meet the needs of chromatographic monitoring. Sensitivity is reduced by a factor of 5–10, and the mass spectral pattern is modified (see page 18). These compromises are considered undesirable. However, a good ion monitor record as well as standard voltage mass spectral scans can be obtained by use of a switching circuit that permits the electron gun to operate at 20 eV for ion monitoring and at 80 eV for mass spectral scans. The elec-

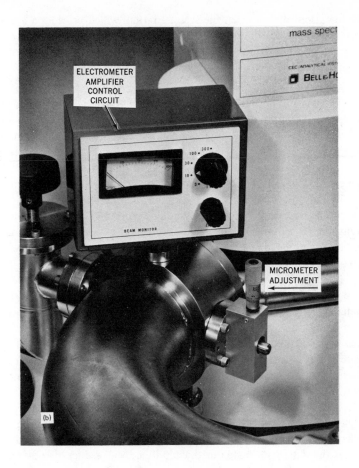

FIGURE 6.16. (b) Photograph of ion beam monitor located after electric sector (DuPont Instrument Company, Model 21–492).

trical changes are fast enough that no abnormality is obtained in the mass spectrum, and the chromatographic trace is taken at a sensitivity compatible with the GCMS run.

In a double-focusing mass spectrometer, the ion-focusing voltages and electrostatic field plate voltages can sometimes be adjusted so that only a small fraction of helium ions pass through the electric sector. This empirical adjustment results in a stable base for the ion beam monitor, but the level of helium ion signal may be affected by changes in the fringe magnetic field. A sudden jump in current occurs when the scan is started, which provides a convenient scan marker, but this jump makes the monitor system impractical for monitoring in a cyclic scan mode since the signal is obtained on a very broad sweeping base line.

Auxiliary ion chambers. To avoid the inconvenient He^+ current observed with the total ion monitor, a second ion chamber can be constructed in the mass spectrometer vacuum envelope. Strictly speaking, this is not a form of total ion monitor but rather a separate chromatographic vacuum ionization detector. However, in function, the separate chamber gives a representation of the conditions in the ion chamber and hence indicates a loss of chromatographic separation or sample decomposition the same as a beam monitor does.

The advantage of the auxiliary ion chamber is that the second electron beam can be operated independently at an energy of 20 eV, thus eliminating the current due to helium. When this ionization source is used with a conventional strip chart recorder, the electrometer output can be amplified to give a sensitivity comparable to a hydrogen flame detector.

Three possible configurations for the auxiliary chamber are given in Figure 6.17. In (a), the sample stream is split in a preselected ratio, and the chromatographic detection is completely independent of the mass spectrometric ion chamber. The stream split may be from 1/10 to 1/2 the total, but the decrease in the mass spectral signal is seldom significant in either case. In case (b), all the sample flows through the mass spectrometer ion chamber first and then a fraction through a small orifice to the detector chamber.[257] Again, the quantity going to the second chamber will be 1/10 to 1/2 the total. The bulk of the sample leaves the first chamber by the ion exit slit and electron beam slit. Superficially, it might be expected that method (b) will have greater sample utilization, but this is not so. The increased conductance of the ion chamber due to the additional leak reduces the effective sample pressure by an amount equivalent to the split mode as in (a). (See page 126 for discussion of ion chamber conductances.) Both (a) and (b) give excellent results, and the total ion chromatograms are comparable to the chromatograms obtained using a flame ionization detector.

In the third alternative, a vacuum ion gauge circuit is modified for operation with 20 eV bombardment.[258] The ion gauge can be connected to a "T" on the entrance line and pumped separately, or it can be connected to the ion source pump if convenient. This ionization detector can be added to most existing mass spectrometers, but it has not been widely used, in part because the modification is not commercially available. When used in the presence of excessive column bleed, the electron filament is quickly poisoned and the gauge must be frequently replaced.[259]

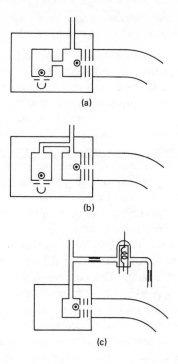

(a)

(b)

(c)

FIGURE 6.17. Possible configuration of auxiliary ionization detectors for chromatographic monitoring in GCMS.

Total ion collection in a time-of-flight mass spectrometer. The commercial time-of-flight mass spectrometer is not set up for direct measurement of the total ion beam but the unique characteristics of the time-of-flight output permit easy selection of any or all masses from each individual pulse. The gating system used for scanning the mass spectrum was discussed on page 49. The principle is also used to collect the total ion current from a selected number of cycles, generally every tenth cycle, on one of the gates.[260] (See Figure 2.26.) The current due to helium ions is rejected by one of two methods. The time span of the total ion monitor gate pulse is set so that it starts after a preselected mass and hence collects only masses from, say, 34 to end of cycle. Alternately, a voltage pulse coincident with the arrival of mass 4 is impressed on an electrode in front of the electron multiplier, thus rejecting the helium ion current.[260]

This method of chromatographic monitoring gives excellent results and does not interfere with the normal mass spectral scan or oscilloscope display.

A comparison of the chromatograms taken by a total ion monitor and by conventional hydrogen flame is given in Figure 6.18.

Total ion collection in a quadrupole mass spectrometer. A beam monitor placed before the quadrupole lenses can be used for a total ion chromatographic record from a quadrupole mass spectrometer in exactly the same manner as it is used with a magnetic spectrometer. The advantages and disadvantages with respect to the helium ion current are identical.

Two other techniques are possible. In one, the quadrupole voltages are adjusted to permit all ions within a specified range to pass to the electron multiplier. Figure 2.29 illustrated the oscillation stability of ions of specified mass with respect to the dc voltage U and the rf voltage V_0. By selecting a low value for the ratio U/V_0, a wide range of ions can be made to have a stable oscillation through the quadrupole system. In this way, a total ion record can be continuously obtained during the GCMS run. When a mass spectrum is scanned, the U/V_0 ratio switches to give the necessary mass resolution, and the spectrum is recorded by varying the values of U and V_0 at the constant U/V_0 ratio. This method gives a good chromatographic record but prohibits the possible use of a simultaneous oscilloscope display and requires a complex switching system. Consequently, it is seldom used.

The preferred procedure for total ion monitoring with a quadrupole mass spectrometer is very similar in principle to the process described for the time-of-flight mass spectrometer. The instrument is adjusted for conventional operation (i.e., resolution about 500) and scanned with a fast cyclic scan at the rate of 10–30 cycles/sec. The scan rate is too fast for most spectral recorders, but the complete ion signal can be integrated by a relatively slow amplifier/recorder system to give a good chromatographic trace.[262] This mode of monitoring has the important advantage that the voltage parameters can be easily set to cover any desired span, thus eliminating the ion current from helium, water, and air. Furthermore, the spectrum from the fast cyclic scan is easily displayed on an oscilloscope for simultaneous visual monitoring of the mass spectrum. When a spectrum is recorded, a single switch changes from the fast cyclic scan to a single scan at a conventional rate (2–6 sec/decade).

The rapid scan/total collection method can also be applied on a magnetic sector mass spectrometer using an electrostatic scan,[181] but magnetic scan is not fast enough for good results. Electrostatic scan has the serious disadvantage of mass discrimination as discussed on page 38, but the discrimination is not significant for chromatographic recording. However, if electrostatic cyclic scan is used, then magnetic cyclic scan is completely ruled out. Since magnetic cyclic scan is important for computer data acquisition with

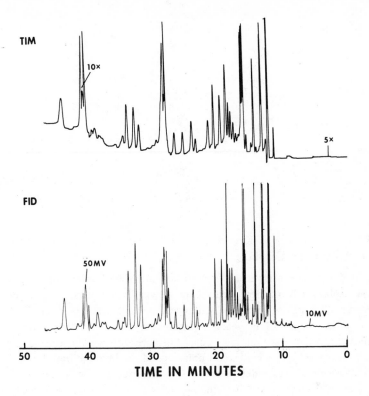

TIM

10×

5×

FID

50MV

10MV

50   40   30   20   10   0

TIME IN MINUTES

FIGURE 6.18. Comparison of chromatograms obtained using a flame ionization detector and the total ion monitor method of the Bendix time-of-flight mass spectrometer.[193, 261]

many magnetic sector instruments, the rapid scan/total collection method has not been popular except with the time-of-flight and quadrupole spectrometers.

Computer regeneration of the chromatogram. If cyclic scan mass spectral data have been stored in a computer system, all the necessary information is available for the computer to output a total ion chromatogram. The sum of the ion intensities over any selected range (e.g., mass 34–600) for each scan cycle provides a point on the chromatogram and, in addition, relates that point or mass spectral scan number to a position in the chromatogram. A typical example of a computer-regenerated chromatogram is given in Figure 6.19. The advantages, options, and output methods are discussed in detail in Chapter 7.

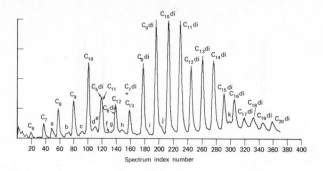

FIGURE 6.19. Computer-regenerated chromatogram from analysis of the methyl ester obtained from a green river oil shale extract.[148]

3. Oscilloscope Display

One of the most controversial "necessities" in GCMS hardware is the oscilloscope display. Historically, oscilloscope viewing was not common in mass spectrometry until the time-of-flight spectrometer became popular.[14, 82] Many mass spectrometrists of 10 years' experience or more do not consider the oscilloscope to be essential and bypass this accessory when specifying hardware needs. However, anyone who has used a good oscilloscope monitor for more than a few months considers it to be very important. Apparently, an oscilloscope attachment is something you can do without if you have never had it, but you must have it if you are used to it.

Oscilloscope display can be obtained from almost any mass spectrometer scan. The oscilloscope has a fast and sensitive response, and the only necessary instrument modification is a trigger signal to correlate the start of the oscilloscope scan with the start of the mass spectral scan.

When an oscilloscope trace is obtained from a magnetically scanned sector instrument, the scan cycle is 2–6 sec. The oscilloscope trace passes across the viewing face, generating the mass spectrum (Figure 6.20a). The complete spectrum is viewed for only a second at most at the end of the scan, a flash occurs, and another scan starts. This constant on-off effect with each cycle is not bothersome when the sample is not changing rapidly, and a persistent phosphor or storage scope provides a useful visual monitor. If the pattern is changing rapidly, as for example, when viewing inhomogenous chromatographic peaks, the precise nature of the changes is obscured by the on-off flash effect in the scope display.

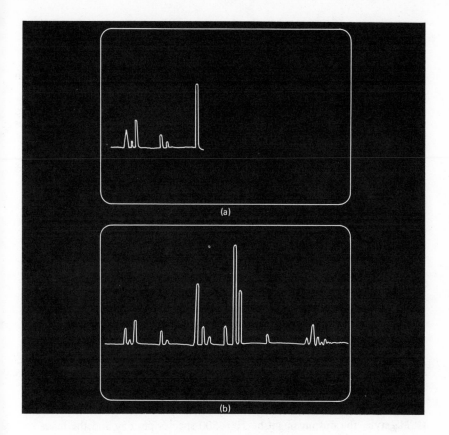

(a)

(b)

FIGURE 6.20. Mass spectral oscilloscope display. (a) After
2 sec of a 4-sec magnetic scan. (b) Continuous display of
10 cps electrostatic scan or TOF fast scan.

With the time-of-flight mass spectrometer an oscilloscope spectrum can
be obtained on every pulse cycle of the instrument.[14, 82] Since the common
repetition rate is 10,000 to 20,000 cps, the pattern represents a continuous
view of the conditions in the ion chamber (Figure 6.20b), and even minute
pattern changes that occur for a few seconds are easily discerned. This class
of monitor is highly desirable for any GCMS runs that are not automated.

High-quality oscilloscope display is an integral part of the time-of-flight
spectrometer, but it is not unique to that system. A 10-cps scan rate is used
for an oscilloscope display of the quadrupole spectra and gives a good visual
presentation. Electrostatic scan of a magnetic sector instrument at a 10 cps

scan rate also gives an acceptable pattern, but mass discrimination effects are noticed in the high molecular weight range as previously discussed.

A split-screen storage oscilloscope can be used for convenient GCMS monitoring of magnetically scanned spectra.[263] The bottom half of the screen stores a background spectrum that can be manually erased and replaced with the next spectrum at the operator's discretion. The top half of the screen is used to display the spectrum on-line at that scan point. Thus, even though the top spectrum is constantly being written, comparison with the stored background spectrum facilitates detection of minor spectral changes.

An oscilloscope output can also be generated by a computer system in either on-line or off-line mode.[264] Strictly speaking, the on-line process is not performed in real time, but rather the computer outputs the normalized, mass-marked spectrum taken in the previous scan cycle so that it is viewed a few seconds after the scan. The display is useful as a general monitor and for previewing data, but it does not function as a signal monitor to establish the start of a scan. The computer/oscilloscope relation will be discussed further in Chapter 7.

F. MANUAL TREATMENT OF GCMS DATA

One of the consequences of the widespread use of GCMS analysis has been a horrendous output of mass spectral data. A mass spectrometer used for qualitative analysis on batch samples might put out 10–20 mass spectra per day. The process of reducing the data is well within the capability of one man, along with other normal duties. If the same mass spectrometer is operated in GCMS style, the output might be 100–500 spectra per day and the mass-marking process would be a boring full-time job for one man. (See Table 7.1, page 279.)

Before long, the computer will take over the processing of GCMS data, but at the present time, well over 90% of this duty is still performed manually. Virtually all of these spectra are taken with an oscillographic recorder using 2–4 galvanometers at different attenuations. A typical trace was shown in Figure 2.38.

Before the chemist can begin to think of a mass spectrum in chemical terms, the raw data peaks from the oscillograph must be expressed in terms of mass numbers. In some cases, quantitative information beyond a visual inspection of peak height may be needed. Ideally, each spectrum of interest should be reduced to a tabular or graphical form as shown in Figure 6.21. Unfortunately, it is very impractical to perform manually such a complete data reduction on each GCMS spectrum. The process of mass marking takes 1–2 minutes per spectrum. The process of measuring peak heights, normalizing

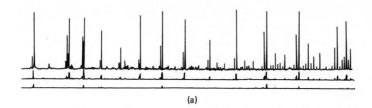

(a)

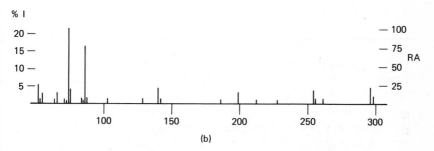

(b)

m/e	RA	m/e	RA	m/e	RA	m/e	RA
55	27	71	8	88	7	255	13
56	54	73	2	101	6	256	3
57	20	74	*100*	129	6	267	6
58	1	75	20	130	2	298	21 p
59	7	76	1	143	21	299	4
67	4	83	11	144	2		
68	2	84	4	199	9		
69	16	85	5	200	1		
70	3	87	76	241	3		

(c)

FIGURE 6.21. Comparison of different forms of mass spectral data. (a) Raw unmarked oscillographic chart. (b) Normalized bar graph with most abundant peak equal to 100. (c) Tabulated relative abundance mass spectral data.

intensity data, and plotting the spectrum takes 15–60 minutes per spectrum depending on the thoroughness and fastidiousness of the individual worker. Thus, it could take 50–100 hours for the simple process of reducing data from a single GCMS run to the preferred usable form.

It is easy to see why initially, most spectroscopists refine most raw GCMS data only to the mass-marking stage. Eventually, selected spectra may be measured and plotted for report purposes. In addition, the spectra of new compounds may be extracted from the huge bulk of data and submitted to public or private spectral files. However, both these important processes are performed after the GCMS data have been interpreted and are not, strictly speaking, part of the manual data processing method but rather, are part of data storage or presentation.

1. Mass Marking Raw Spectra

The most common method of manual mass marking is to identify a specific peak in the trace and proceed to count the peaks in their respective order. The usual starting point is the mass 28 peak due to N_2^+. Even a mass spectrometer with good vacuum seals (background pressure $< 10^{-7}$ Torr) will have a discernible peak at mass 28 and another of about 1/4 the intensity at mass 32 (O_2^+). In a GCMS system, the carrier gas elutes residual air from the chromatographic components, and vacuum seals in the interface are not usually as rigorously tightened as those in the mass spectrometer. Thus, the mass 28 and 32 peaks are always prominent and easily recognized.

Ion peaks do not necessarily appear at each mass, but the operator can make a visual extrapolation and continue with the counting. For systems operating with a nonlinear scan function, the extrapolation may be subject to error, but in the lower mass range (below 200), the marking process is reasonably accurate.

Above mass 200, the frequency of peaks in a typical spectrum decreases, and a manual count is often very difficult. Visual extrapolation over 20–30 mass units at mass 300 may easily be out one or two mass units, and the mass assignment must be interpreted with additional caution. In this range, the natural linear scan function of a quadrupole is highly advantageous, and frequently, the exponential scan function of a magnetic spectrometer will be modified to be linear. The expedience thus attained for marking masses above 200 often overrides the modest disadvantages discussed on page 245.

One advantage of a linear scan function is the ease of preparing and using a calibration chart. Actually, a calibration chart can be set up for any scan function, but accuracy of use often diminishes at higher mass ranges when the mass increments are very close together (Figure 6.11a). Nevertheless, much of the uncertainty involved in marking higher mass numbers can be reduced by using a calibration curve, particularly if it is made from the data on the lower mass numbers of the same chart.

Often, one typical spectrum from a GCMS run will be mass marked and then used as an overlay calibration spectrum for all the other charts. Actually, it is easier and more convenient to obtain a calibration chart of a known material right after completion of the GCMS run. This chart, rather than one of the unknown spectra, is then used for the overlay comparisons. The spectrum of perfluorokerosene is frequently employed for this purpose because of the considerable mass range and convenient periodicity of calibration peaks (Figure 6.22). A paraffinic hydrocarbon mixture or perfluorotributyl amine is also a useful mass-marking material. In addition, the background peaks due to column bleed may provide useful guides or calibration checks. For mass spectrometers with a mercury diffusion pump, the several mercury isotope peaks provide a convenient mass check in the limited range of mass 198–204.

Many mass spectrometers can be equipped with a mass calibration system that gives a simultaneous galvanometer trace with mass units marked automatically. An example is shown in Figure 6.23. Mass counting is further facilitated by accentuation of every tenth and every hundredth mass mark. Most automatic mass markers used on magnetic mass spectrometers are based on a Hall probe voltage developed in the magnetic field (Figure 6.24). Since this voltage is directly proportional to the magnetic field, digitization by an analog-to-digital convertor provides a convenient relationship between magnetic field (time) and mass.

Some of the mass markers used a few years ago gave unreliable mass assignments, and user complaints were frequent. Improvements in these devices have taken care of such problems, and mass markers are now reliable even at fast GCMS scan rates. The automatic mass marker gives considerable reduction in data processing time, but more important, mass assignment in higher mass ranges (300–500) is made with complete confidence. Mass markers are very useful, but unfortunately, usually cost $7,000–$9,000, which may strain a limited budget.

2. Normalization of Raw Spectra

For a good comparison of an unknown spectrum with a catalogued spectrum, quantitative measurement of peak intensities and normalization to a standard base should be performed. For GCMS runs, the refinement is seldom made because of the excessive time needed to treat the large number of spectra. Instead, an "eyeball" technique is applied in which the estimated peak height of four or five peaks is compared with the values of authentic spectra in a card catalogue or spectral compilation. This procedure amounts to little more

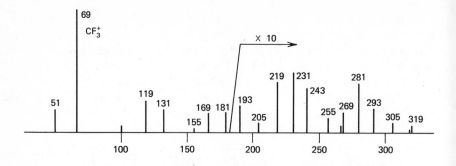

FIGURE 6.22. The mass spectrum of perfluorinated kerosene (PFK) showing the useful repetition of mass peaks every 100 mass units.

than a casual "small, medium, or large" assignment of peak intensity, but confident identifications are possible because of the large amount of specific information given in each mass spectrum.

This simple approach is a real timesaver when several hundred spectra must be examined. In any GCMS run, many of the chromatographic peaks are from simple, well-known compounds and a quick inspection of the mass data is all that is needed to make a safe assignment.

For important identifications, the peak intensities should be tabulated for a more precise comparison. Preferably, the known spectrum should be obtained with an authentic sample on the same mass spectrometer. As was previously pointed out (page 107), all new and important identifications should be confirmed by other methods whenever possible.

Two systems are used for normalization of mass spectral data. In the more common process, the most intense peak in the spectrum is used as a base for the peak comparisons and assigned the value of 100. The intensity of the other peaks is then expressed as a percent fraction, or relative abundance, compared to this base peak. In the other normalization method, the sume of the intensities of all peaks is used as a base, and individual peak intensities are expressed as a percent fraction of the total ionization. The graphical comparison of the two methods is given in Figure 6.21. As would be expected, the two processes have the same pattern and differ only in the value of the assigned numbers.

The percent of total ion current process is a more fundamental system and is preferred in basic studies of mass spectral fragmentation. For analytical

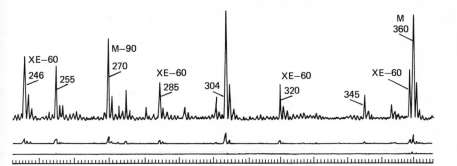

FIGURE 6.23. Raw oscillographic mass spectral chart with mass marker scale. Format shows elongated line every 10 mass units (LKB Instrument Corporation).

purposes, relative abundance has advantages. For one thing, it is not necessary to measure the whole spectrum to compare the intensity of only four or five peaks. The presence of an impurity does not affect every peak in the spectrum. The pattern changes caused by an impurity depend on the nature of the two spectra, but may not even be significant if the two compounds are quite different. When percent ion current is used for normalization, the relative intensity for every peak is reduced proportional to the concentration of the impurity.

3. Interpretation of GCMS Data

There are many fine text books devoted primarily to the interpretation and understanding of mass spectra, and the reader will want to refer to these volumes for a proper account of ion fragmentation processes.[4-11] Basically, there is no difference between the interpretation of a single spectrum obtained from a probe sample and the interpretation of several hundred spectra obtained from a GCMS run. Nevertheless, time is again an important element. A spectroscopist might well devote several hours to the spectrum from a single valuable sample. When faced with several hundred spectra each week, shortcuts are necessary, and trivial spectra must be quickly eliminated. The purpose of this section is to suggest procedures that can be used to process the large quantities of spectra with minimum effort.

FIGURE 6.24. Location of Hall probe on magnetic sector (DuPont Instrument Products Division).

Comparison of unknown spectra with catalogued spectral files. Positive identification by any spectroscopic method usually demands an accepted match of the unknown spectrum with a known one. In mass spectrometry, there are several sources of catalogued spectra,[265-269] and available compilations of spectra[270,271] in form that permits rapid search and comparison. A mass spectrometry laboratory should also develop a file of their own spectra. After a few years of operation, the in-house file becomes the most valuable prime source of spectra since it is made up of the type of compounds encountered in that laboratory. Needless to say, if the information can be released, these private spectra should be made available to the literature or commercial compilations.

The number of spectra in most listings is around 10,000 to 20,000. Because these data are taken from different sources, duplication occurs, and the exact number of compounds is less than the total number. However, the files provide a significant source of data for GCMS search purposes and at least one of these compilations should be available in a GCMS laboratory.

A typical compilation will be arranged with four to six different index listings: molecular weight, compound name, order of ion intensity, and empirical formula are most common. Examples of the molecular weight listing and the ion intensity listing are given in Figures 6.25 and 6.26.

If the molecular weight of the unknown spectrum can be ascertained, the molecular weight listing provides the easiest search process. Suppose, for example, that an unknown spectrum has a molecular weight of 148. Compounds of this molecular weight are catalogued as a group. A secondary order listing within the molecular weight 148 group is made according to the mass of the more intense ions. The relative abundance of each listed ion is also tabulated. Comparison of the unknown spectrum with the several catalogued spectra at that molecular weight permits a quick assignment or denial.

The ion intensities in the unknown spectrum do not usually need to be measured. If an ion is listed in the catalog as 40% relative abundance, a visual comparison with the intensity of that ion in the GCMS spectrum should indicate it to be between 20–60%. This crude degree of accuracy is sufficient for most preliminary identifications.

One of the unfortunate banes of the mass spectrometrist's existence is his frequent inability to observe directly the molecular weight. Many compounds break down so rapidly in the ion chamber that only fragment ions are collected, and the molecular weight must be established by deduction. When determination of the molecular weight is not possible, the unknown mass spectral pattern can be searched using the ions of highest intensity (Figure 6.26). The base peak is used for the first comparison. Unfortunately, different mass spectrometers do not always give the same base peak. If identification is not made by the end of the "most abundant ion" listing, the "second most abundant ion" list is examined, and sometimes examination of the "third most abundant ion" list may be necessary.

The above process should be sufficient to search the files. In some instances, the second most abundant ion in the unknown spectrum must be used to find a match, but as a rule, the base peak of the unknown is expected in the first three ion abundance tabulations of the catalogued spectra.

These search methods apply to almost any extensive set of GCMS spectra. The average time needed to peruse the files for each spectrum is about 5 min. The procedure is particularly effective for the types of compounds encountered in fruit volatiles, petroleum mixture, or essential oils. For more esoteric compounds such as are found in body fluids, metabolite studies, or pesticide residues, the catalogued spectra are often inadequate due to the fact that sufficient compounds of that type have not yet been circulated in the commercial files. This deficiency is being rapidly overcome, particularly through specific but limited files that list all expected poisons, drugs, pesticides, and metabolites.[272-274] For example, to search for a pesticide or its metabolic

COMPOUND NAME	MASS SPECTRUM	C	H	D	Br	Cl	F	I	N	O	P	S	Si	Di	MOL. WT.
1 4 DIISOPROPYLBENZENE	147 119 162 91 43 105 41 148 39 27 12 18														162
N# 978 1 M=162 B=1.84	204 252 173 160 143 136 171 107 99														
3 6 DIMETHYL 1 THIAINDENE	162 161 28 39 147 29 41 27 80 45 10 10 . . . . 1 . .														162
N# 1231 1 M=162	890 470 290 290 250 230 220 170 140														
1 3 5 TRIETHYLBENZENE	147 133 162 105 91 27 29 39 15 117 12 18														162
N# 1175 1 M=162 B=1.17	80A 578 253 197 187 172 137 133 127														
3 PHENYLHEXANE	91 2 119 133 162 41 92 39 29 77 12 18														162
N# 1189 1 M=162	28 214 141 114 101 77 72 68 47														
3 PHENYL 3 METHYLPENTANE	91 133 55 27 29 105 39 92 77 162 12 18														162
N# 1190 1 M=162	646 106 149 149 145 89 77 72 70														
1 PHENYL 2 ETHYLBUTANE	92 43 91 29 71 27 162 41 39 65 12 18														162
N# 1191 1 M=162	769 643 251 205 201 179 147 143 118														
3 , 7 METHYLPHENYL . PENTANE	105 133 162 41 91 27 39 79 119 43 12 18														162
N# 1192 1 M=162	586 313 309 282 197 143 128 106 100														
3 . 4 METHYLPHENYL . PENTANE	105 133 41 162 27 42 91 39 29 134 12 18														162
N# 1193 1 M=162	969 279 236 181 165 158 147 137 103														
1 4 DI N PROPYLBENZENE	133 91 28 162 27 29 134 41 104 39 12 18														162
N# 1194 1 M=162	30V 228 195 172 121 107 91 85 83														
1 3 DIMETHYL 5 TERT BUTYLBENZENE	147 162 41 119 107 91 39 148 27 15 12 18														162
N# 1195 1 M=162	250 217 210 283 131 129 124 106 87														
1 3 5 TRIETHYLBENZENE	147 133 162 105 29 27 91 39 28 15 12 18														162
N# 1196 1 M=162	843 499 272 254 221 204 159 133 128														
2 METHYL 2 PHENYLPENTANE	119 91 41 27 105 39 120 162 29 77 12 18														162
N# 1446 1 M=162	379 236 134 105 99 92 85 72 68														
4 ISOPROPYLACETOPHENONE	147 43 91 162 51 41 39 77 119 148 11 14 . . . . . 1														162
N# 1749 1 M=162 H=1.30	909 234 219 185 172 161 161 125 110														
1.4 DIACETYLBENZENE	147 43 91 162 119 50 76 77 148 51 10 10 . . . . . 2														162
N# 1753 1 M=162 B=1.14	573 240 226 186 156 129 110 100														
2 PHENYLHEXANE	105 91 27 162 106 29 77 41 28 39 12 18														162
N# 1767 1 M=162	131 123 106 103 86 71 67 62 61														
2.3 DIMETHYL 3 PHENYLBUTANE	119 91 41 27 40 39 118 120 105 77 12 18														162
N# 1768 1 M=162	365 244 132 127 121 106 104 87 86														
1 2 4 TRIMETHYL 5 ISOPROPYLBENZENE	147 162 148 41 39 91 27 119 15 105 12 18														162
N# 1769 1 M=162	221 115 103 91 87 85 75 59 55														
HEXAMETHYLBENZENE	147 162 27 148 39 91 41 161 163 77 12 18														162
N# 1770 1 M=162	580 121 111 105 95 98 68 67 60														
CYCLODODECATRIENE	54 39 41 67 79 80 93 53 91 66 12 18														162
N# 42A02 2 M=162	513 506 473 423 314 271 154 139 102														
2 N HEXYLPYRIDINE	93 27 39 134 29 120 41 28 106 65 11 17 . . . . . 1														163
N# 190 2 M=163	163 135 133 114 102 98 96 90 78														
ETHYL PHENYL ACETATE	91 29 31 164 27 92 65 39 45 43 10 12 . . . . . . 2														164
N# 43A02 1 M=164	390 146 139 114 111 119 91 65 52														
ETHYLPHENYL ACETATE	91 29 164 65 92 39 27 28 63 89 10 12 . . . . . . 2														164
N# 117A02 1 M=164 B=1.30	253 155 117 108 71 49 45 38 33														
6 TERTIARYBUTYL M CRESOL	149 121 164 91 77 150 107 108 115 82 11 16 . . . . . 1														164
N# 64A2 M=164 H=1.01	484 265 184 140 119 118 102 94 80														
TETRACHLOROETHYLENE "	166 164 129 131 168 47 94 35 96 133 2 . . . . . 4														164
N# 698 2 M=164	786 686 655 484 415 398 352 256 210														
EUGENOL	164 15 27 76 148 38 90 54 50 130 10 12 . . . . . . 2														164
N# 1408 2 M=164	973 632 377 326 311 246 263 241 232														
BENZYL PROPIONATE	91 108 57 29 90 27 65 79 39 51 10 12 . . . . . . 2														164
N# 141R 2 M=164	833 462 428 274 220 178 159 152 152														
PROPIONATE OF BENZYL	91 108 29 57 164 27 90 65 39 10 12 . . . . . . 2														164
N#A 3AC3 M=164 B=g.55	802 456 427 241 259 255 188 152 150														
BETA PHENYLETHYL ACETATE	104 43 91 105 65 39 51 77 103 78 10 12 . . . . . . 2														164
N#A 8AC3 M=164 B=g.43	820 210 128 90 87 79 70 58 57														
1 PROPYL BENZOATE	105 77 123 122 59 164 51 43 106 41 10 12 . . . . . . 2														164
N#7 24AC3 M=164 B=g.56	320 310 200 140 130 120 110 110 80														
1 PROPYL BENZOATE	105 77 123 51 43 59 122 27 41 164 10 12 . . . . . . 2														164
N#A 20AC3 M=164 B=g.63	412 272 262 250 215 187 1A0 116 115														
2 4 6 TRIMETHYL BENZOIC ACID	146 164 119 147 118 91 77 105 91 10 12 . . . . . . 2														164
N#7 5AC3 M=164 B=g.28	870 420 400 330 190 140 120 100 90														
P ISOPROPYL BENZOIC ACID	149 164 105 119 77 79 91 150 51 . . 12 . . . . . . 2														164
N#7 6AC3 M=164 B=g.44	470 330 320 190 140 110 110 100 60														
N HEXYL BROMIDE	43 135 85 41 55 57 27 29 56 42 6 13 . 1														164
N#R 51aC3 M=164 B=g.38	500 490 3A0 340 220 180 180 170 140														
1 METHYL AMYL BROMIDE	43 85 41 55 57 27 29 56 69 28 6 13 . 1														164
N#R 52aC3 M=164 B=g.61	900 270 150 150 120 100 50 40 20														
1 ETHYL BUTYL BROMIDE	43 85 41 57 27 29 42 56 28 69 6 13 . 1														164
N#R 53aC3 M=164 B=n.41	870 270 150 120 120 100 70 50 50														
3 METHYL AMYL BROMIDE	57 85 55 41 84 29 43 69 56 27 6 13 . 1														164
N#A 54aC3 M=164 B=n.46	620 450 440 280 260 230 200 190 150														
3.3.3 TRIFLUORO 1.2 DICHLOROPROPANE	129 164 131 69 166 85 145 95 75 114 3 . 1 . . 2 3														164
N# 485Nw4 M=164 B=n.31	492 338 322 312 169 129 115 104 94														
2.TERT.BUTYL.M.CRESOL	149 121 164 150 91 77 175 116 109 41 11 16 . . . . . 1														164
N# 1113Nw4 M=164 B=n.43	343 310 113 96 60 59 54 51 49														
4.TERT.BUTYL.O.CRESOL	149 164 121 150 109 41 91 77 165 39 11 16 . . . . . 1														164
N# 1113Nw4 M=164 B=n.51	272 167 111 97 60 58 50 32 31														
CHLORAL.HYDRATE	18 82 17 84 83 111 47 29 85 113 2 . 3 . . 3 . . . 2 . . .														164
N# 1482Nw4 M=164 B=t.02	223 219 143 84 79 73 73 55 52														
2.1.PROPENYL.6.METHOXYPHENOL	164 149 131 121 103 77 91 165 132 144 10 12 . . . . . . 2														164
N# 1A23Nw4 M=164 B=n.45	271 231 226 221 200 143 113 89 88														
P METHOXYPROPIOPHENONE	135 164 77 92 107 136 64 63 50 29 10 12 . . . . . . 2														164
N# 1299Nw4 M=164 B=n.80	163 148 91 94 88 49 32 20 19														
2.BROMO.3.PENTANONE	57 29 27 56 107 109 28 137 135 164 5 9 . 1														164
N# 2244Nw4 M=164 B=n.42	451 161 121 112 110 77 44 45 35														
P.A.4.TRIMETHYL PHENETHYL.ALCOHOL	106 59 91 105 107 149 43 77 31 79 11 16 . . . . . 1														164
N# 2533Nw4 M=164 B=n.69	542 470 182 105 104 95 79 55 50														
3.BROMOHEXANE	43 85 41 55 57 27 29 42 39 56 6 13 . 1														164
N# 2902Nw4 M=164 B=n.41	869 274 217 150 119 117 109 95 67														
P .1.1 DIMETHYLPROPYL . PHENOL	135 107 164 135 95 41 149 77 91 55 11 16 . . . . . 1														164
N# 2575Nw4 M=164 B=t.23	260 114 100 75 59 54 49 41 37														
1.BROMO.3.METHYLPENTANE	57 85 55 41 84 29 43 69 56 27 6 13 . 1														164
N# 3238Nw4 M=164 B=t.61	620 454 439 283 262 229 199 192 151														
2.BROMOHEXANE	43 85 41 57 55 27 29 42 39 46 6 13 . 1														164
N# 3239Nw4 M=164 B=n.61	900 273 151 147 121 102 96 90 58														
P.TERT.BUTYLANISOLE	149 164 121 150 109 135 91 77 39 41 10 12 . . . . . . 2														164
N# 3554Nw4 M=164 B=n.98	233 161 137 112 88 77 60 60 33														
P.ISOPROPYLBENZOIC.ACID	149 164 105 119 77 79 131 79 103 91 150 10 12 . . . . . . 2														164
N# 356ANw4 M=164 B=n.44	466 325 317 189 141 136 131 126 112														

59 B

FIGURE 6.25. Specimen page from "Compilation of Mass Spectral Data" showing molecular weight listing.[270]

COMPOUND NAME	MASS SPECTRUM	C	H	D	Br	Cl	F	I	N	O	P	S	Si	Di	M/E
BENZOYL CHLORIDE	105 77 51 50 38 106 74 37 140 39	7	5	.	.	1	.	.	.	1	.	.	.	.	105
N= 244 2 M=140 B=1.05	654 352 238 107 78 75 70 47 46														
BENZOYL CHLORIDE	105 77 51 50 52+106 74 37 38+ 27	7	5	.	.	1	.	.	.	1	.	.	.	.	105
N= 598 2 M=140 B=0.79	674 377 240 107 92 79 78 59 52														
HIPPURIC ACID	105 77 51 76 122 117 50 161 44 106	9	.	.	.	.	.	1	3	.	.	.	.	.	105
N= 84H2 M=179	782 416 323 295 258 235 231 121 90														
2.CHLORJACETOPHENONE	105 77 51 106 50 91 78 74 39 76	8	7	.	1	.	.	.	.	1	.	.	.	.	105
N= 1986Dw4 M=154 T=0.88	407 113 79 47 37 38 18 18 17														
A.A.DICHLORO.W.HYDROXYPROPIOPHENONE	105 77 51 106 78 50 29 30 28 76	9	8	.	.	2	.	.	.	2	.	.	.	.	105
N= 2112Dw4 M=218	348 84 77 42 32 31 25 24 16														
ISOBUTYROPHENONE	105 77 51 106 148 78 50 27 41 43	10	12	.	.	.	.	.	.	1	.	.	.	.	105
N= 2676Dw4 M=148 T=1.90	291 83 79 74 24 22 21 20 19														
ACETOPHENONE	105 77 51 120 43 50 78 28 39 74	8	8	.	.	.	.	.	.	1	.	.	.	.	105
N= 121AD2 M=120 B=0.43	826 305 247 177 128 89 55 52 44														
ETHYL BENZOATE	105 77 51 122 150 27 50 29 106 28	9	10	.	.	.	.	.	.	2	.	.	.	.	105
N= 194C3 M=150 B=0.45	481 244 257 178 141 124 122 92 53														
METHYL BENZOATE	105 77 51 136 50 106 29 78 74 39	8	8	.	.	.	.	.	.	2	.	.	.	.	105
N=6 18AC3 M=136 B=0.33	677 376 284 183 79 58 52 48 45														
METHYL BENZOATE	105 77 51 136 50 106 78 74 39 76	8	8	.	.	.	.	.	.	2	.	.	.	.	105
N= 1752 1 M=136 B=1.27	655 343 322 179 81 50 47 44 42														
ALPHA BROMOMETHYL 2 METHYL PHENYL	105 77 79 51 106 103 184 104 78 39	8	9	. 1	.	.	.	.	.	.	.	.	.	.	105
N=8 191AC3 M=184 T=0.72	110 110 100 100 90 80 60 50 40														
ALPHA BROMOMETHYL 4 METHYL PHENYL	105 77 79 51 106 103 184 104 91 78	8	9	. 1	.	.	.	.	.	.	.	.	.	.	105
N=8 193AC3 M=184 T=0.74	100 100 90 90 80 70 50 40 40														
A.PHENYLPROPIONALDEHYDE	105 77 79 134 104 103 91 51 78 39	9	10	.	.	.	.	.	.	1	.	.	.	.	105
N= 1442 M=134 T=1.20	178 160 139 104 92 71 78 40 37														
PHENYLBENZOATE	105 77 106 51 50 65 78 107 93 52	13	10	.	.	.	.	.	.	2	.	.	.	.	105
N=7 30AC3 M=198 T=0.97	290 80 70 20 20 20 06 05 04														
ALLYL BENZOATE	105 77 106 51 162 41 117 78 50 76	10	10	.	.	.	.	.	.	2	.	.	.	.	105
N=7 23AC3 M=162 T=1.37	280 80 80 60 40 30 20 20 10														
ACETOPHENONE	105 77 120 51 43 50 78 106 39 74	8	8	.	.	.	.	.	.	1	.	.	.	.	105
N= 767Dw4 M=120 T=0.88	745 249 229 120 91 77 77 33 33														
BENZOIC ACID	105 77 122 51 50 39 38 74 95 78	7	6	.	.	.	.	.	.	2	.	.	.	.	105
N= 1442 M=122	851 735 597 392 148 133 122 111 99														
ALPHA HYDROXYACETOPHENONE	105 77 122 51 106 18 50 78 44 28	8	8	.	.	.	.	.	.	2	.	.	.	.	105
N= 2674Dw4 M=136 T=0.70	561 146 162 84 64 60 44 32 30														
ETHYL.BENZOATE	105 77 122 150 51 106 50 78 29 27	9	10	.	.	.	.	.	.	2	.	.	.	.	105
N=8 3673Dw4 M=150 T=0.82	359 359 272 120 111 48 32 31 30														
I PROPYL WEAZOATE	105 77 123 51 43 59 122 27 41 164	10	12	.	.	.	.	.	.	2	.	.	.	.	105
N=6 20AC3 M=164 B=0.63	412 272 242 226 215 187 160 116 115														
I PROPYL BENZOATE	105 77 123 122 59 164 51 43 100 41	10	12	.	.	.	.	.	.	2	.	.	.	.	105
N=6 24AC3 M=164 T=0.56	322 310 200 190 130 120 110 110 60														
ISOPROPYL.BENZOATE	105 77 123 122 59 164 51 43 106 41	10	12	.	.	.	.	.	.	2	.	.	.	.	105
N= 3609Dw4 M=164 T=0.56	316 388 204 189 132 118 113 108 57														
PROPIOPHENONE	105 77 134 51 106 50 78 27 135 29	9	10	.	.	.	.	.	.	1	.	.	.	.	105
N= 1086Dw4 M=134 T=1.40	433 191 126 77 40 34 23 19 19														
PHTALIDE	105 77 134 51 133 50 76 106 74 78	8	6	.	.	.	.	.	.	2	.	.	.	.	105
N= 5152Dw4 M=134 T=0.78	441 345 128 122 106 102 78 122 41														
N.BUTYROPHENONE	105 77 146 51 120 106 78 50 27 39	10	12	.	.	.	.	.	.	1	.	.	.	.	105
N= 2477Dw4 M=148 T=1.10	383 163 107 86 79 38 28 25 20														
BENZOPHENONE	105 77 182 51 106 183 181 28 50 76	12	22	.	.	.	.	.	.	1	.	.	.	.	105
N= 2773Dw4 M=182 T=0.97	567 542 173 78 75 68 52 49 43														
6.6.DIMETHYLBICYCLO.3.1.1.2.HEPTENE 2.ETHANOL	105 79 91 93 92 41 77 106 122 133	11	18	.	.	.	.	.	.	1	.	.	.	.	105
N= 3125Dw4 M=166 T=0.68	203 107 176 166 139 123 103 99 78														
2 VINYLPYRIDINE	105 79 104 51 78 106 77 78 26 39	7	7	.	.	.	.	.	1	.	.	.	.	.	105
N= 41 2 M=105	931 558 468 422 340 213 247 170 140														
SULFUR OXYTETRAFLUORIDE	105 86 67 32 107 89 48 51 19 70	.	.	.	.	.	4	.	.	1	.	1	.	.	105
N= 158C2 M=124 B=0.96	126 93 52 47 37 27 25 18 17														
2 PHENYLHEXANE	105 91 27 162 106 29 77 41 28 39	12	18	.	.	.	.	.	.	1	.	.	.	.	105
N= 1767 1 M=162	131 123 106 103 86 71 67 62 61														
BETA ETHYL PHENYL IODIDE	105 91 77 79 103 106 51 78 104 39	8	9	.	.	.	.	1	.	.	.	.	.	.	105
N=8 203AC3 M=232 T=0.45	180 160 140 110 90 90 60 60 40														
BENZYL BENZOATE	105 91 77 106 51 90 107 65 167 92	14	12	.	.	.	.	.	.	2	.	.	.	.	105
N=7 31AC3 M=212 T=1.21	410 210 90 80 70 70 60 60 45														
BENZYL.BENZOATE	105 91 77 212 51 106 65 90 194 107	14	12	.	.	.	.	.	.	2	.	.	.	.	105
N= 1765Dw4 M=212 T=0.76	445 259 212 104 86 82 73 68 64														
2 PHENYLDODECANE	105 91 184 41 43 246 104 77 77 59	18	30	.	.	.	.	.	.	.	.	.	.	.	105
N= 1743 1 M=246 B=5.68	148 133 108 77 56 44 38 38 33														
2 PHENYL OCTANE	105 91 106 91 41 77 39 79 104 103	14	22	.	.	.	.	.	.	.	.	.	.	.	105
N= 890 1 M=190 B=2.06	132 120 83 77 59 47 46 41 29														
I.METHYTBUTYL.BENZENE	105 91 148 119 106 77 79 103 41 92	11	16	.	.	.	.	.	.	.	.	.	.	.	105
N= 2578Dw4 M=148 T=1.29	401 216 171 102 93 65 49 49 49														
CINNAMYL.PROPYL.ETHER	105 92 176 117 91 115 133 43 78 134	12	16	.	.	.	.	.	.	1	.	.	.	.	105
N= 3467Dw4 M=176 T=1.37	751 656 504 364 349 350 338 235 214														
ERYTHRO 1.2 DIPHENYLETHYL D? BENZOATE	105 104 91 51 180 78 27 77 141 1A2	43	92	21	17	1	.	.	.	2	.	.	.	.	105
N= 4135Dw4 M=303 T=0.08	606 517 492 400 295 214 190 185 179														
3 VINYL PYRIDINE	105 104 51 52 78 50 39 79 77 27	7	7	.	.	.	.	.	1	.	.	.	.	.	105
N= 828 2 M=105 B=0.90	572 429 364 303 176 158 152 132														
3 VINYLPYRIDINE	105 104 51 52 78 50 39 79 77 27	7	7	.	.	.	.	.	1	.	.	.	.	.	105
N= 24A2 M=105 B=0.90	572 429 368 310 303 176 158 152 132														
I DEUTEROMETHENYLBENZENE	105 104 77 78 51 79 50 52 103 106	8	7	1	.	.	.	.	.	.	.	.	.	.	105
N= 1289 1 M=106 B=1.20	121 109 100 100 99 96 94 92														
2 DEUTEROETHENYLBENZENE	105 104 78 51 79 103 50 52 39 77	8	7	1	.	.	.	.	.	.	.	.	.	.	105
N= 1790 1 M=107 B=1.11	394 215 203 201 167 128 114 108 102														
4 DEUTERO I ETHENYLBENZENE	105 104 79 78 51 52 50 106 103 77	8	7	1	.	.	.	.	.	.	.	.	.	.	105
N= 1791 1 M=105 B=1.38	372 300 175 170 132 98 89 82 70														
ALPHA BROMOETHYL PHENYL	105 104 103 77 51 79 106 78 50 63	8	9	. 1	.	.	.	.	.	.	.	.	.	.	105
N=8 104AC3 M=184 T=0.71	160 140 103 77 40 40 30														
2 METHYL BENZYL 2.METHYLBENZOATE	105 104 119 91 77 106 79 103 78 89	16	16	.	.	.	.	.	.	2	.	.	.	.	105
N= 544A2 M=240	916 364 240 145 86 79 73 58 45														
I PHENYL I CYCLOHEXYLETHANE	105 106 41 39 27 104 91 55 51 188	14	20	.	.	.	.	.	.	.	.	.	.	.	105
N= 584 1 M=188 B=0.46	862 435 385 365 322 304 286 232 106														
2 PHENYLEICOSANE	105 106 43 41 57 55 104 358 79 26	46	.	.	.	.	.	.	.	.	.	.	.	.	105
N= 589 1 M=358 B=1.85	149 128 90 70 68 59 39 34														
2.PHENYLPROPANOL.D.	105 106 79 77 136 92 91 137 103 107	9	11	1	.	.	.	.	.	1	.	.	.	.	105
N= 3022Dw4 M=137 T=1.10	299 258 212 124 110 74 43 46 45														
I .ALPHA METHYLBENZYL. CYCLOHEXENE I	105 106 80 41 91 79 91 77 104 44 106	14	18	.	.	.	.	.	.	.	.	.	.	.	105
N= 1916Dw4 M=186 T=0.75	394 310 275 205 183 157 143 142 141														
2 PHENYLDODECANE	105 106 91 246 104 55 77 79 57 103	18	30	.	.	.	.	.	.	.	.	.	.	.	105
N= 1772 1 M=246	129 102 77 43 34 30 29 20 18														

230 D

FIGURE 6.26. Specimen page from "Compilation of Mass Spectral Data" showing most abundant ion listing.[270]

residue, it is not necessary to go through the many thousands of known compounds and file search of a few hundred compounds will give identification or denial. The reduced file is also more amenable to automatic computer file search. (See Costello, page 376.)

Compound identification by analogy. Every spectrometrist knows that he will not find all unknowns in the spectral catalog. Fortunately, the members of homologous series of compounds always fragment with essentially the same reactions. Thus, the mass peaks from one member of a group often occur in the spectra of close homologues displaced 14, 28, etc. mass units. If an ion fragment does not contain the element of additional chain length, then that mass fragment will appear in the spectra of all the homologues. Spectral correlations of this type are of tremendous value in extending the usefulness of spectral files. Many significant papers have been published on correlative relationships in many classes of compounds. (See, for example, the basic texts, references 4–11.) By studying the spectra of homologues, fragmentation mechanisms can be deduced that permit a reasonable suggestion of compound identity. These complimental relationships also play an important role in fundamental research on mass spectral fragmentation.

The scope of organic mass spectrometry has become so extensive in recent years that there are literally hundreds of detailed studies of various compound types.[2,3] The basic principles of organic mass spectrometry are given in the several available textbooks. The novice can soon become familiar with the spectral characteristics of simple compounds (esters, ketones, aldehydes, acetals, thioles, amines, halogens, etc.) in the aliphatic and aromatic series. More exotic series such as heterocyclic compounds, peptides, carbohydrates, alkaloids, and steroids[275] necessitate greater depth of understanding, and after an initial study of basic principles, the chemist must concentrate in his own field. He is soon gratified to realize that, within his narrow realm of application, his specialized knowledge of mass spectral fragmentation exceeds the general knowledge of many experienced mass spectrometrists.

Manual interpretation of GCMS data is greatly aided by use of our mental catalog of mass spectral features. During the mass-marking procedure, the experienced spectroscopist will search his mind for correlative traits. The molecular weight, rearrangement ion peaks, strong bond-break ions, appearance of unsaturation or aromaticity, all and any of these spectral properties can be quickly linked together to suggest a possible structure. In this way, spectra from commonly encountered compounds are interpreted as fast as the mass-marking process can take place and the spectral file search is facilitated by the prior suggestion of a possible compound.

If an easy hit is not obtained from the file search or from simple correlative relations, the literature spectra must be explored in depth to find any possible spectral analogues. This "in depth" literature examination is quite obviously a duty for the chemist, but the surfeit of mass spectral data now available makes the procedure very time consuming. The task is facilitated by the excellent services provided in periodical reviews, but several hours may be used on a good manual literature search on one difficult spectrum.

If the number of spectral charts is small, an extensive search poses little problem. The difficulty is encountered in GCMS studies that generate as many as 20–30 important unmatched spectra in a single day. Ideally, all these should be identified, but too often, the time investment cannot be made. It is imperative that the chemist learns to select the important chromatographic peaks either by chemical reactivity, aroma, biological response, abundance in the mixture, or whatever specific properties are important to the total problem. If a selection is not made, the excessive data produced by GCMS methods will result in unwarranted expenditure of time on spectra of no immediate value. Unfortunately, the danger of ignoring important spectra is always present.

Thus, GCMS analyses create a paradox. If the data are thoroughly interpreted, time may be wasted on trivial spectra. If the data are selectively interpreted, valuable spectra may be overlooked. Without doubt, the answer to the dilemma is to use computer-automated data interpretation.

7

The Role of the Computer in GCMS

A. WHY USE A COMPUTER?

Today, everybody wants to use a computer. Probably no implement in the history of man has been so fashionable since the discovery of the wheel, and computers are now widely used in the business world, government agencies, and all branches of science and engineering. It is even becoming voguish to own a small hand-size computer for personal calculations. Application of the computer to GCMS has been a natural outcome of this enthusiasm.

Use of the computer in organic mass spectrometry became important with the extensive use of high-resolution mass spectrometers.[276] It was soon apparent that the extensive amount of data generated by a GCMS system was also amenable to computer acquisition, reduction, and interpretation.[96, 98, 241, 242, 277-283] Several general articles review these applications both in GCMS and high-resolution mass spectrometry, as well as in other analytical systems.[10, 284-287] In GCMS, the value of a computer is realized two ways: in a saving of time (hence money) and in increased efficiency.

1. Time Saving

Anytime a chemist must perform a large number of relatively routine operations, he should use a computer. In mass spectrometry, a high rate of data output is routinely attained, and the mass spectrometrist is continually overloaded with repetitive data processing. For high-resolution mass spectrometric data, the large number of measurements, calculations, and value comparisons for each spectrum makes computer processing a necessity for proper data analysis. In GCMS, the large number of spectra generated during a run makes computer acquisition and processing highly desireable. Although most GCMS data are still treated manually, use of computers is rapidly expanding and will soon be considered routine in most laboratories.

The GCMS system in a large, well-organized laboratory may perform analyses for 4–5 hours during a typical 8-hour day. Depending on the type of samples involved, this might be 3–5 runs or as many as 8–10 runs. In any event, it is highly improbable that less than 100 spectra would be generated and quite possibly the number of spectra would exceed 300–400.

These charts must be mass marked and processed with at least a cursory search or spectral comparison. A complete data treatment involves measurement, normalization, and plotting of data followed by an intense file search and further deductive consideration. The approximate time needed to carry out these procedures manually is suggested in column 2 of Table 7.1. Assuming an operation that puts out 150 spectra per day, 250 days per year, reasonable estimations can be given for the total daily processing time and the annual cost for treatment of mass spectral data.

The figures given in Table 7.1 are realistic, but care must be taken to avoid misinterpretation. For one thing, the average GCMS output level of 150 spectra/day, 250 days/year would by typical for about 15 project chemists using the GCMS facility. (Whether the data come from more than one GCMS system is immaterial when assessing the value of a computer.) For fewer groups, the output and costs are correspondingly reduced, and a single group may use about 2 1/2 man moths per year manually processing 2,500 GCMS spectra. The laboratory time used per group is about 50–60 hours/year of GCMS operation, exclusive of time taken to establish the spectra of standard compounds.

Estimated annual cost based on $100/day is probably low for most overheaded laboratory accounts. The apparent cost depends upon the accounting system, but choosing a low value only helps emphasize the considerable expense encountered and the savings that are possible.

In setting up Table 7.1, it was assumed that 7–8% of the spectra will be measured and normalized for spectral files or publication. Another 2–4% will be given a thorough analysis by a chemist using spectral correlations and deductive reasoning. A higher conversion rate would be preferred, but even this low level of complete interpretive treatment makes up more than 50% of the data processing cost.

Computer systems can be designed to perform any or all of the functions listed in Table 7.1, but the cost to install and operate each procedure must

TABLE 7.1. Estimate of time and cost for manual processing of GCMS data in a laboratory with 15 projects using the GCMS facility.

Process	Time per Spectrum	Probable average time per day (15 project chemists)	Man days per year	Approx. annual overhead cost, $ ($100/man day)[1]
Mass mark	1 min	150 spec, 2 hr	60	6,000
Search files	4 min	150 spec, 10 hr	300	30,000
Measure, normalize, plot	20 min	12 spec, 4 hr	120	12,000
Correlation or deductive analysis	2 hr	5 spec, 10 hr	300	30,000
			Total estimated cost	$ 78,000

[1] Depends on the organization or business. For example, graduate students are usually a little cheaper than the suggested price.

balance against the profits. It would not be practical to use a computer for the sole purpose of mass marking, but on the other hand, a considerable convenience and time saving can be realized if the spectra are also normalized and plotted.

Currently, mass marking, normalization, and plotting are the services most frequently performed by the computer in GCMS data processing. Yet these constitute only about 25% of the total work. The spectra are all presented in a form that is easily handled and compared with standard spectra. In a large laboratory, approximately $20,000 time saving will be realized each year, and in addition, there will be a saving of $5,000–$10,000 for oscillographic chart paper. Purchase of a $60,000 computer system is more than warranted for an operation of this size.

The more expensive data treatments listed in Table 7.1 are file search, and correlation and deductive analysis. Each of these comprises about 35% of the total time. Available commercial file search programs have not been widely tested, and the various spectral interpretation programs have been used only in advanced research laboratories. Because of the considerable potential gain by automatic search or data analysis, a computer system purchased today for data acquisition and reduction should be adaptable in the future for conversion to data interpretation programs if possible.

For some time it has been recognized that a computer system is essential for thorough analysis of conventional high-resolution mass spectral data. Even for one mass spectrum, accurate mass measurement of several peaks and proper comparison of exact masses with tabulated values may take a few hours. To reduce the excessive data obtained in a high-resolution GCMS run, a large, well-programmed computer system must be available. The considerable cost has discouraged widespread use of high-resolution GCMS. Even the few highly specialized laboratories that can afford this luxury do not use it routinely.

2. Increased Efficiency

From the previous discussion it is clear that proper use of a computer results in a considerable saving of time and money. Even when used only for data reduction and presentation, the purchase is readily justified. In addition to the economic factor, computers also provide improved quality and increased efficiency in many of the data acquisition and reduction processes. Some of the ways that a computer can serve us beyond the call of Mammon are listed below.

(1) Cyclic scan on a 4–5-sec cycle gives a record of virtually all available information. Operator decisions and/or fatigue errors are avoided.

(2) Consecutive jump scan (page 250) is possible on electrically scanned mass spectrometers. The computer thus sits on a peak for a longer time period and attains increased sensitivity.

(3) Mass marking is more reliable in the high molecular weight ranges.

(4) A computer will normalize and print all spectra of interest, not just a few that are selected niggardly to avoid excess time expenditure. The spectra are clearly read and easily handled.

(5) A computer program can subtract the background, or an adjacent spectrum, as requested.

(6) The computer will plot mass chromatograms for any selected masses, thus giving clarity for interpretation and presentation of data.

(7) An efficient search of mass spectral files can be performed by the computer. A good manual search of a file that exceeds 10,000 entries is very difficult, and if a large number of spectra must be processed, operator fatigue may lead to errors of omission, sometimes even by deliberate short-cutting. If the file is in the range of 30,000 spectra or more (not yet attained in mass spectral catalog), a thorough manual search is virtually impossible for a large number of unknown spectra.

(8) When an unknown spectrum cannot be matched in the files, cor-relative or deductive analysis of the data is necessary. Although man does a good job on this aspect, he uses only information he knows personally and can recall at that moment. A computer uses all the information that has been pro-vided.

(9) For high-resolution mass spectral data, a computer is needed for accurate mass determination of all the useful mass peaks in the spectrum. Without a computer treatment, accurate mass assignment is usually limited to a few selected peaks, and important data is overlooked for expediency. As a corollary of this statement, one notes that for high-resolution GCMS, an expen-sive computer system is essential.

In spite of these obvious merits, it is important not to overlook some weaknesses or disadvantages that can occur when using automatic data pro-cessing. A few examples are listed below. They are not always critical, and methods can be devised to obviate the difficulties, but in assessing the value of computer systems, the disadvantages must be weighed with the advantages.

(1) If a total scan cycle in a cyclic scan process is as long as 5−8 sec, im-portant data may be lost or distorted on fairly sharp peaks (see page 98).

Similarly, in some data systems, the computer uses every second cycle to perform a specific data manipulation, such as outputting the previous spectrum to an oscilloscope. If the total scan cycle is a reasonable 3–5 sec, the gap of 3–5 sec for every second scan may result in a loss of important data.

(2) Although computer mass calibration is generally quite reliable, low-resolution systems cannot use an internal chemical standard. In the event that some minor instrumental malfunction causes the calibration to be in error, the whole GCMS run may be lost.

(3) If spectra are plotted on a slow digital printer, the complete GCMS-computer system may be tied up for an unacceptable period of time. This often necessitates plotting the spectra overnight. If it is decided subsequently that additional spectra should be plotted, a further delay is encountered. Fast plotting methods are available (i.e., a few seconds per spectrum) and should be obtained whenever possible.

(4) The plotted spectra are conveniently normalized as relative abundance, and minor peaks less than 1% RA are not displayed. Frequently, in GCMS, these minor peaks indicate an important trace component. The operator can request a multiplication of any part of the spectrum, but of course, this is done only when some unusual low-intensity peaks are observed. Valuable data points are often lost in this way that would be observed in raw oscillographic traces.

(5) Since the spectra are normalized with the most abundant ion equal to 100, the intensity of ions from small chromatographic peaks may be less than the background. The unknown spectra are thus not properly presented. Special instructions must be given to the computer for subtraction of background which thus necessitates additional operator attention.

(6) The absolute intensity of each scan is usually printed at the top of the plot so that comparisons can be made with adjacent spectra. This is not as convenient as comparing the oscillograph spectra from adjacent scans, particularly if the spectra are taken with two or more components eluting very close together.

(7) Most mass spectral file search programs in current use require operator computer conversation. Completely automatic systems are much more expensive and often are limited to searching a small number of spectra from specific types of compounds. (For example, poisons and drugs, see page 377.)

(8) Similarly, programs which perform correlation studies have been developed only on relatively large, expensive systems and are not yet accessible to the average chemist who wishes to purchase a computer system for around $60,000.

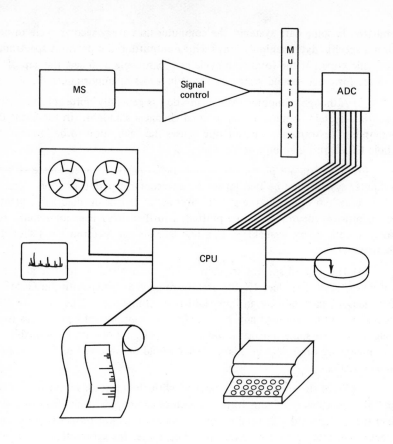

FIGURE 7.1. Hardware configuration in a typical on-line GCMS-Computer system.

B. THE BASIC COMPUTER SYSTEM

A typical GCMS computer hardware configuration is shown schematically in Figure 7.1. Many variations are possible depending upon the operations of the laboratory and the investment that can be made in computer hardware and software. For a small laboratory, it may be economically necessary to store raw analog or digital data directly on the tape unit for off-line processing. In a large, well-equipped laboratory, additional flexibility is obtained by using the facilities of a large central computer system along with a small dedicated mini-computer.

On-line GCMS data acquisition imposes several demands on the specifications of the computer data processing equipment. It is important to relate available hardware capabilities with the needs of the laboratory, particularly since the cost of options may exceed the economic factor for a specific operation.

1. Signal Controller

If a mass spectrometer is not designed for use with a specific computer system, the mass spectral signal may not be properly rated for a given analog-to-digital convertor (ADC). Some of the problems that arise are: (1) the amplifier output impedance may not match the input impedance of the converter, (2) the signal voltage level from the mass spectrometer amplifier may be too high for the convertor, and (3) high-frequency noise from the amplifier ground system may exceed the desired threshold voltage. To remove these obstacles, a differential amplifier is used between the mass spectrometer analog output and the analog-to-digital convertor.

The output impedance of the analog amplifier may vary from 1,000 ohms to as high as 100,000 ohms. The input impedance of most A/D convertors will be 1,000 ohms. The differential amplifier is chosen to match the specific values of these two components.

The maximum signal level output of a mass spectrometer will be somewhere in the range of 10–100 volts. The maximum input signal to the A/D convertor will most likely be 1 volt or 10 volts. The appropriate match of these two signals is easily obtained by adjusting the gain of the differential amplifier.

Perhaps the most important signal control function is to eliminate the high-frequency noise inherent in most analog amplifiers. If the mass spectrometer amplifier and A/D convertor have the same ground connection, the ground loop signal is often several millivolts and can be as high as 100 mV or more. Since threshold signal in the ADC is usually set in the range 2–5 mV, it is essential that common ground noise is eliminated.

2. Multiplexor

If a computer is being used solely for recording a mass spectral signal, a multiplex switching device is not needed prior to the A/D conversion. In some arrangements, the computer is also used to store auxiliary data such as temperature, total ion current, or the voltage from a Hall magnetic field probe. In highly sophisticated systems, the computer may control the mass spectrometer by monitoring operational parameters. To effect these refinements, the various input signals must be time-selected by a multiplexing unit.

Solid state multiplexors are readily obtained for operation with most A/D convertors. GCMS operation seldom uses more than eight channels, and the timing for each switch is easily selected by the computer program.

3. Analog-to-Digital Convertor

Before a computer can operate on mass spectral data, the continuous analog signal from the electrometer amplifier must be converted to a digital form. Figure 7.2 shows a simplified explanation of this transformation. At the top of the diagram, a mass spectral peak is divided into eight separate sample regions. The average voltage developed during a sample period is stored on a "sample and hold" amplifier (primarily a charged condenser), and then sampled by the analog-to-digital convertor. The discharged current passes through a resistive network and produces a voltage across many switch points. If the voltage drop produced in a particular resistor exceeds a reference voltage stored in the convertor system, a switch is activated and a positive signal is received in that register. The computer thus receives a series of digital signals indicating changes in the nature of the input analog voltage. The "shape" of the digital pattern recorded by the computer is shown on the right in Figure 7.2 and a probable digital sequence for the first four samples is shown below the pattern.

For convenience, a 5-bit ADC is shown in Figure 7.2, but in practice, a 12- or 14-bit convertor is commonly used in GCMS. The number of unique digital levels obtainable from a specific system is given by $2^n - 1$ where n is the number of bits in the convertor. Thus, a 12-bit convertor can represent 4,095 different voltages. If the convertor has a signal level of 0–10 volts, each digit represents 10 volts/4,095, or 2.4 mV. The value is close to the noise level encountered from most analog amplifiers, so that a 10-volt, 12-bit A/D convertor is a compatible specification for most GCMS purposes.

In Figure 7.2, eight voltage samples were taken across the base of the analog mass peak. The digital representation is seen to be slightly distorted and, depending on the nature of the unfiltered analog signal, it can be worse (Figure 7.3). As a rule, 8–10 digital samples per mass peak is considered a reasonable minimum, and 16–20 samples may be used when the scan rate and sensitivity considerations permit.

The time taken to traverse a mass peak at resolution 500 and at a scan rate of 3 sec/decade is close to 2×10^{-3} sec. A digitization rate of 8,000 KHz will give about 16 samples per mass peak which satisfactorily represents the peak shape. At resolution 1,000, the A/D conversion rate would have to be 16,000 KHz to have 16 windows across the peak.

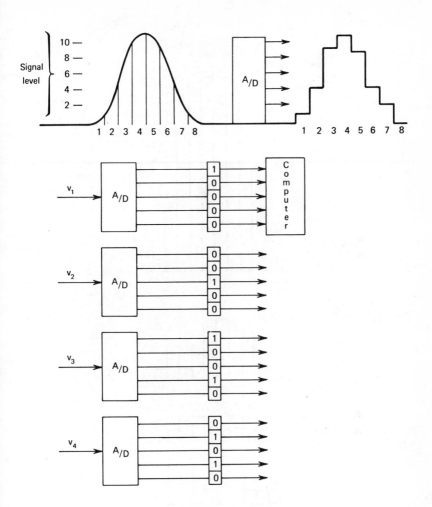

FIGURE 7.2. Schematic illustration of analog-to-digital signal conversion.

For high-resolution GCMS (M/ΔM = 10,000), the digitization rate at 3 sec/decade would exceed a prohibitive 10^5 voltage samples/sec for 16 samples per peak, and a compromise is necessary in the scan time. As a rule, 10 sec/decade is accepted, and a conversion rate of 25,000 samples/sec gives 7–8 windows across the peak. The number is sufficient to obtain a reasonable representation of the peak shape, even at low ion current levels (see Figure 7.3).

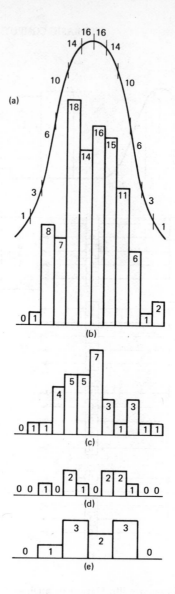

FIGURE 7.3. Illustration of the number of ions collected during the scan of a GCMS peak. (a) Analog peak, 3 x 10^{-14} amp at maximum Average of 100 ions collected in 10^{-3} sec. (b) Digital peak, 3 x 10^{-14} amp at maximum. 99 ions sampled in 12 windows at 12 KHz A/D. (c) Digital peak, 1 x 10^{-14} amp at maximum. 32 ions sampled in 12 windows at 12 KHz A/D. (d) Digital peak, 3 x 10^{-15} amp at maximum. 9 ions sampled in 12 windows at 12 KHz A/D. (e) Digital peak, 3 x 10^{-15} amp at maximum. Same 9 ions sampled in 6 windows at 6KHz A/D.

Faster A/D conversion is not practical for sensitivity reasons, and slower scan rates should be used whenever possible.

Conventional mass spectral recording on an oscillograph is restricted in scan rate primarily by the response characteristics of the galvanometers. For computer acquisition of data, the digitization rates of available A/D convertors are sufficient to meet any GCMS recording needs, but it is also important that the electrometer amplifier band-pass is increased to match the high sample rate at the computer interface. For a sample rate of 10,000 KHz, the electrometer input resistor should be about 10^7 ohms. With a multiplier gain of 10^6, a single ion collected in 10^{-4} sec will give a voltage pulse of 1–10 mV, which is within the detection limit of the computer and comparable to the noise level of the system. However, the sensitivity limit is still determined by the statistical arrival of ions at the multiplier (Figure 7.3).

An ion beam of 1.2×10^{-14} amp peak intensity will deliver about 170 ions in 10^{-3} sec at the maximum point. If a mass peak of this intensity is scanned at 3 sec/decade with a resolution of 1,000, the time taken to traverse the total peak will be slightly greater than 10^{-3} sec. At a digitization rate of 12 KHz (about 12 sample windows across the peak), the peak maximum will have an average of only 16 ions in one sample period and significantly less on the sides of the peak. The total number of ions collected in the peak will average about 100 (Figure 7.3). Consequently, the digital peak shape will be influenced by the statistical variation in the average number of ions collected during each digital sample (i.e., $\pm\sqrt{n}$). One of the many expected digital shapes for collection of 100 ions is shown in Figure 7.3b. At this intensity level (1.2×10^{-14} amp at peak), the 100 ions gives an excellent representation of the analog peak shape. Even at one-third of this current level, an average statistical distribution of the 33 ions still gives a useful signal profile which is easily recognized as a mass peak and for which an accurate centroid point can be established (Figure 7.3c). However, as shown in Figure 7.3d, a peak current of 10^{-15} amp would have less than 10 ions across the 12 digital windows, and many of the data would be rejected as not representing a mass peak.

For low-resolution GCMS, the shape of the digital peak is of no consequence provided the computer can determine that a true ion signal has arrived. The location of the center of gravity of the ion beam will not vary significantly insofar as a unit mass assignment is concerned. Threshold and signal continuity criteria are usually met when 12–15 ions are collected.

For high-resolution GCMS, the statistical fluctuations in the digital peak shape can limit the accuracy of mass measurement. Green et al.[239] found that an accuracy of ±10 ppm could be obtained from a signal containing 10–20 ions. When the beam produces fewer ions than this, spurious stray ions or noise spikes are unduly weighted in the peak histogram and cause errors as high as 50 ppm. For an assured accuracy of 5–6 ppm, the collected beam should contain 30 ions or more.

The illustrations in Figure 7.3 show that the limit of current sensitivity for computer acquisition of GCMS data is comparable to that suggested for conventional oscillographic recording. In both cases, the limitation is due to the statistical fluctuation or counting errors encountered in sampling a small number of ions. Figure 7.3d and e also show that at the limiting current level, a better peak representation can be obtained by using a slower digitization rate. Careful consideration of this point is particularly meritorious where threshold rules might cause frequent rejection of minimal-size mass peaks of diagnostic importance.

To eliminate spurious ion signals and noise spikes, a GCMS-computer system may have several methods of selecting threshold options. In some cases, the threshold is established electrically at the computer interface, but it is more flexible to use software threshold modes. These may include: (1) the elimination of small digital signals, (2) a specified number of consecutive positive digital samples, (3) a limited number of zero signal samples in peak, and (4) the total number of digital signals must exceed a specific value.

1. Elimination of small digital samples is easily accomplished either by software or hardware methods. Usually, a voltage signal level corresponding to that put out by 1 or 2 ions is suitable. However, care should be taken in setting this level since it severely influences the lower level of ion current sensitivity. As noted in Figure 7.3d, rejection of signals due to one ion may leave very little for mass peak evaluation, and the net result of three or four digital samples could be eliminated by other peak recognition specifications.

2. A specified number of digital samples is commonly requested to avoid acceptance of sharp noise spikes that may be several millivolts or more. The total number will depend on the mass spectral scan rate and the digitization rate. For the example of Figure 7.3, the operator might request that 3–5 digital signals constitute a peak, depending, of course, on the other threshold values in use. For example, if four digital signals are requested and all ion signals due to only one ion are software eliminated, then (c), corresponding to a peak current of 1×10^{-14} amp, is accepted as a peak but (d), corresponding to a peak current of 3×10^{-15} amp is rejected, since only three windows have a signal greater than that due to one ion.

3. By placing a limit on the number of zero signal samples (i.e., those below the threshold voltage) that can occur after a peak has started, erroneous peak detection of two or three consecutive noise spikes is avoided. For example, a noise signal could occur, giving a significant voltage for two

digital windows, followed by no signal for three or four windows, and then followed by another noise spike of two or three windows' duration. By setting a maximum of two open samples between the signal windows, this type of signal is usually eliminated, but a true peak such as (d) would have a good probability of being retained. If this threshold mode is used, it is not necessary to specify the total number of digital samples as well. Each method serves very close to the same function.

4. Specifying the total number of digital samples in a given period that define a peak is primarily a necessary instruction of recognition, but it also serves a threshold function. Unfortunately, the reasonable number anticipated for low-intensity peaks may be only four or five, so that this instruction will not always reject false peaks due to consecutive noise spikes.

The example of Figure 7.3 illustrates the important interplay between scan rate, digitization rate, and threshold methods in establishing a minimum current detection. It is also clear why consecutive jump scan (page 250) or multiple ion detection methods (page 252) are so advantageous when dealing with extremely small samples. Careful consideration should be given to optimize these operations when recognition of small signals is important. A careless threshold setting or indiscriminate use of fast digitization may lead to a loss of a factor of 2–3 in sensitivity.

4. The Computer

The operational processes of a computer in analytical instrumentation are twofold. First, the computer acquires data from the basic instrument, and second, it performs a series of logic operations which prepare the data for further interpretations. The basic components of the computer that perform these duties are logic elements, memory elements, and a control unit that transfers instruction out of the memory. These instructional functions, or software, permit the logic and memory elements to operate on input data in alternate pathways and thus establish the data treatment.

Most automatic GCMS data processing is performed by a dedicated mini-computer with limited ability for performance outside the field of dedication. A variety of computer systems are commercially available. Some of these are designed and sold for one specific mass spectrometer. Others can be purchased and adapted to different types of mass spectrometer. The size of the computer and the peripheral hardware options greatly affect the ability of the total data system, and the user should be sure that the speed of data processing and the output options and formats meet his expectations. It is also important that the vendor accepts complete system responsibility since it may be difficult to determine whether unsatisfactory performance is the fault of the computer system or the mass spectrometer.

The core size of the computer used for GCMS data reduction may be as small as 4,096 12-bit words or in some cases as large as 16,384 16-bit words. Smaller systems offer limited initial economy, but have a slower processing rate and inefficient storage. A 16-bit word computer is often preferred for this reason. Flexibility for other applications is also facilitated, and with reasonable selection of input/output devices, an 8K 16-bit word computer can serve other functions without disturbing the dedication to GCMS.

Speed of data processing is important in GCMS and may significantly influence the mode of operation. If raw data from a complete scan is converted to mass and intensity in one second or so, the data can be output to storage in that form. In some cases, the conversion takes many seconds and necessitates storage of raw data. Subsequent processing of a few hundred spectra is time consuming, especially if the data access is slow. A 16-bit word computer greatly facilitates a fast conversion, but efficiency in the software program and proper selection of auxiliary hardware are also essential.

It is important that the storage capacity of the GCMS-computer system can retain the total number of scans anticipated during a run. If automatic cyclic scan is used, this may mean around 700 scans per hour, and for general utility, the storage should be at least that high. The number of scans that can be stored depends not only upon the storage capacity, but also on the scan mode and whether or not the computer has time to refine the data prior to placing it in storage. Consecutive jump scan, for example, is considerably more economical from the computer point of view in that the mass-to-voltage calibration is performed prior to a run. Only one intensity point per mass is read during a scan, hence the total data in a scan is only 300–500 points. For a magnetic scan digitized at around 10 KHz, more than 20,000–30,000 intensity samples are taken which must be reduced to a few hundred time-intensity or mass-intensity points for convenient storage.

If the computer cannot perform partial data reduction in the second or so that is available between scans, raw data must be stored and the total number of scans that can be taken in a run is drastically reduced. Automatic data acquisition is no longer practical, but the operator can utilize the computer by manually actuating the scan (or recording function if cyclic scan is used) at a selected point on the chromatographic peak. The number of scans stored during a run is thus only around 30–200 depending on the complexity of the chromatogram. If the storage of the computer package is not sufficient to meet these demands, use of the complete system may be severely restricted.

Most GCMS-computer systems store the mass spectral scans on digital tape. For reasons of economy, disc or core storage is seldom practical. Even a suitable tape unit may cost from $8,000–$16,000 which, for a small laboratory, is a considerable additional cost for peripheral hardware. However, good storage is a necessity for practical computerization of GCMS. Operationally, the disadvantage of tape storage is limited access. This is not important when the computer systematically reduces the mass spectral data from one run. On the other hand, if two or three spectra, each from several different runs, are to be selected at a later date, the computer spends considerable time searching the tape for these spectra, during which time it is not available for other work.

5. Input/Output Devices

All GCMS-computer systems are equipped with a Teletype or similar keyboard for addressing the computer. The keyboard is necessary for activation of the GCMS scan program, for setting up the calibration parameters prior to acquisition of data, and to instruct the computer to take the GCMS data when ready. In addition, the keyboard is used to request the various output options at the end of a run. Sometimes the Teletype is also used to plot the mass spectra if a suitable digital plotter is not available. This form of output is not preferred, however, in that the printing time is excessive, the process is noisy (if other work must be performed nearby), and the output, though readable, is not attractive. When so much money is invested into a GCMS-computer setup, an additional amount for a suitable plotter should be considered essential.

Hard copy GCMS output is usually obtained from some form of digital plotter. The popular Cal-Comp plotter, or equivalent instruments, have been most commonly used, but these suffer a considerable disadvantage in the length of time needed to plot a single chart. Depending on the size and other specifications, the time to plot a single spectrum may be 2–4 minutes and hence, in the range of 3–5 hours for the spectra from a complete run. If several runs are performed in a day, the computer must be fully instructed prior to an all-night printing session. Subsequently, examination of the data may indicate the need for additional spectra or replotting with multiplication factors, and a further annoying delay is encountered.

Several fast electrostatic digital printers are available that provide tremendous time saving when a large number of plots must be made. The time used to obtain a complete spectrum is reduced to a few seconds, and as a rule, the complete run can be plotted while the chromatograph or mass spectrometer is reset for the next run. Most fast printers give a high-quality output suitable for insertion into reports or publications. The cost is not significantly greater than the slower ink printers, and the advantage of having the data printed immediately after the run is extremely valuable. A slow plotter is a tremendous disadvantage in a GCMS-computer system and should be avoided.

Microfilm photography is also used for obtaining records of the GCMS data by photographing the spectrum as it is displayed on an oscilloscope.[297] The method is fast and the data can be easily stored. Inspection or reading of the spectra is reasonably convenient with a microfilm reader, but if copies are desired for reports or general circulation, an additional photographic printing process must be performed.

Use of an oscilloscope display for monitoring the GCMS run has been extolled (page 264). An oscilloscope output is also a very valuable part of the peripheral computer hardware and can be used to select or reject spectra prior to obtaining a printed copy. The operator can request display of the selected spectrum and, by visual inspection, can determine whether the spectrum should be printed as is, or whether background corrections or multiplicative factors are needed. If the presence of two or more compounds is indicated, the operator may select a different scan that will give a cleaner spectral record. Thus, the final plots are obtained in the best form and repeat printing with different instructions is avoided. Oscilloscope examination greatly reduces the number of times that a tape must be reloaded and searched for a few spectra.

C. COMPUTER ACQUISITION AND REDUCTION OF GCMS DATA

1. Calibration

The computer acquires mass spectral data as a series of peaks that occur periodically at specific times or specific voltages. To relate these time or voltage parameters to mass, the computer must have calibration data from a known compound. For high-resolution mass spectra, internal calibration is attained with a standard compound entering the mass spectrometer from a batch inlet at the same time as the unknown. The calibration compound will usually be a perfluorokerosene, or other fluorinated material. Since fluorine has a large mass defect, the fluorinated ion peaks are well separated at 10,000 resolution from those of most unknowns. For low-resolution mass spectrometry, and particularly GCMS, internal calibration is not practical since the mass peaks of the fluorocarbon are not separated from those of an unknown. Hence, a calibration run is necessary prior to taking GCMS data.

The calibration process generally involves two stages. First, a background scan is taken and the computer prints out time-intensity or voltage-intensity relations for observed peaks. The operator must recognize a familiar peak, such as mass 28 due to N_2^+, and relate the time or voltage back to the computer along with acceptable tolerance factors. When the computer clock is thus set, the calibration compound is introduced and another scan is taken. The spectrum of the calibration compound must be known to the computer so that it can seek the most abundant peak and relate the observed time or voltage to that stored for mass 28. Fluorine compounds are not necessary for low-resolution calibration, but are often chosen because the spectrum has a convenient base peak at mass 69 due to CF_3^+. The dispersion of the mass spectrometer is thus established by using the time or voltage values for mass 28 and 69 in the appropriate scan function (page 244).

The correctness of fit for the scan function is checked by searching for other known peaks in the spectrum of the calibration compound. If most of these are found within the allowed tolerance limits, consistent with the spectral intensity and threshold parameters, the calibration is accepted. If the important calibration peaks are not found, the operator must reject the run and determine whether tolerance parameters, threshold parameters, or instrumental variations have caused the failure. In some programs, minor instrumental fluctuations are corrected by using a dynamic scan function which is corrected by the computer whenever a peak appears significantly outside of the accepted limits.

The reliability of the calibration is easily checked by requesting another scan on the calibration compound while it is still in the mass spectrometer. A similar check should be performed if the system has been unmonitored for several hours since instrumental drift could cause complete failure of a GCMS run. When the scan mode is electrically controlled, the calibration data is usually quite reliable for 24 hours and often for several days. Magnetically scanned instruments are more susceptible to drift, and a new calibration may be necessary each day. A periodic test is essential for some of the older sector mass spectrometers, but improved magnet quality has increased the reliability of the newer magnetic instruments close to that expected from the electrically controlled systems.

2. On-Line Acquisition of Data

A typical GCMS data acquisition system was shown in block form in Figure 7.1. Digital information comes to the computer as a series of intensity and time or voltage points. The computer must determine that a peak exists, establish the absolute intensity for the peak, establish the precise time or voltage, and relate that value to a mass as given by the calibration. Several different algorithms are used, each of which appears to give satisfactory results. One possible process is outlined in Figure 7.4.

Recognition of the occurrence of an analog peak was discussed under digitization, and various threshold methods were suggested. The statistical

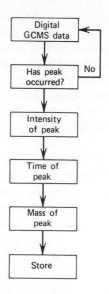

FIGURE 7.4. Schematic representation of computer treatment of digitized GCMS data.

nature of ion collection was emphasized, and as was shown, low-intensity mass peaks (current less than 3×10^{-15} amp) were often established with 3–4 ions collected during the digital sample. In effect, these ions constitute a digital signal arriving at the electron multiplier. The multiplier enhances the signal which is converted to an analog pattern by the electrometer amplifier.

Since an electron multiplier pulse width is only about 10^{-9} sec, that signal can be used directly as digital data by interfacing the computer with a pulse counter. Each output event is counted as one ion. The pulse counter is preset to count for a specified period (typically around 10^{-4} sec), and if the number of counts is above a requested threshold, the information is stored. A peak is defined when the total number of consecutive "stores" meets the operator's specification (generally 3–5).

The pulse-counting method has the advantage that a signal threshold can be set corresponding to the output pulse due to one ion. Low-level noise currents are thus eliminated and a cleaner output signal is obtained. It has the disadvantage that at high ion current levels around 3×10^{-12} to 10^{-11} amp, the counter can no longer discriminate individual pulses.

Peak intensity determination can be made either from the maximum intensity occurring in the several digital samples of the peak,[241] or from the total area of all samples in the peak.[288] If the maximum intensity of any single sample is used, intensity measurements may be in error of 5–10% for ion currents around 10^{-14} amp, and as high as 40% for ion currents of 4×10^{-15} amp or less. Incorporation of a mathematical smoothing operation substantially improves the accuracy of the intensity determination, but use of total area is usually preferred. The total area value is determined by summation of the signals measured in each digital sample. Use of height times half-width for an analog peak is an equivalent process. However, the method of total area can be applied only if the natural scan function of the mass spectrometer is used (page 244). For example, for a linear up-scan with a magnetic instrument, the sampling time is much greater for the high mass peaks, and consequently, the relative intensities would be distorted (Figure 6.11). When used with the natural scan mode, total area determination gives a more accurate intensity measurement with less computer involvement.

Several methods have been proposed to determine the center position of the peak within the several digital samples. These are: (1) center of gravity of digital peak,[288] (2) point of maximum intensity determined from digital point smoothing,[74] and (3) center of the peak area or median.[74] For high-resolution measurement of exact mass, accurate determination of peak position is essential, and the various methods have been appraised by several investigators. The center-of-gravity methods are usually preferred.[288, 289, 290] For low-resolution GCMS, the algorithm used to establish the peak position is not critical, and mass assignment can vary as much as ± 0.3 amu without affecting the ultimate interpretation. Consequently, for computer reduction of low-resolution data, the only critical factor is stability of the scan function since an internal standard is not practical.

Scan function stability is far more reliable when electrical control of the scan is practical. Many existing magnetic spectrometers suffer minor fluctuations in field reproducibility due to hysteresis effects in the magnet, and precise reproduction of the constants in the scan function is not always possible. However, the magnet can usually be conditioned by keeping it in a state of constant cycle, and reproducible GCMS cans are obtained that can be related to the calibration. The restriction of cyclic operation does not impose any inconvenience unless the scan cycle is too slow to catch the mass spectrum at an optimum position on the chromatographic peak. Since computerized GCMS should be directed toward automatic data acquisition, cyclic scan is a necessary condition and should not be regarded as an inconvenience that must be suffered by magnetic instruments.

Many of the newer magnetic mass spectrometers are fabricated with superior magnetic materials and improved control circuits. Scan reproducibility has been demonstrated without a cyclic mode, and stability of scan function parameters is achieved for at least 24 hours.

3. Direct Storage of GCMS Data on Magnetic Tape

Occasionally, a GCMS laboratory will have access to a large time-shared computer system and will wish to record GCMS data in a form that can be processed by the master facility. Acquisition of data directly on FM magnetic tape for computer processing has been adequately demonstrated, and there are no basic technical difficulties that have not been overcome. However, in most situations, turnaround time is usually 1–3 days. Communication with the data processing crew is minimal, and often a modified data plot is desired a posteriori that necessitates another turnaround delay. For the most part, the intermediate stop process (i.e., direct recording on tape) is considered undesirable and will be avoided if at all possible.

The main incentive to direct tape recording is economy, but since a high-quality recorder must be used, it is questionable whether there is any saving in the long run. If the peripheral hardware of the time-sharing facility is properly used, the cost of a commercial fully programmed mini-computer could be less than the cost of a good FM tape recorder plus additional in-house software development. Interface devices would be approximately the same in both cases. Exact cost figures cannot be given for a specific laboratory until all available hardware is specified, but it is imperative to examine the merits of both systems and to be certain that the disadvantages of off-line processing will truly be offset by a capital saving. When considered with the total capital and operating expenditures for a GCMS laboratory, the saving is usually an insignificant percentage.

4. Output Options

Several different output functions can be performed by the computer to provide data in a form that facilitates interpretation. Some of these output options are:

(1) Plot total ion chromatogram

(2) Plot normalized mass spectrum in bar graph form

(3) Print partial or complete mass spectrum in digital form

(4) Display normalized mass spectrum on oscilloscope

(5) Plot mass chromatogram

(6) Correct mass spectrum for background

(7) Present difference mass spectrum by subtracting any specified scan

(8) Enhance selected sections of spectrum by a chosen multiplication factor.

Most of these programs are incorporated into even a small GCMS computer system since at the time the output is performed, the computer has no other duties. The operator can select the options according to the type of information desired from the data and consistent with the available peripheral hardware.

Total ion chromatogram. If data storage of the computer system is sufficient to contain all the scans taken during the chromatographic process, the computer can regenerate a total ion chromatogram after the run. Summation of the ion current for each scan is made and each sum is plotted as a point on the chromatogram. Prominent background ions such as $18(H_2O^+)$, $28(N_2^+)$, $32(O_2^+)$, $40(A^+)$, and $44(CO_2^+)$ are usually omitted from this summation, and He^+ (mass 4) is seldom in the scan range. The computer can be requested to plot the chromatogram in a size appropriate to the length of the chromatographic run. Sometimes a multiplying factor may be introduced to enhance smaller peaks in certain regions. An example of a typical regenerated total ion chromatogram is shown in Figure 7.5.

The total ion chromatogram provides an important monitor record of the chromatographic process for comparison with the chromatograms obtained previously or simultaneously on the flame ionization detector. The specific sensitivities of the two modes of recording are not the same for all compounds, but are close enough (10–15% difference) for direct comparison. The two chromatograms should be effectively the same. However, sometimes the regenerated total ion chromatogram will fail to show separation of peaks observed in the flame ionization record. The discrepancy arises because the computer-regenerated chromatogram is a digital record from a scan cycle of several seconds; hence, there are several seconds of blindness between each point. If a minimum between two fused peaks occurs in the blind region, the digital record fails to show the attained separation. A conventional total ion chromatogram (page 256) is necessary to be sure that separation has not been lost, but for most purposes, a minor inconsistency is not important. A comparison of a flame ionization chromatogram and of a regenerated total ion plot is given in Figure 7.6.

For automatic data acquisition in cyclic scan mode, the regenerated chromatogram is essential to identify the scan number with a position in the chromatogram. The operator must inspect the chromatogram and select scan numbers

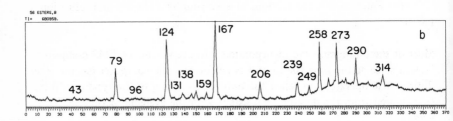

FIGURE 7.5. Regenerated total ion chromatogram of branched cyclic ester fraction from Black Sea core sample.[288]

corresponding to specific peaks. Usually, two or three scans per peak will be chosen. These scan numbers are then relayed to the computer in the request for mass spectra.

Mass spectral plots. When the decision has been made regarding the desired spectra, a plot request is directed to the computer. The operator specifies the size of the plot and any peaks that should not be used for relative abundance normalization. Several convenience options are available, such as multiplying part of the spectra for stronger display, but as a rule, without having any guiding information, the first request will be for an untouched relative abundance plot.

Two examples are shown in Figure 7.7. The first (Figure 7.7a) was plotted on a Cal-Comp plotter. A permanent high quality printout is obtained which is suitable for direct reproduction without further touch-up or modification. Sometimes hard-copy spectra are plotted on the Teletype, but this output mode gives poor quality spectra, and the disadvantage is not offset by a significant savings in capital expenditure or time. The current trend is toward use of fast plotters that give a complete spectrum within a few seconds (Figure 7.7b). Although the photographic or electrostatic processes may result in some loss of quality, the time saving is extremely valuable when a large number of spectra must be plotted from a long GCMS run. Overnight plotting sessions are virtually eliminated, and the spectra can often be obtained right after the run while the chromatograph is being reconditioned for the next sample.

Frequently, examination of the spectra suggests a modified mode of

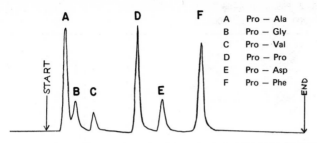

GLC CHART RECORD OF TFA—DIPETIDE METHYL ESTERS

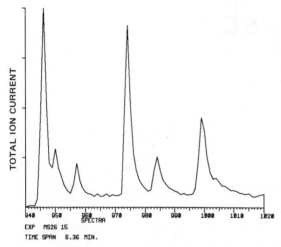

A COMPUTER PLOT OF THE SUMMED PEAKS (IONS) OF EACH SPECTRUM

FIGURE 7.6. Comparison of computer-generated chromatogram and flame ionization chromatogram for dipeptide methyl ester mixture.[280]

plotting. Higher molecular weight ions must often be enhanced by a multiplying factor. Perhaps the background should be subtracted from weak spectra. In some cases, the spectrum of a trace component can be obtained by subtracting two close or adjacent spectra. Such options are available in most programs, but the necessity of reloading the tape and recalling several spectra can be quite time consuming. If the plotting mode is slow, the process may interfere with on-line mass spectral work or have to be postponed. The need for a fast plotting system is greatly emphasized when these important replots are desired.

87   ØR–2 P55–7 8–4–69
x   1 > 200   186 =   7008

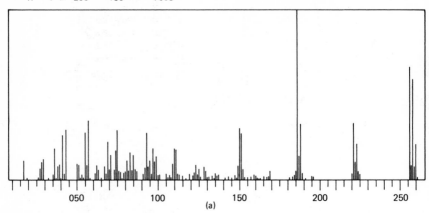

(a)

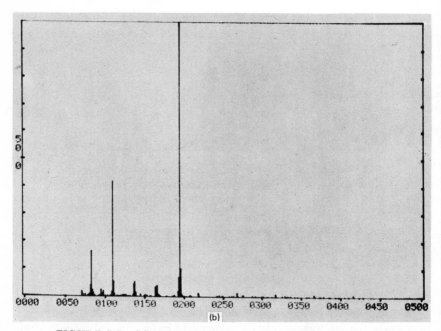

(b)

FIGURE 7.7. (a) Computer-plotted mass spectrum of a trichloro-
biphenyl contaminant from Apollo 12 York mesh monitors.[242, 291]
Plot obtained on Cal-Comp plotter. (b) Computer plotted mass
spectrum of caffeine. Plot obtained using Tektronix Model 4010–1
Visual Display Unit and Model 4610 Hardcopy Unit with AEI Data
System DS–30.

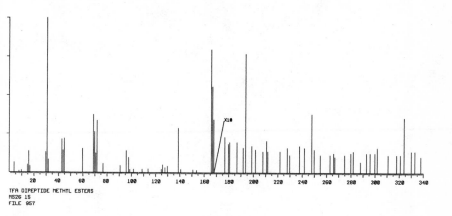

TFA DIPEPTIDE METHYL ESTERS
MS26 15
FILE 957

FIGURE 7.8. A computer-plotted mass spectrum in which the computer has multiplied all peaks above mass 168 by a factor of 10. The computer also edited out small peaks by printing only three peaks every 14 mass units.[280]

Figure 7.8 shows an example of a replotted spectrum in which the ions above mass 168 have been multiplied by 10. This multiplied display permits closer examination for the possible presence of a trace compound or for structural features often revealed in low-intensity, high molecular weight ions. In this plot, the computer was asked to print only the three most abundant peaks every 14 mass units (i.e., in ranges 1–14, 15–28, etc.).

Oscilloscope display. The selection of spectra and decisions regarding plotting options are greatly facilitated if the spectra can be displayed on an oscilloscope for preliminary examination. A large oscilloscope face is preferred (Figure 7.9a), but even a more economical small display can be useful by splitting the spectrum onto several lines (Figure 7.9b). This preliminary inspection enables the operator to choose the best direct print or modified print of the several spectra taken on a chromatographic peak. Extraneous copies are avoided, and it is not necessary to replay the tape for replots.

The oscilloscope display option is very convenient when used in conjunction with either slow or fast digital plotters, but it is particularly valuable when the operator can examine the spectra, modify to attain the best display, and then immediately obtain the result on a fast digital recorder or micro-

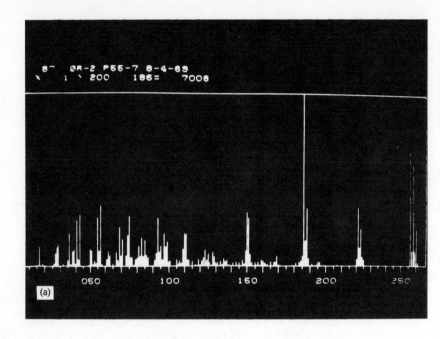

FIGURE 7.9. (a) Large-screen oscilloscope display of computer-normalized spectrum.[242]

film camera.[297] Such sophistication is important if maximum utility is to be achieved from a GCMS-computer system.

Mass chromatograms. An organic mass spectrum contains a wealth of information relating to the structure of the molecule. However, for many experiments all the information is excessive and may even be confusing when presented to co-workers who are not experienced in mass spectrometry. Often the important points can be obtained from the intensity of one specific ion or of a few selected ions. Such singular information is especially valuable to search out particular types of compounds from a complex chromatogram. For example, if the principal interest is in methyl esters, a search for mass 74 ions may be sufficient. The presence of chlorinated pesticides or specific drugs and their metabolites can often be clearly presented with the signal due to a few prominent ions. Such data are contained in a GCMS-computer run, and the most convenient and definitive way to present them is in the form of a mass chromatogram.

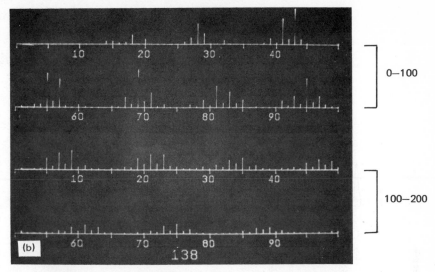

FIGURE 7.9. (b) Small-screen split oscilloscope display showing pattern from mass 0–200.[293]

The computer can plot a single ion current chromatogram in the same way as a total ion chromatogram is given. Each spectrum of the run is searched for the specified ion, and the intensity of that ion is plotted in the same form as the regenerated total ion chromatogram. An example is shown in Figure 7.10 which gives the total ion plot, the mass 74 plot, and the mass 98 plot for the analysis of a methyl ester mixture. Mass 74 is specific for most methyl esters. Mass 98 is prominent in the spectra of dicarboxylic methyl esters. Thus, as is noted in the latter part of the chromatograms, predominance of mass 98 plot over the mass 74 establishes the peaks as being due to dicarboxylic esters.

Another example is given in Figure 7.11. The top chromatogram (a) is the total ion trace from a GCMS run on a mixture of chlorinated biphenyls. To separate the peaks due to 4, 5, or 6 chlorine substitutions, the computer is asked to plot the mass chromatogram for the molecular ion at masses 290, 324, and 358. The mass chromatogram for mass 290 (b) clearly shows which peaks are due to tetrachlorinated species. The tremendous power of the specific data is further shown in Figure 7.11c and d. The total ion chromatogram of a Los Angeles sewage extract (c) shows very little of interest. The mass chromatogram for mass 290 gives a very strong suggestion that tetrachlorobiphenyls are present. Additional mass chromatograms will confirm the suggestion.

Mass chromatogram plots give the same data as are obtained by multiple ion detection methods (page 252). The multiple ion detection techni-

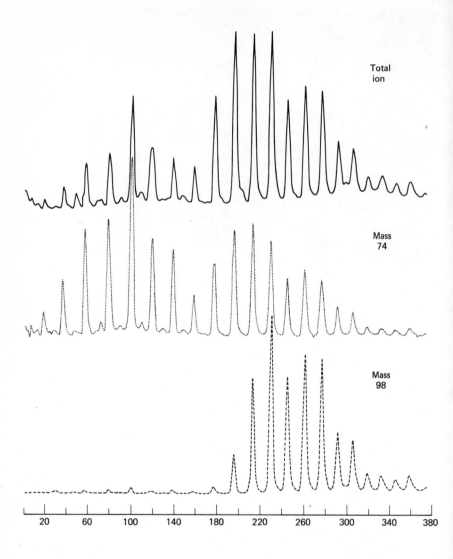

FIGURE 7.10. Total ion chromatogram and mass chromatograms for masses 74 and 98. Data obtained from a methyl ester fraction of a Green River oil shale extract.[241]

ques have the advantage that the slow scan over a peak permits higher sensitivity. On the other hand, the operator must select the desired ions before the run and is restricted to 3–8 ions in most systems. If other masses are desired a posteriori, the run must be repeated.

D. COMPUTER IDENTIFICATION OF GCMS DATA

1. File Search

File search is one of the more time-consuming functions that must be performed on GCMS data. As was suggested in Table 7.1, a manual file search of the spectra from a typical GCMS run might take around 10 hours or 35–40% of the total data processing time. Furthermore, a manual file search becomes less efficient as the number of spectra in the catalog increases. A good manual file search of 100 unknown spectra is possible using a catalog of 3,000–5,000 spectra; it becomes difficult and fatiguing with a catalog of 8,000–10,000 spectra; it is nearly impossible to perform a good manual search on a catalog of 20,000 spectra or more.

Computer searching of mass spectral data would be relatively straightforward if a compound always gave exactly the same spectra on different instruments and with different operating modes. For GCMS data, the possibility of excessive concentration changes during the scan imposes further discrepancies. Consequently, spectra must be coded and searched in a manner that will take these aberrations into account.

Each mass spectrum contains a great deal of information regarding both the mass and intensity of all the fragment ions. To store all the data for 10,000–20,000 compounds would use several hundred thousand bits of computer storage, and in addition, the average computer time needed to search one unknown spectrum would probably be several minutes. Fortunately, a successful file match of an unknown does not need every bit of information in the spectrum, and storage and search problems can be greatly reduced by using an abbreviated mass spectral code.

Some of the methods proposed for selecting mass spectral information for storage and file search purposes are summarized in Table 7.2. The variations range from selecting a few of the most intense peaks and encoding with intensity data, to selecting all peaks above a threshold level, but without differentiating with respect to intensity. The latter process permits binary encoding (i.e., yes or no) and can be programmed to reduce the total search time significantly. In tests, most methods have shown a reasonable degree of success, and it is difficult to see a clear preference of one over the other, insofar as file search is concerned. However, as will be discussed, the method suggested by Biemann and co-workers[297] has the advantage of retaining most data that might contain structural information while rejecting trivial peaks of no apparent value.

It is unrealistic to expect a search procedure to find a perfect match between a standard and an unknown spectrum when variations in the experi-

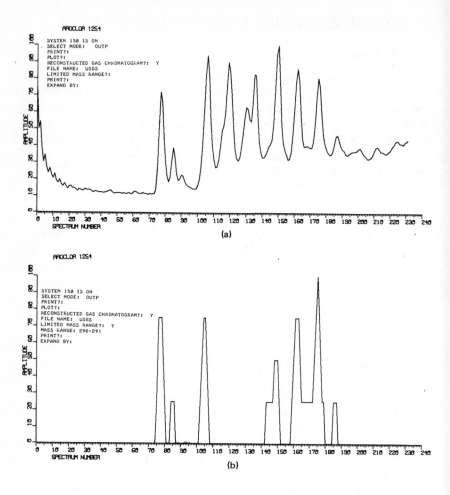

FIGURE 7.11. Total ion chromatograms and mass chromatograms of polychlorinated biphenyl mixtures. (a) Aroclor 1254 (a commercial PCB mixture) total ion chromatogram. (b) Aroclor 1254 mass 290 chromatogram.

mental conditions are known to impart significant variations to the collected data. To cope with these differences, the matching process must measure a "closeness of fit" or mismatch of the spectra. All the successful processes incorporate a method of numerically relating the match of the unknown to the cataloged spectra. The printed list includes not just one hit per spectrum,

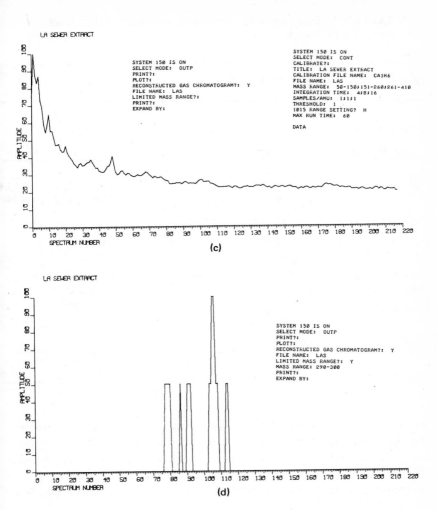

FIGURE 7.11. Total ion chromatograms and mass chromatograms of polychlorinated biphenyl mixtures. (c) Los Angeles sewer extract total ion chromatogram. (d) Los Angeles sewer extract mass 290 chromatogram.[294]

as would be obtained from a perfect match system, but a numerical comparison of several (usually ten) of the best fits. The different modes of encoding spectral data have led to different algorithms for the mismatch values and

TABLE 7.2. Some methods of selecting mass spectral data from computer file search.

Peak Selection	Mode of Intensity	Reference
5 intense peaks	relative intensity	Abrahamsson[281]
6 intense peaks	relative intensity	Crawford and Morrison[295]
"n" intense peaks	mass only, no order of intensity	Knock, Smith, Wright Ridley, and Kelly[296]
"n" intense peaks	in order of intensity	ibid.
"n" intense peaks in mass range "m"	intensity ordered within range only	ibid.
2 intense peaks in mass range 14	mass and intensity	Hertz, Hites, and Biemann[274, 297]
all peaks	binary intensity (yes or no)	Grotch[298, 299]
352 mass positions	binary intensity	Wangen, Woodward, and Isenhour[300]

hence, the relative success of the various methods cannot be readily evaluated. The significant fact that appears foremost is that all processes do extract the correct compound as one of the first three or four compounds listed, provided of course, that the unknown is in the spectral catalog. This condition is of paramount importance since a system that fails to identify the true unknown as one of the best 10 hits would be unacceptable.

The process of listing several of the better spectral matches provides an important consequence when the spectral file does not have the unknown. Frequently, one or more homologues or similar compounds are printed in the list, and these serve as a useful guide to further data interpretation. Since mass spectral catalog files are limited (currently around 10,000–20,000 compounds), the spectra of unknowns have an annoying habit of not being present in the file. Thus, the suggestion that the unknown may be structurally similar to some of the printed hits provides an important key for further deductive interpretation.

The ultimate aim of a file search system on GCMS data is to be able to go through a run from sample injection to computer identification with no operator intervention. At the present time, this is somewhat impractical except for limited specialized cases. In a cyclical GCMS run, several hundred scans may be recorded, but perhaps only 50–60 of these will be on peaks or shoulders. The operator does not want to waste computer time searching spectral files with base line data, nor does he want to waste his own time later in selecting the pertinent search reports from this plethora of output. Thus, in the absence of a very elaborate computer instruction program, the operator prefers to inspect the regenerated total ion chromatogram and instruct the computer regarding the spectra of interest. However, it is very important at this point that the computer can take the GCMS data, encode it in a suitable format, and search the spectral file without further communication. Most search methods under development have been directed towards this objective.

Details of the computer search developed by Biemann and co-workers illustrate the basic approach inherent in most systems. The mass peaks are selected for coding purposes by taking the two most intense peaks every 14 mass units starting with the interval 6–19. Experience has shown that this procedure retains the maximum amount of useful information. As an example, Figure 7.12 compares the complete spectrum of 10-ethyl, 10-n-propyl docosane, the abbreviated spectrum obtained as suggested above, and the spectrum obtained by selecting the six most intense ion peaks. As is noted, it is primarily only trivial mass peaks that are eliminated by selecting two peaks in a 14 mass unit interval. Important data such as the molecular ion and the molecular ion minus CH_3 are thus not discarded. This mode of peak selection is a special case of the method suggested by Knock et al.[296] in which n = 2 and m = 14.

The abbreviated spectra are searched by the computer first with a presearch designed to eliminate file spectra that are grossly different from the unknown, and then with a detailed point-by-point comparison of the spectrum with each remaining file member. A presearch mode is necessary to reduce the total computer time for searching each compound. Some of the simpler presearch methods that were tried are molecular weight selection and comparison of abbreviated spectra with binary (yes-no) intensity data. However, better success was attained with a more complex algorithm that required (1) comparison of base peaks, each at least 25% of the other, (2) the mass ranges of the two spectra must not differ by more than a factor of three, and (3) the total ion abundance of homologues series ions must be similar (i.e., the sums of the ions 1 + 14n, 2 + 14n, ..., 14 + 14n). Even using this relatively elaborate presearch method, the average computer time per spectrum for a search of 7,600 spectra is around one minute.

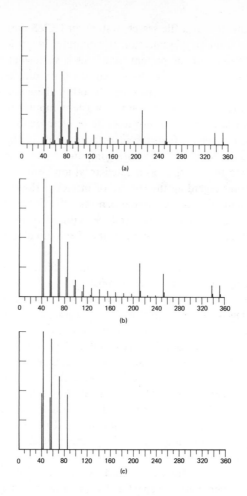

FIGURE 7.12. Comparison of a complete mass spectrum with two methods of abbreviating spectra. (a) Complete spectrum of 10-ethyl 10-n-propyl docosane. (b) Two most intense peaks every 14 mass units. (c) Six most intense peaks.[297]

Use of gas chromatographic retention indices would also provide a very efficient presearch method, and a catalog of 10,000 mass spectra could be rapidly reduced to about 100. Unfortunately, this technique would eliminate members of homologues or compounds which were similar to the unknown, and the valuable feature of finding "like" compounds in the computer search would be lost.

The algorithms for the detailed comparison that follows the presearch must give a useful numerical value for intercomparison. The merit of the search can be based on similarities or dissimilarities. In Biemann's search program (SERCH), a similarity index is calculated and the 10 compounds that give the best agreement are printed out with the numerical value of the index. Briefly, the similarity index is calculated as follows.

(1) The ratio R of the intensity of each peak in the unknown to that peak in the known is calculated. If a peak is absent in either spectra, R is set to zero.

(2) A very large or very small ratio can be due to different compounds (true misfit) or distortions in the spectrum due to GCMS concentration variations. To distinguish these two cases, R is divided by the average R of all peaks over 10%. For a good fit, this average R will be close to unity.

(3) The reciprocal of R is taken if $R > 1$.

(4) Abundant peaks are weighted more than less abundant peaks by the formula: $> 10\%$, weighting factor = 12; $1-10\%$, weighting factor = 4; $< 1\%$, weighting factor = 1.

(5) The average weighted ratio is calculated by summing the weighted ratios and dividing by the weighted number.

(6) The fraction of unmatched peaks (i.e., peaks present in one spectrum only) is determined.

(7) The similarity index is given by the following formula:

$$\text{similarity index } = \frac{\text{average weighted ratio}}{\text{fraction unmatched spectra} + 1}$$

The index number thus calculated can have a maximum value of 1.000 for a perfect fit and will be close to 0.000 for dissimilar spectra.

Figure 7.13 shows the computer-regenerated chromatogram of a test mixture used to evaluate the computer search process. The arrows indicate the spectra chosen for the search, and the results are given in Table 7.3. Although the computer is instructed to print the best 10 finds, only two are used for illustration here.

Most of the compounds that are correctly identified have similarity indices ranging from about 0.4–0.8. In some cases, such as vanillin spectrum number 178, the similarity index is low due to the fact that the compound was not separated by the chromatographic process. A low degree of similarity is expected if the chromatographic peaks are broad or have shoulders. Computer

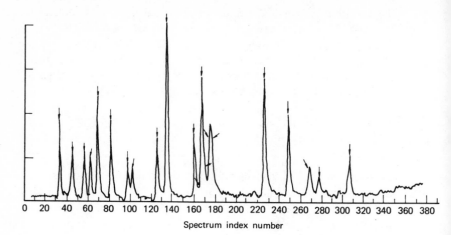

0   20   40   60   80   100 120 140 160 180 200 220 240 260 280 300 320 340 360 380
Spectrum index number

FIGURE 7.13. Computer-regenerated total ion chromatogram for synthetic test mixture.[297]

software options, such as subtracting one of the previous spectra, will give a much better fit.

The spectra numbered 103 and 161 demonstrate the importance of having a program that extracts similar compounds when the unknown is not in the search file. For spectrum 103, the similarity indices for p-hydroxy-acetophenone (SI = 0.428) and m-hydroxyacetophenone (SI = 0.367) are in a satisfactory range even though the unknown compound was o-hydroxyaceto-phenone. This level of agreement is often found in the spectra of geometric isomers, particularly the ortho-meta-para isomers of aromatic compounds, and the operator must always examine each identification from a jaundiced point of view. Confirmation by retention indices is very useful at this stage.

For spectrum 161, the similarity index of 0.24 suggests that the unknown is similar but not identical to the finds. The operator is thus alerted to inspect the spectrum closely and also is given information regarding the general chemical classification that might be anticipated.

When the similarity index is less than 0.10, it is unlikely that the unknown is in the computer file, provided of course, that the spectrum is not that of a mixture. Structural similarities will be rather superficial, and the operator must examine the spectra for possible correlative or deductive suggestions. However, the time-consuming process of searching for absent compounds has been eliminated.

The success of these search programs is exhilarating, but unfortunately, the program algorithms require a large computer with considerable storage capacity in both disc and core. Search programs with the speed and efficiency described above are not amenable to run on a mini-computer and, for this phase of data processing with small economical systems, one must endure many compromises. Often the spectral file can be searched only in small blocks (e.g., 25 spectra transferred to core at one time), and the search process is thus quite slow. Usually, the small computer cannot encode the spectra directly from a tape, and the operator must communicate chosen search peaks to the computer. This is a severe disadvantage in that is uses valuable operator time both in preliminary encoding and in instruction, but the personal involvement also permits a much more flexible search program. For example, the operator can request a successful search based on four or five peaks that have been chosen because the operator recognized structural relevance. Consequently, although spectral search on a small computer system involves a lot of operator time, it may be faster than a manual search and very much more efficient. These compromise methods contribute an important stage to the ultimate goal of complete automation.

Recently, the Division of Computer Research and Technology of the National Institutes of Health, in cooperation with the Aldermaston Mass Spectrometry Data Center, England[271] has implemented a time-shared mass spectral information system.[301] Communication with the computer system can be made via a standard Teletype, an acoustic coupler, or similar telecommunication device. Several optional programs enable the user to search a relatively large file (currently over 9,000 spectra) with a method tailored to each specific unknown spectrum. Thus, although significant operator time is involved, the efficiency of the search is high, and continuous operator-computer dialogue quickly narrows the search to a few plausible finds.

As an example, consider a search routine on the spectrum shown in Figure 7.14. Intuitively, the operator would sense that the higher molecular weight peaks contain more specific information and would set up a search based on these. As is shown in Figure 7.15a, the search was narrowed to two compounds by specifying only the first three highest molecular weight peaks. On the other hand, if the peaks were selected according to relative abundance, the search still contained seven possible compounds even after six peaks were specified (Figure 7.15b). Of course, the same result is expected from the automatic system. However, at the present time, such sophistication demands a large initial capital outlay, while the time-shared system requires only minimal communication hardware.

Other options are available in the NIH system. A search can be made by ion peaks as illustrated, or by molecular weight, molecular formula, or a com-

TABLE 7.3.

Spectrum Index No.	SERCH Find No. 1 Name	Sim. Ind.	SERCH Find No. 2 Name	Sim. Ind.	Compound in mixture
34	Isopropylbenzene Dow 444	0.695	Isopropylbenzene (Cumene) API 311	0.678	Isopropylbenzene
46	Octanone 3	0.509	2-Ethylbutyraldehyde Dow 249	0.290	3-Octanone
57	n-Butylbenzene Dow 623	0.590	nor-Butylbenzene API 494	0.463	n-Butylbenzene
63	Cycloheptanone Dow 364	0.817	Cycloheptanone ASTM 1421	0.713	Cycloheptanone
70	nor-Dodecane API 1598	0.770	n-Dodecane API 23	0.575	n-Dodecane
82	Methyl benzoate Dow 669	0.647	Methyl benzoate MCA 88	0.469	Methyl benzoate
98	3-Acetylpyridine	0.460	2-Ethylpyridine	0.306	3-Acetylpyridine
103	p-Hydroxyacetophenone Dow 685	0.428	m-Hydroxyacetophenone Dow 684	0.367	o-Hydroxyacetophenone
126	Quinoline MCA 105	0.499	Quinoline API 625	0.497	Quinoline
135	1-Methylnaphthalene Dow 738	0.658	2-Methylnaphthalene Dow 753	0.581	1-Methylnaphthalene
161	ar-Methoxybenzaldehyde Dow 675	0.242	p-Methoxybenzaldehyde	0.241	p-Methoxy methyl benzoate[1]
166	o-Fluorochlorobenzene	0.064	1-Chloro-2-fluoro-benzene Dow 562	0.051	3-Methylindole[2]

168	Methyl dodecanoate	0.726	Methyl tridecanoate ASTM 1905	0.441	Methyl dodecanoate
172	Propiophenone Dow 633	0.268	Phthalide Dow 656	0.255	Phthalide
175	Dihydrocoumarin Dow 826	0.514	Anethole MSDC 61	0.096	Dihydrocoumarin
178	Vanillin ASTM 1045	0.258	p-Methoxybenzoic acid ASTM 1188	0.117	Vanillin
227	Ethyl myristate	0.602	Ethyl caprylate ASTM 1068	0.211	Ethyl tetradecanoate (myristate)
249	9-Fluorenone Dow 1297	0.664	Fluorenone ASTM 1435	0.583	9-Fluorenone
268	Carbazole MSDC 62	0.197	Carbazole MCA 106	0.178	Carbazole
278	Caffein MSDC 63	0.091	10,10-Dimethylacridane ASTM 1367	0.027	Caffein
307	4-Bromobenzophenone Dow 1872	0.053	ar-Methyl-ar-chloro-phenyl phenyl ether Dow 1693	0.039	Triphenylcarbinol[1]

[1] The spectrum of this compound is not in the reference collection.

[2] This compound represented a small shoulder on a previous peak and the spectrum searched was obtained by subtracting the spectrum of the previous component (see text).

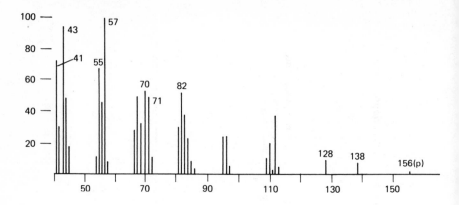

FIGURE 7.14. Mass spectrum of n-decanal.

bination of any two of these. In addition, the complete spectrum of a compound is printed out on request, thus providing an easy source of data when the find is not present in the user's shelf file. (Although spectra are searched using the Biemann code of two peaks every 14 mass units, the complete spectra are stored in the file.)

In the future, the system will be transferred to a commercial time-sharing network. Depending on the price set at that time, the system promises to be extremely useful with a high dollar value. The effectiveness will increase as the number of compounds in the file is increased. Ultimately, the system may be extended to take GCMS data directly and perform an automatic search using a search mode similar to the one described by Biemann.

2. Computer Interpretation of Mass Spectra

It was pointed out in Table 7.1 that correlative or deductive processes consume about 35% of the total data reduction and interpretation time spent on GCMS data. In any given set of spectra there are always a few that are not recognized or identified by the file search process, and for these, the chemist must apply his knowledge of mass spectrometry, chemistry, and sample history to deduce some formula. Here is where the chemist has the most fun, but also where he can make the most glaring errors.

Several methods are under investigation for computer interpretation of mass spectra.[302-310] The objective is to have the computer duplicate the logic processes carried out by a chemist during a correlative or deductive analysis of data. Although easy to state, the objective is not easy to accomplish, primarily because a good analysis of this type falls into the same category as artistic creativity. The outstanding data analyst will perceive alternatives that are not obvious to others, but the actual processes of mentally sorting through previous knowledge are not easily defineable.

Various processes have been developed to apply the computer to the problem of interpreting mass spectral data. One of the simpler approaches deals with the identification of the molecular class or functionality. More elaborate methods use the computer as a learning machine or as a source of artificial intelligence. All the reports show that satisfactory answers are obtained 5–10 times faster than answers obtained by knowledgable graduate students. Unfortunately, most methods have been demonstrated thus far only on large computer systems, often with as much as 50K–100K words of core memory. Hence, the techniques are not yet useful to the GCMS operator who has access only to a small mini-computer with 4K–8K words. Nevertheless, the encouraging preliminary results suggest that computer interpretation will be generally available within a few years.

The work of Smith[309] serves as an interesting illustration of functional identification by computer. It has been observed that for members of a homologous series, a constant sum is obtained for the total percent ion current due to all ions of a specific degree of saturation. Thus, for any chemical homologue, the normalized sum of all ions at the molecular weights corresponding to C_nH_{2n}, C_nH_{2n-1}, etc., are constants. This fact is shown graphically in Figure 7.16, which plots these ion series sums for normal alkanes and for mono-n-alkyl benzenes.

The average value for each ion series as obtained above provides a means of fabricating a reduced mass spectrum that is characteristic of each functional group. The reduced mass spectrum or ion series spectrum thus consists of a total of 14 peaks corresponding to the sums of ions from $2n + 1$ to $2n - 11$. Typical examples of ion series spectra are shown in Figure 7.17.

Ion series spectra have been calculated by Smith for approximately 50 different molecular types. Each reduced spectrum is uniquely characteristic and permits functional identification. A computer program was developed that calculates the reduced spectrum for an unknown and compares that with the ion series spectra in the computer file. A mismatch algorithm permits evaluation of the fit. Although the system must be refined to distinguish geometric isomers, the ability to distinguish the various compound classes has been thoroughly demonstrated.

Although limited in scope, functional identification can be very valuable in many research problems. The ion series are also useful as a subroutine in other interpretations or searches, especially when used as a presearch mode (page 311).

```
PROGRAM: MASS SPEC PEAK AND INTENSITY SEARCH
USER:  INTENSITY RANGE FACTOR FOR THIS SEARCH IS: 4

TYPE PEAK, INT
CR TO EXIT, 1 FOR ID #/NAMES

USER: 156, 1
                # REFS     M/E PEAKS

                  128         156

      USER: 138, 6
                # REFS     M/E PEAKS

                   18        156 138

      USER: 128, 9
                # REFS     M/E PEAKS

                    2        156 138 128

      USER: 1

        ID#            NAME


        2540      GAMMA-NONALACTONE
        3726      N-DECANAL
                          (a)
```

FIGURE 7.15. Search results with the NIH mass spectral search
system. (a) Using three highest molecular weight peaks. (Computer
print-out superfluous to the illustration has been deleted.)

```
PROGRAM: MASS SPEC PEAK AND INTENSITY SEARCH

USER:   INTENSITY RANGE FACTOR FOR THIS SEARCH IS: 4


TYPE PEAK,INT
CR TO EXIT, 1 FOR ID #/NAMES
USER:57,1ØØ
                # REFS      M/E PEAKS

                  126Ø          57

USER:43,95

                # REFS      M/E PEAKS

                  823         57  43

USER:41,77
                # REFS     M/E PEAKS

                  723        57  43  41

USER:55,65
                # REFS     M/E PEAKS

                  54Ø        57  43  41  55

USER:7Ø,55
                # REFS     M/E PEAKS

                  1Ø9        57  43  41  55  7Ø

USER:82,52
                # REFS     M/E PEAKS

                   8         57  43  41  55  7Ø  82

USER:71,5Ø
                # REFS     M/E PEAKS

                   1         57  43  41  55  7Ø  82  71
USER:1
   ID#           NAME


  3726     N-DECANAL
```
(b)

FIGURE 7.15. Search results with the NIH mass spectral search system. (b) Using seven most abundant peaks.[301] (Computer print-out superfluous to the illustration has been deleted.)

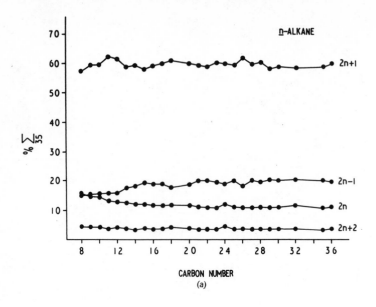

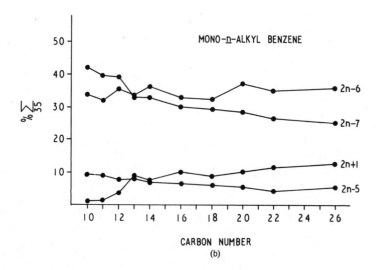

FIGURE 7.16. Ion series plots giving the percent total ionization as a function of carbon number. (a) Series 2n + 2, 2n + 1, 2n, and 2n − 1 for n-alkanes. (b) Series 2n + 1, 2n − 5, 2n − 6, and 2n − 7 for mono-n-alkyl benzenes.[309]

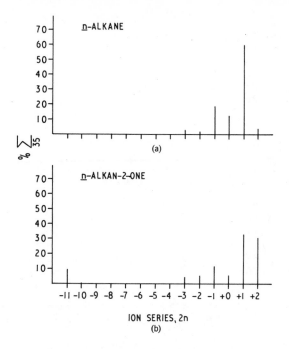

FIGURE 7.17. Mass spectral data of compound class reduced to ion series mass spectra. (a) n-Alkane. (b) n-Alkan-2-one.

One of the most advanced applications of computer technology to the interpretation of mass spectra is the work of the Stanford University Mass Spectrometry Group.[310] In their method, the computer is asked to select the best structure from every possible structure consistent with an unknown mass spectrum. Superficially, this basic, straightforward approach appears simple, but because of the unlimited variations that can occur in molecular structures, extensive development was necessary to obtain a program (DENDRAL) that would perform the calculations and comparisons necessary to establish identity via known spectral correlations. Currently, the DENDRAL algorithm can formulate all possible acyclic and cyclic structures corresponding to a given empirical formula.

The overall DENDRAL method is illustrated by the example given in Figure 7.18. The input data includes the empirical formula as well as the mass spectrum. In the first step (Preliminary Inference Maker), the theory of organic mass spectra is used to suggest the best possible functionality for the unknown.

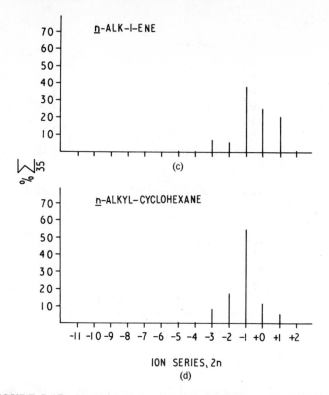

FIGURE 7.17. Mass spectral data of compound class reduced to
ion series mass spectra. (c) n-Alk-1-ene. (d) n-alkyl-cyclohexane.[309]

The computer then builds every possible structure consistent with constraints
imposed by the input mass spectrum. In the example, 1,936 possible acyclic
structures were suggested, but of this number, only 82 were ketones. The
Predicator predicts spectra for each of the 82 Plausible Candidates and compares these with that of the unknown. Inconsistent spectra (e.g., an important peak appearing in one but not in the other) are eliminated and a Scoring
Function is applied to rank all remaining candidates. In the example, all
structures were eliminated except $(CH_3)_2CHCOCH(CH_3)CH_2CH_2CH_3$ (the
correct structure) and $(CH_3)_2CHCOCH(CH_3)CH(CH_3)_2$, which were rated
equally.

The DENDRAL program has been successfully demonstrated on a considerable number of compound types. A typical set of results is shown in
Table 7.4. The effectiveness of each stage of inspection (Figure 7.18) in reducing the field is quite apparent and, as noted in the last column, the correct structure is always given a high rating. For two of the examples listed,

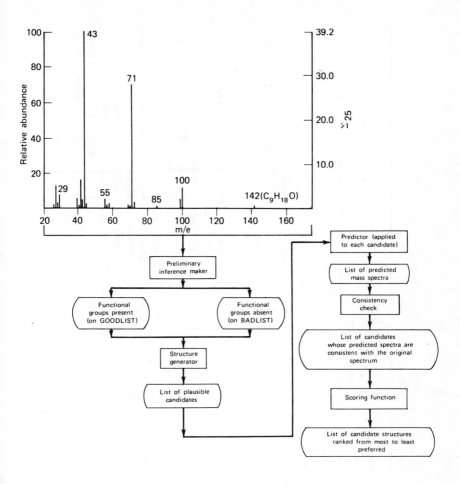

FIGURE 7.18. Flow diagram for structure determination of an unknown spectrum using heuristic DENDRAL Program (Reference 310, Paper No. II).

the consistency check eliminated the correct compound because an expected rearrangement peak was absent. The program is continually modified to eliminate these failures. More recent examples have been even more successful. When tested against mass spectrometry graduate students and postdoctoral fellows, the computer obtained comparable results in one-tenth the time.

TABLE 7.4. Typical results of structure elucidation of aliphatic ketones using the heuristic DENDRAL Program.[1]

Compound	No. of aliphatic Isomers	Ketones	No. of candidates from Structure generator	Consistency check	Ranking of Candidates
2-Butanone	11	1	1	1	1st, 2-butanone
3-Pentanone	14	3	1	1	1st, 3-pentanone
3-Hexanone	91	6	1	1	1st, 3-hexanone
2-Methylhexan-3-one	254	15	1	1	1st, 2-methylhexan-3-one
3-Heptanone	254	15	2	2	Tie for 1st, 3-heptanone and 5-methylhexan-3-one
3-Octanone	698	33	4	4	1st, 3-octanone
4-Octanone	698	33	2	1	1st, 4-octanone
2,4-Dimethylhexan-3-one	698	33	4	3	Tie for 1st, 2,4-dimethylflexan-3-one and 2,2-dimethylhexan-3-one
6-Methylheptan-3-one	698	33	4	4	1st, 3-octanone; tied for 2nd, 6-methyl-heptan-3-one, 5-methylheptan-3-one, and 5,5-dimethylhexan-3-one
3-Nonanone	1936	82	7	7	1st, 3-nonanone
2-Methyloctan-3-one	1936	82	4	3	Consistency check eliminated correct structure because no McLafferty + 1 peak was present in original mass spectrum
4-Nonanone	1936	82	4	0	Consistency check eliminated all candidates since no peak was present at m/e 114 (Mc-Lafferty rearrangement) in original mass spectrum

[1] Reference 310, paper No. II.

E. THE FUTURE OF COMPUTERS IN GCMS

It is apparent from the previous sections that technically, it is now possible for a computer to perform all stages of data acquisition, reduction, and interpretation in GCMS. The success of even the most embryonic programs has established the feasibility of completely computerized systems in which the operator's only responsibility is to inject the sample. Unfortunately, cost must still be reckoned and most laboratories cannot consider the tremendous capital investment needed for such an advanced program. In fact, no complete automated system for all these functions has yet been assembled, and the most advanced GCMS-computer operation terminates the computer automation with a search of a highly specific file of 300–400 drugs and poisons. (See Costello, page 377). This is truly a remarkable advance, but it also serves to emphasize the limitations of computerization to the small GCMS user and suggests that for some considerable time in the future, operator interaction will still play an important role.

It is always difficult to predict future developments, especially in the field of computers. Nevertheless, such predictions are more than just parlor games and must be considered seriously for present and future planning of a GCMS laboratory.

An attempt to summarize the role of the computer in GCMS is given in Figure 7.19. The various manual functions and the approximate time consumed are shown on the right-hand side and the computer processes on the left. It is noted that only a small part of the total manipulation is considered to be "in practice," and the important computer contributions of file search and structural deductions are classified as "research." Matching or searching systems are available to the average user today (page 315), but operator input is necessary for virtually every peak of every spectrum. As just described, artificial intelligence is being used for structural determinations, but the size of the computer required to implement the program is prohibitive to most GCMS laboratories. These functions are therefore considered in the future.

The value of this type of prediction lies in planning for current and future equipment purchases. Thus, a proposed GCMS-computer system should include a data acquisition and reduction process. Accommodation for a manual file search (probably time-shared) should be made with serious consideration for automation within 2–3 years. It does not seem feasible at the present to expect the more elaborate interpretive processes to be generally available for at least five years or more.

Ultimately, the computer will handle all operations in the GCMS laboratory. Highly trained personnel will be released for important creative duties, and menial laboratory functions will be performed by unskilled technicians (Figure 7.20).

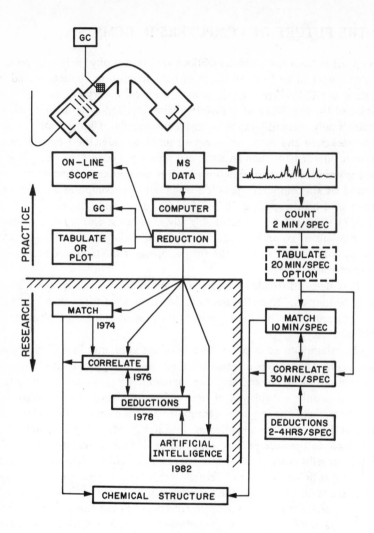

FIGURE 7.19. Current status and future role of the computer in GCMS.

FIGURE 7.20. Ultimate objective of computerization of the GCMS laboratory: to release highly trained personnel for creative duties.

8

Applications
of GCMS

Less than a decade ago, it was difficult to find more than two or three fields of chemical research that could provide interesting applications of GCMS. Today, it is impossible to do justice to all the important applications, and many significant works must go unmentioned. Virtually every branch of chemistry now makes extensive use of the powerful GCMS techniques.

Insofar as is practical, the following illustrations have been chosen because they show a specific point or technique, because they are of general interest, or because of the important impact that the GCMS method will have in that field. The accounts have been written by the original research worker to maintain the freshness and excitement of discovery that is easily lost in secondary reports.

Editorial comments are added after most of these mini-papers. The comments are certainly not criticisms of the work but rather, are intended to guide the tyro to note important features as well as unavoidable shortcomings in each application.

A. FLAVORS AND FRAGRANCES

1. The Flavor of Bell Peppers

R. G. Buttery
Western Regional Research Laboratory
Agricultural Research Service
U. S. Department of Agriculture
Berkeley, California 94710

In the normal dehydration of foods, significant amounts of the volatile flavoring compounds are lost along with the water. This is of small importance with bland products such as potatoes but becomes more important with products known for their characteristic flavor. Bell peppers are such a product. With the aim of improving the flavor of dehydrated bell peppers, we undertook to study which compounds were responsible for their natural flavor.[311]

The concentration of volatile oil in bell peppers is quite low (ca. 1 ppm) and is usually isolated using a steam distillation continuous extraction apparatus under vacuum. The oil obtained in this way was analyzed by gas chromatography using a 300-meter 0.075-cm ID stainless steel capillary column coated with Silicone SF96[100] containing 5% Igepal CO–880. The chromatographic flow was 10 cm^3 atm/min. The column was temperature programmed from 70–170°C at 1/2°C per minute and held at the upper limit. The analysis showed over 100 components (Figure 8.1). For GCMS analysis, a 50/50 splitter was attached to the end of the capillary. This was simply a Swagelok tube "T" with equal lengths (ca. 30 cm) of 0.05-cm ID stainless steel tubing leading to a flame ionization detector and to a Llewellyn-Littlejohn type[192] silicone membrane separator (see page 191). A modified Consolidated 21–620 cycloidal mass spectrometer was used without additional monitoring. Because the cycloidal mass spectrometer does not have a geometry suitable for an electron multiplier, scan times were relatively long (10–20 sec) but were usable with the slow (1/2°C per minute) chromatographic temperature program rate which gave fairly broad peaks. The silicone rubber membrane separator, having its outlet at atmospheric pressure, allowed the odor of the effluent to be evaluated continuously.

Chromatography of the steam volatile bell pepper oil separated a peak which possessed a potent odor characteristic of bell peppers. The mass spectrum obtained is shown in Figure 8.2. From its odor, it was suspected to contain nitrogen, in which case an even molecular weight of 166 meant that the molecule would contain two nitrogen atoms (or some multiple of two). Enough of the compound was isolated by packed-column gas chromatography to obtain ultraviolet absorption and micro-infrared absorption spectra. These spectra indicated an aromatic compound, possibly a pyrazine. Sufficient information was now available to guess that the compound was a butylmethoxypyrazine but a number of other structures were possible and more information was needed. Additional sample (ca. 100 μgm) was collected for a micro-proton magnetic resonance spectrum (PMR) using computer noise averaging techniques. The data showed an isobutyl group and a methoxy group attached to the same side of a pyrazine ring. At the same time, results from a high-resolution mass spectrum carried out on a Consolidated 21–110 B double-focusing instrument

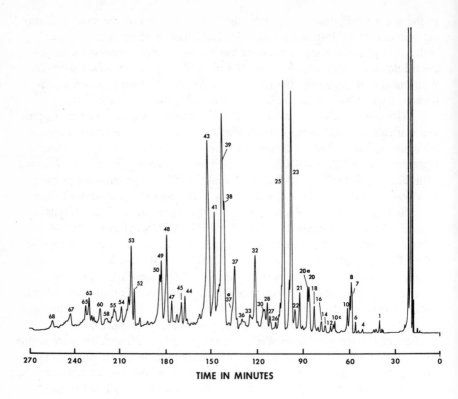

FIGURE 8.1. Chromatogram of bell pepper oil. Active aroma component, peak 39.[311]

showed a molecular weight of 166.1102, consistent with the empirical formula, $C_9H_{14}N_2O$.

The combined spectral evidence pointed to the identity of the compound as 2-isobutyl-3-methoxy-pyrazine, a compound not previously reported in the literature. It was readily synthesized by condensation of the amide of the amino acid leucine with glyoxal, followed by methylation of the resultant 2-isobutyl-3-hydroxypyrazine with diazomethane. The mass, infrared, PMR, and UV spectra of the synthetic compound were all identical to that of the unknown from bell peppers.

The compound has proven to be one of the most potent odorants known and is detectable at a concentration of two parts in 10^{12} parts of water. One drop (0.05 ml) would be sufficient to impart bell pepper flavor

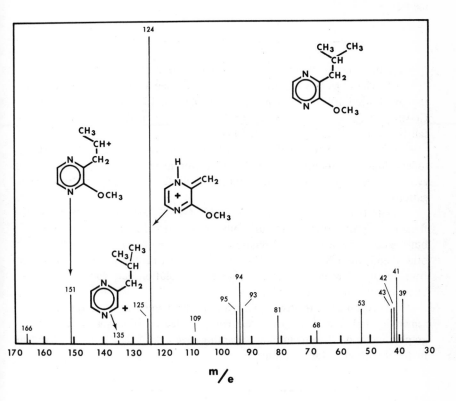

FIGURE 8.2. The mass spectrum of bell pepper aroma component, peak 39.

to a large city swimming pool. The compound has been accepted by the food industry and is now widely used in commercial flavoring, particularly to improve the flavor of products containing dehydrated bell peppers.

It is interesting to note that since publication of this work, a number of related alkylmethoxypyrazines have been found in other foods including peas,[312] coffee,[313] and galbanum. In at least one of these cases,[313] the publication of the mass spectrum of the 2-isobutyl-3-methoxypyrazine provided the key to the elucidation of the mass spectra of the related compounds which were present in such small amounts that GCMS was the only practical spectral method.

Editor's note. Several important points are illustrated by this work.

(1) The objective was to identify the bell pepper aroma. The chromatogram showed 100 components or more that could be mass analyzed by GCMS. To do so would waste valuable time off the main purpose. This pitfall is particularly dangerous to the beginner.

(2) The spectrum of this type of compound is difficult to interpret unless similar compounds are catalogued in the mass spectral literature. Whenever possible, additional sample should be collected for other spectral investigations.

Note that in this case, high-resolution GCMS data might have been sufficient for most structural features but would not define the structure of the butyl side chain or its vicinal relationship to the methoxy group. On the other hand, the UV, IR, and PMR data combine with the low-resolution spectrum to give all essential features. The high-resolution data gave only valuable confirmation in this application.

(3) Although a relatively old mass spectrometer was used, the low sensitivity and slow scan speed were compensated by adapting the chromatographic method. The membrane separator gave a sufficient transfer of organic material without a fast pumping system.

---

2. Computer Data Acquisition and Processing of Low-Resolution GCMS Data in Flavor Research

Paul Vallon
Givaudan Corporation
100 Delawanna Avenue
Clifton, New Jersey 07014

For several years the volume of work for the flavor chemist has steadily increased. Today, more and more samples must be analyzed with greater accuracy and lower limits of detection. The experimental sophistication needed to solve research problems is constantly rising. Combined GCMS analysis is perfectly suited to the types of samples that the flavor chemist encounters. It allows continuous sampling of a gas chromatographic effluent, and spectral information is obtained on each chromatographic peak. Losses due to sample manipulation, contamination, and decomposition are minimized.

Because of the necessity for computer acquisition and processing of high-resolution mass spectral data, few chemists would consider purchasing a high-resolution mass spectrometer without computer facilities. Yet many operate low-resolution GCMS systems without the benefit of automatic data processing. Even though it is possible to operate a low-resolution GCMS system manually, maximum utilization requires continuous recording under variable scan conditions. A large volume of data is generated and it can take weeks to organize and analyze thoroughly a fraction of all the data produced in a single day. The spectra must be sorted, counted, interpreted, and compared with known spectra. Determination of the correct mass of each peak may be difficult under fast scan (3 sec/decade or less) conditions. Unambiguous mass determination is, of course, a necessity since an error of one mass unit will result in an incorrect interpretation. A computer rapidly analyzes all of the GCMS data generated, and allows more efficient use of the GCMS system as well as increased accuracy and manipulation of the data.[241]

A small dedicated computer system suitable for GCMS data acquisition is shown in Figure 8.3. The function of the computer is to acquire GCMS data, to process it, and to present it in a useful and meaningful form to the chemist. This allows the chemist to make maximum use of his time and talent on sample analysis. The system structure allows real-time interaction of the computer, mass spectrometer, and its human operator. Although the operator is the slowest and most careless element in the system, he is also the most flexible. He has the greatest range of decision-making ability, and the system must take full advantage of this ability to economize on hardware and software components.

There are five phases in the operation of an on-line instrumental system: hardware data acquisition, mass calibration, software data acquisition, data manipulation, and data presentation. In hardware data acquisition, the computer samples the continuous electrical analog signal from the electron multiplier at precise time intervals. To calculate the mass for any unknown, a mass scale must be set up. Mass calibration of the spectra is accomplished by taking the position of each peak centroid vs. time and comparing the time to a preset known time vs. mass table. In software data acquisition, the system acquires the spectra (intensity and peak position in time) and stores it on some permanent storage device. All manipulation of data is a non real-time operation. The data is brought into core and processed for data presentation.

Figure 8.4 shows a chromatogram of an expressed lime oil. The components were separated on a 62-meter 0.05-cm ID Carbowax (20 M) capillary column, programmed from 70–180°C at 2.5°C per minute. The chromatogram was recorded on the total ion current monitor of a Perkin Elmer 270 mass

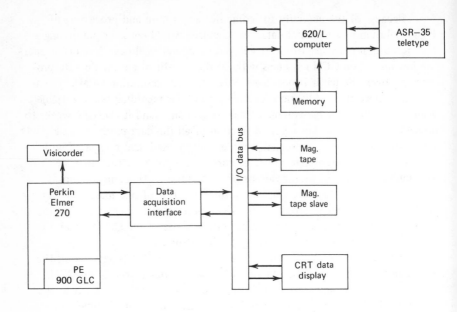

FIGURE 8.3. Block diagram of GCMS computer system.

spectrometer which was interfaced to a Varian mass spectral data system.

Peak A was identified as limonene. Table 8.1 compares the spectrum tabulated from the visicorder trace to the spectrum given in the computer printout. Both have the same sensitivity, and for most purposes, the two printouts are equivalent. The important difference is that the tabulated visicorder spectrum requires operator time whereas the computer output is automatically printed on request. The time for manual manipulation of the mass spectrum varies, but may average 10–20 minutes per spectrum. If we assume 200 spectra (a low figure) per week are processed, then approximately 2500 man-hours per year can be saved by automatic data processing. The resultant savings in time and capacity make the application of a computer to on-line GCMS data acquisition, storage, and processing almost mandatory. Furthermore, some errors are unavoidable when such quantities of data are processed manually and these errors are often difficult to detect and to correct.

Table 8.2 compares part of the spectrum of peak C from the lime oil isolate with the spectrum of caryophyllene from the reference library. The identification was confirmed by gas chromatographic retention data on two columns. Total time for searching through a library of 6000 known spectra

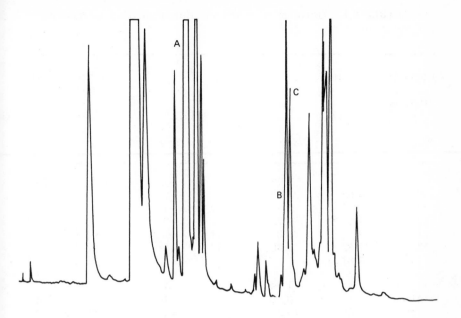

FIGURE 8.4. Chromatogram of expressed lime oil. (Total ion monitor.)

was less than five minutes.

Table 8.3 shows two spectra that were taken on peak B. The spectrum of 4-terpineol was taken on the shoulder at the very beginning of the peak while the spectrum of the C_{15} sesquiterpene was taken at the top of the peak. This illustrates that even small shoulder peaks can be picked out and processed by the computer to give useful information.

Another interesting example was observed in a spectrum obtained from a tomato isolate (Table 8.4). Since it seemed that two components were present, a known sample of 1-borneol was run and the spectrum was subtracted out by the computer, leaving just the unknown spectrum. This spectrum was later identified as due to α-terpineol by comparison with an authentic sample. In both cases, search routines made the appropriate identifications on the computer-subtracted spectra.

By combining the GCMS system with on-line computer analysis and searching of an encoded reference library, it is possible to identify quickly a large number of simple compounds that are expected in a flavor isolate. This frees the mass spectroscopist to concentrate his efforts on the more difficult and interesting structural problems presented by unknown compounds.

TABLE 8.1. Visicorder vs. computer printout of the mass spectrum of limonene

Computer Spectrum					Tabulated Visicorder Spectrum	
*Date Base 4414						
*/Spec # Sum 22741				117/LP/	Limonene	
Peak	I/Base, %	I/Sum, %	Mass	Test	Mass	%
7	15.38	2.98	27	−1	27	15
8	144.63	28.07	28	−2		
9	9.37	1.82	29	−1	29	8
10	24.53	4.75	32	−2		
15	24.76	4.80	41	−2	41	24
22	22.67	4.40	53	−2	53	23
24	7.81	1.51	55	−1	55	6
					57	7
27	6.63	1.28	65	−1	65	6
28	54.19	10.51	67	−1	67	54
29	100.00	19.40	68	−1	68	100
30	6.68	1.29	69	−1		
31	13.32	2.58	77	−2	77	12
33	23.47	4.55	79	−1	79	23
34	9.01	1.75	80	−1	80	8
35	8.67	1.68	81	−1	81	8
37	12.61	2.44	91	−1	91	12
38	14.88	2.88	92	−1	92	15
39	46.55	9.03	93	−0	93	47
40	17.05	3.31	94	−0	94	17
43	3.37	0.65	105	−1	105	2
45	11.71	2.27	107	−0	107	11
46	3.39	0.65	108	−1	108	3
50	11.78	2.28	121	−1	121	11
52	9.92	1.92	136	−1	136	9

TABLE 8.2. Comparison of the computer tabulation of the spectrum of Peak C with that of authentic Caryophyllene

Peak C					Caryophyllene				
*Date Base 885					*Date Base 6552				
*/Spec # Sum 10600			127/LP/		*/Spec # Sum 68498			179/LP/	
Peak	I/Base, %	I/Sum, %	Mass	Test	Peak	I/Base, %	I/Sum, %	Mass	Test
8	25.73	2.07	27	−1	7	23.19	2.21	27	−2
9	327.67	240.47	28	−1	8	193.25	18.48	28	−2
10	45.61	3.67	29	−1	9	22.19	2.12	29	−2
12	327.67	50.52	32	−1	10	23.58	2.25	32	−3
13	37.19	3.00	39	−2	11	32.73	3.13	39	−3
14	50.99	4.11	40	−3	12	8.21	0.78	40	−3
15	100.00	8.06	41	−2	13	100.00	9.56	41	−3
16	6.54	0.52	42	−2					
17	22.33	1.80	43	−2	15	13.06	1.24	43	−3
18	44.09	3.55	44	−3					
21	25.02	2.01	53	−2	20	24.14	2.30	53	−3
22	3.74	0.30	54	−2					
23	37.66	3.03	55	−1	22	38.70	3.70	55	−2
26	14.85	1.19	65	−1	27	11.88	1.13	65	−2
27	4.79	0.38	66	−1					
28	30.87	2.49	67	−1	29	31.01	2.96	67	−2
29	9.12	0.73	68	−1	30	6.73	0.64	68	−2
30	63.97	5.16	69	−1	31	77.38	7.40	69	−1
33	29.12	2.34	77	−1	34	26.44	2.53	77	−3
34	8.42	0.67	78	−2	35	7.32	0.70	78	−3
35	52.74	4.25	79	−1	36	51.06	4.88	79	−3
36	13.21	1.06	80	−1	37	14.71	1.40	80	−3
37	29.23	2.35	81	−1	38	33.21	3.17	81	−3
40	51.57	4.16	91	−1	41	48.13	4.60	91	−1
41	16.14	1.30	92	−1	42	16.26	1.55	92	−2
42	73.09	5.89	93	−1	43	73.19	7.00	93	−1
43	16.02	1.29	94	−0	44	14.07	1.34	94	−1
44	16.49	1.33	95	−1	45	15.38	1.47	95	−1
46	33.80	2.72	105	−1	49	33.92	3.24	105	−2
47	18.12	1.46	106	−1	50	19.42	1.85	106	−1
48	26.31	2.12	107	−1	51	31.04	2.96	107	−2
53	25.73	2.07	119	−1	58	22.95	2.19	119	−1
54	21.98	1.77	120	−1	59	25.10	2.40	120	−1
55	18.36	1.48	121	−1	60	17.18	1.64	121	−1
59	42.57	3.43	133	−1	64	49.26	4.71	133	−1
60	13.09	1.05	134	−1	65	12.14	1.16	134	−2
					69	14.78	1.41	147	−2
63	14.26	1.15	148	−1					
64	16.72	1.34	161	−1	71	16.98	1.62	161	−2
65	4.44	0.35	175	−4					
66	8.42	0.67	189	0	74	8.69	0.83	189	−1
67	4.09	0.33	204	0	75	5.29	0.50	204	0

TABLE 8.3. Two spectra taken on Peak B. 4-Terpineol spectrum taken at beginning of peak, C_{15}-sesquiterpene spectrum taken at top

4-Terpineol					C_{15} Sesquiterpene				
*Date Base 989					*Date Base 4820				
*/Spec # Sum 11731			123/LP/		*/Spec # Sum 44989			124/LP/	
Peak	I/Base, %	I/Sum, %	Mass	Test	Peak	I/Base, %	I/Sum, %	Mass	Test
8	30.94	2.60	27	−1	10	327.67	69.42	28	4
9	327.67	266.23	28	3	11	23.60	2.52	29	−1
10	86.85	7.32	29	−1	14	327.67	38.58	32	−1
13	327.67	145.49	32	−2	19	29.31	3.14	40	−3
17	30.43	2.56	39	−2	20	63.65	6.81	41	−2
18	141.45	11.92	40	−3	21	3.87	0.41	42	−1
19	61.27	5.16	41	−2	22	17.73	1.90	43	−2
20	9.80	0.82	42	−2	23	26.68	2.85	44	−2
21	73.40	6.18	43	−2	30	33.46	3.58	55	−1
22	111.52	9.40	44	−3	31	10.18	1.09	56	−1
27	18.50	1.55	53	−1	38	12.11	1.29	67	−1
28	36.70	3.09	55	−1	39	4.58	0.49	68	−1
36	19.71	1.66	67	−1	40	47.28	5.06	69	−0
37	18.90	1.59	68	−1	46	25.10	2.68	77	−2
38	34.47	2.90	69	−1	47	6.41	0.68	78	−2
39	9.20	0.77	70	−1	48	27.01.	2.89	79	−2
40	100.00	8.43	71	−1	49	7.19	0.77	80	−2
43	25.17	2.12	77	−2	50	12.05	1.29	81	−1
45	20.42	1.72	79	−2	56	36.28	3.88	91	−1
47	16.27	1.37	81	−1	57	15.20	1.62	92	−1
49	13.04	1.09	83	−1	58	100.00	10.71	93	−1
52	23.55	1.98	86	−1	59	13.63	1.46	94	−1
55	36.80	3.10	91	−1	65	23.44	2.51	105	−1
56	14.56	1.22	92	−1	66	5.29	0.56	106	−1
57	84.63	7.13	93	−1	67	32.59	3.49	107	−2
58	10.31	0.86	94	−1	76	87.63	9.38	119	−1
59	11.93	1.00	95	−1	77	13.60	1.45	120	−1
65	10.51	0.88	107	−2	78	10.78	1.15	121	−1
69	34.27	2.88	111	−1	85	5.91	0.63	132	−0
73	48.83	4.11	119	−0	86	6.49	0.69	133	−1
75	16.27	1.37	121	−1	87	6.20	0.66	135	−1
79	12.63	1.06	136	−2	95	5.47	0.58	161	−2
80	10.71	0.90	154	−1	99	2.55	0.27	189	−1
					101	2.24	0.24	202	−1
					102	3.21	0.34	204	1

TABLE 8.4. Mixed spectrum from tomato isolate

				112/LP/
*Date Base 2960				
*/Spec # Sum 33458				
Peak	I/Base, %	I/Sum, %	Mass	Test
9	26.62	2.35	27	−1
10	327.67	86.71	28	−1
11	27.43	2.42	29	−1
13	11.35	1.00	31	−1
14	150.00	13.27	32	−1
17	33.98	3.00	39	−2
18	17.36	1.53	40	−3
19	53.81	4.76	41	−2
20	5.33	0.47	42	−2
21	74.56	6.59	43	−2
22	14.69	1.30	44	−3
27	21.35	1.88	53	−2
29	29.79	2.63	55	−1
33	100.00	8.84	59	−1
40	36.45	3.22	67	−1
41	26.92	2.38	68	−0
42	13.71	1.21	69	−0
44	14.02	1.24	71	−0
46	21.55	1.90	77	−1
48	30.74	2.71	79	−1
49	10.60	0.93	80	−1
50	40.20	3.55	81	−1
56	24.45	2.16	91	−1
57	21.41	1.89	92	−0
58	90.40	7.99	93	−0
59	13.98	1.23	94	−0
60	71.41	6.31	95	−0
66	14.52	1.28	107	−0
73	11.28	0.99	119	0
74	77.43	6.85	121	−0
75	7.22	0.63	122	−1
78	45.74	4.04	136	−1
80	8.44	0.74	139	−1

B. GEOCHEMISTRY

1. Analysis of the Saturated Fraction of Green River Shale Extract

Emilio J. Gallegos

Chevron Research Company

Richmond, California

Organic geochemistry is concerned not only with the origins of petroleum, oil shales, and coal, but also of organic material found in sedimentary rocks and meteorites. The economic importance of petroleum, coal, and shale oil, and the growing awareness and intrinsic need of man to probe his own origin, have combined with the increased sophistication of analytical instrumentation in recent years to create a spectacular upsurge in the investigation of organic substances in sediments, fossils, and meteorites.[315, 316]

The biological origin theory of organic material in sediments is supported by optical activity, C^{12}/C^{13} ratios, and the presence of biological marker compounds. Optical activity is generally associated with molecules of biological origin, and many crude oils and shales display this property. C^{12} enrichment in living matter through biological activity is also observed in organic material from sediments. Isotope effects through fractionation over geological time during the formation of organic material in sediments must be considered in defense of the biogenetic origin theory.

Probably the most important argument for the biological origin of organic sediments is the discovery of organic compounds which have skeletal features closely related to compounds produced by living systems. These have been aptly named "biological marker" compounds.[317] Intact or modified paraffins, isoprenoids,[318, 319] alcohols, amino acids, ketones, carboxylic acids,[320, 321] steroid carboxylic acids,[322] terpanes, steranes,[323–328] and porphyrins[329] have been isolated from the soluble fraction of organic material on many sediments and fossils of various geological ages.

The unique mass spectra of many isoprenoid paraffins, steranes, terpanes, and carotenes found thus far in shales and crude oils make these compounds particularly amenable to mass spectral analysis. The combination of a gas chromatograph and a mass spectrometer provides a very powerful technique for rapid separation and analysis of volatile organic components. This technique is now applied by many investigators in the analysis of organic components in sediments and meteorites. The analysis of the saturate fraction of Green River shale by GCMS serves to illustrate the tremendous amount of information that can result from this type of investigation.

Much work has been done on the isolation and identification of individual components in the branched cyclic hydrocarbon extract from Green River shale. Cummins and Robinson[330] reported the uneven distribution of n-alkanes and the presence of large proportions of C_{16}, C_{18}, C_{19} (pristane) and C_{20} (phytane) isoprenoid alkanes. Burlingame et al.[323] reported the analytical isolation and skeletal identification of C_{27}, C_{28}, and C_{29} steranes and a pentacyclic triterpane. McCormick and Eglinton[331] have identified cholestane ($C_{27}H_{48}$), ergostane ($C_{28}H_{50}$), sitostane ($C_{29}H_{52}$), and perhydro$-\beta-$carotene ($C_{40}H_{78}$). Robinson and Cummins isolated a crystalline solid from a benzene extract of Green River shale which they identified, in collaboration with Hills et al.,[332] as gammacerane ($C_{30}H_{52}$). Recently, Henderson, Wollrab, and Eglinton,[324] using GCMS analysis, analytically isolated $5\beta-$ and $5\alpha-$ cholestane ($C_{27}H_{48}$), a stigmastane isomer ($C_{29}H_{52}$), and gammacerane ($C_{30}H_{52}$). In addition, they identified two ambreanes ($C_{30}H_{56}$), ergostane ($C_{28}H_{50}$), and hopane ($C_{30}H_{52}$), and indicated the presence of a pentacyclic nortriterpane ($C_{29}H_{50}$) and a pentacyclic triterpane ($C_{31}H_{54}$). Anderson et al.[325] isolated a stigmastane isomer ($C_{29}H_{52}$). Very recently, using an extensive isolation technique, Anders and Robinson[327] identified fifty-two cyclic alkanes by mass spectra. Twenty-one have fragmentation patterns similar to tetraalkyl-substituted cyclohexanes and six resembling alkylhexahydroindanes. In addition, eight of the tetracyclic compounds show mass spectra similar to steranes, and six tricyclic and eleven pentacyclic compounds were shown to fragment similar to the pentacyclic terpenoids.

Evidence is presented here from GCMS data for thirty-six compounds in Green River shale, excluding the one- and two-ring cyclic compounds. Twenty-two new components were uncovered as a result of this work.[326]

A Green River shale sample from the vicinity of Grand Valley, Colorado, was crushed to approximately 2–3 mm diameter particle size. The organic material was extracted using 50:50 benzene/methanol, and a saturated fraction was isolated by making several passes through fully active alumina using cyclohexane as the elutant.[326] A 45-meter 0.05-cm ID capillary column coated with 7% OV-17 was coupled directly to the source of an MS-9 mass spectrometer, and used at flow rate of approximately 2 cm^3 atm/min.

Chromatograms of the branched hydrocarbon fraction are shown in Figure 8.5. The total ion monitor trace (bottom) and flame ionization trace (top) show nearly identical resolution. A key to Figure 8.5 is shown in Table 8.5 which gives empirical formula, molecular weight, and relative amounts of components provisionally identified.

Two perhydro$-\beta-$carotenes were identified. Both show nearly identical mass spectra. The isomer present in the largest amount was identified earlier by Eglinton et al.

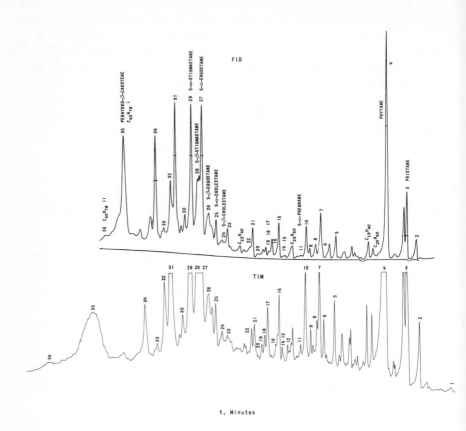

FIGURE 8.5. Chromatogram of the branched hydrocarbon fraction of a Green River shale extract. Top, flame ionization detector chromatogram. Bottom, total ion monitor chromatogram.[326]

The partial mass spectra of 11 tricyclic diterpanes and 5 pentacyclic triterpanes are shown in Figures 8.6 and 8.7. All the tricyclic diterpanes appeared as isomer pairs differing only in retention times. Common features of these mass spectra are a relatively large parent peak, an important parent minus a methyl peak, and a base peak at mass 191. This suggests that the A/B part of the ring is probably identical for all the terpanes observed. The six-membered Ring C in the tricyclic terpanes is chosen because of the prepon-

TABLE 8.5. Key to Figure 8.5 components identified by GCMS in Green River shale.

GC Peak No.	Empirical Formula	Mole Weight	Relative Amount, %
1	$C_{16}H_{34}$	226	0.2
2	$C_{18}H_{38}$	254	0.7
3	$C_{19}H_{40}$	268	2.0
4	$C_{20}H_{42}$	282	9.0
5*	$C_{21}H_{44}$	296	0.8
6*	$C_{20}H_{36}$ (I)	276	0.7
7*	$C_{22}H_{46}$	310	1.9
8*	$C_{20}H_{36}$ (II)	276	0.7
9*	$C_{21}H_{38}$ (I)	290	0.3
10*	$C_{21}H_{38}$ (II)	290	0.1
11*	$C_{21}H_{36}$ 5α	288	< 0.05
12*	$C_{23}H_{42}$ (I)	318	6.0
13*	$C_{23}H_{42}$ (II)	318	6.0
14*	$C_{24}H_{44}$	332	6.0
15*	$C_{25}H_{52}$	352	6.0
16*	$C_{23}H_{40}$ (I)	316	6.0
17*	$C_{23}H_{40}$ (II)	316	6.0
18*	$C_{25}H_{46}$ (I)	346	6.0
19*	$C_{25}H_{46}$ (II)	346	6.0
20*	$C_{26}H_{48}$ (I)	360	6.0
21*	$C_{27}H_{56}$	380	6.0
22*	$C_{26}H_{48}$ (II)	360	6.0
23*	$C_{29}H_{60}$	408	1.2
24	$C_{27}H_{48}$ 5β	372	0.5
25	$C_{27}H_{48}$ 5α	372	1.4
26*	$C_{28}H_{50}$ 5β	386	2.0
27	$C_{28}H_{50}$ 5α	386	5.6
28*	$C_{29}H_{52}$ 5β	400	2.7
29	$C_{29}H_{52}$ 5α	400	7.6
30	$C_{29}H_{50}$	398	2.3
31	$C_{30}H_{52}$ (I)	412	6.6
32	$C_{30}H_{52}$ (II)	412	4.2
33	$C_{31}H_{54}$	426	1.0
34	$C_{30}H_{52}$ (III)	412	2.8
35	$C_{40}H_{78}$ (I)	558	12.9
36*	$C_{40}H_{78}$ (II)	558	0.1

*Those components not previously reported.

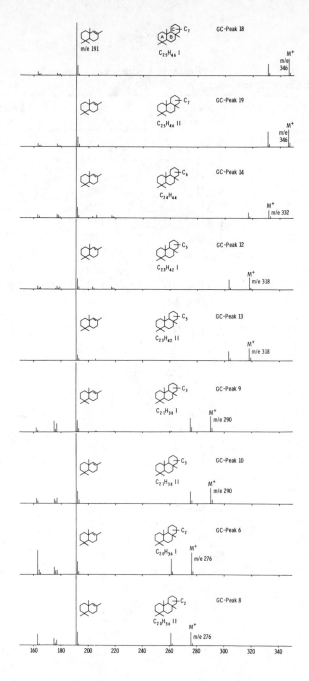

FIGURE 8.6. Mass spectra of C_{20} to C_{25} terpanes from Green River shale extract. Mass spectra obtained with 20 eV electrons.

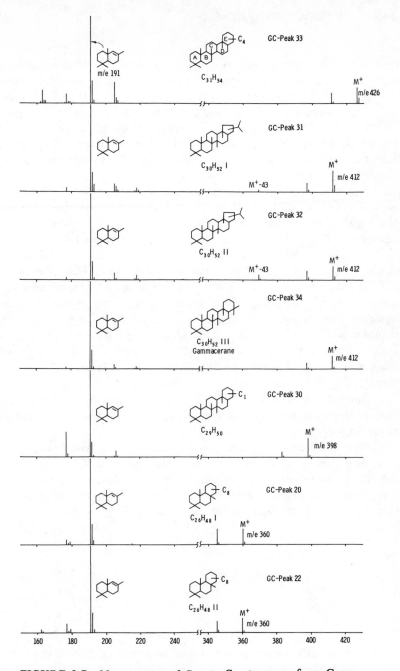

FIGURE 8.7. Mass spectra of C_{26} to C_{31} terpanes from Green River shale extract. Mass spectra obtained with 20 eV electrons.

derance of six-membered rings in terpenoid systems. The same reasoning is used for Ring E for the pentacyclic terpanes except where evidence from mass spectra (e.g., fragments at $M^+ - 43$ in GC peaks 31 and 32) suggests otherwise.

Two components were isolated from Green River shale having a molecular weight of mass 316, $C_{23}H_{40}$. The mass spectra observed can be tentatively explained on the basis of fragmentation expected from the reduced structures of euphane, fusidane, or dammarane. This very tentative speculation has been discussed elsewhere.

Figure 8.8 shows the partial mass spectral patterns of seven steranes isolated and identified by GCMS. The steranes are identified from the presence of a large parent ion peak, an important ion due to parent less methyl, and a base peak at 217 with significant mass 218 and 219 fragments.

All of the α-isomers show a fragment peak at mass 149 which is considerably bigger than the mass 151 fragment peak. The β-isomers show a mass 151 ion peak which is slightly larger than that of the 149 fragment. The spectra of authentic 5α- and 5β- pregnane, bisnorcholane, cholane, cholestane, ergostane, and sitostane show this telltale variation in the mass 149/151 ratio in otherwise identical mass spectra. In addition, in every case the β sterane shows a shorter retention time than that of the α sterane. The relative retention time difference between the β- and the α-isomers is about the same for the C_{29}, C_{28}, and C_{27} steranes.

The ratio of amounts of 5α to 5β steranes is close to 3:1. As a consequence, 5β-pregnane was probably not detected due to the low concentration of the 5α-pregnane.

Nine branched paraffins were also detected. These have been reported.

As a result of this work, a rough estimate can be made of the amount and type of components that are found in the cyclo-hexane extract from Green River shale. The following tabulation includes only those provisionally identified components found in the molecular weight range from 558 to 226.

	%
Isoprenoid paraffins	12.0
Carotenoids	13.0
Terpenoids	20.0
Steroids	20.0

Evidence from GCMS suggests that at least an additional 10% of the components in this fraction are isoprenoids, paraffins, terpenoids, steroids, or carotenoids. The theory of plant and animal contribution to the genesis of this mater-

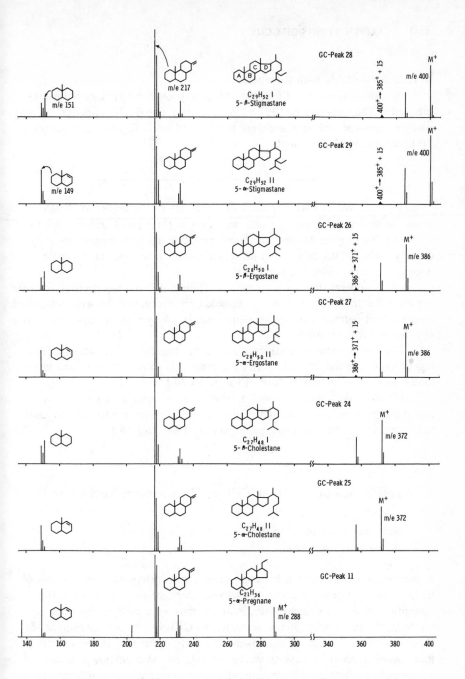

FIGURE 8.8. Mass spectra of C_{27} to C_{29} steranes from Green River shale extract. Mass spectra obtained with 20 eV electrons.

ial rests comfortably with these results.

Acknowledgements. Compounds bisnorcholane and cholane were synthesized at Chevron research by Dr. R. M. Teeter and Dr. W. K. Seifert. Compounds ergostane and sitostane were synthesized by T. C. Hoering, Carnegie Technical Institute, Pittsburgh.

***

Editor's Note. This paper is a fine example of the power of GCMS for complete analysis of complex mixtures. Two or three points merit emphasis.

(1) The original benzene/methanol extract is far too complex to analyze directly. The GCMS data would be very difficult to sort out unless some prior group separations had been used.

(2) It appears that a great deal of definitive information is obtained from the mass spectra. Actually, this is possible only because of the well-established mass spectral correlations for these compounds. As was noted, some authentic samples had to be synthesized.

(3) The chromatogram from the total ion monitor is broadened in the higher molecular weight range. This indicates that some section of the interface is at too low a temperature. (See page 230.)

(4) Note that the mass spectra were obtained using 20 eV electrons. For the type of information that is illustrated in Figures 8.6 to 8.8, the change in the spectral pattern is of no consequence. (See page 18.)

***

2. Applications of the Double-Beam Measurement Technique

J. R. Chapman
AEI Scientific Apparatus Limited
Manchester, England

A portion of a fossilized ichthyosaur vertebrum was submitted for analysis of its lipid and hydrocarbon content. Although straight chain acids from the saponified lipids and straight chain hydrocarbons were expected to make up the bulk of this material, it was subjected to GCMS analysis to maximize the chances of identification of unknowns. To further this aim, it was decided that complete accurate mass measurement data on these unknowns would be advantageous. Because the amounts of sample available were not large, the accurate mass data was obtained using the AEI MS30 double-beam mass spectrometer.[75] (See page 44.)

An integrated Pye 104–MS30 GCMS system with membrane separator[211] was used for the analysis. The MS30 was operated in the double-beam mode. Digital data were collected on-line using the DS30 data acquisition system incorporating a 4K, 12-bit CPU backed by disc store. Subsequent data processing gave accurate mass measurements on all the peaks in each scan although, because of the preliminary nature of the software in use at that time, the mass measurement data was truncated to three decimal places. A visual record was acquired simultaneously on the UV recorder for the examination of metastable data.

The analysis of the hydrocarbon fraction proved to be quite simple. The spectra all compared favorably with standard low-resolution spectra identifying the components as straight chain hydrocarbons. However, analysis of the saponified lipids produced some surprises.

The acids from saponification were methylated with diazomethane and chromatographed under the conditions shown in Figure 8.9. The spectra of the first three peaks in the acid fractions compared favorably with standard mass spectra of methyl myristate (C_{14}), methylpalmitate (C_{16}), and methyl stearate (C_{18}). The next component, labeled 4 in Figure 8.9, showed no molecular ion but could be recognized by comparison with a standard mass spectrum as methyl-9,10-epoxyoctadecanoate (I). It is interesting to note that the most intense metastable peaks in this spectrum correspond to the transitions

m/e 312 → m/e 199 (m* = 127.0) and m/e 312 → m/e 155
(m* = 77.0),

thus linking the major fragment ions to the unobserved molecular ion.

$$CH_3(CH_2)_7 - CH \overset{O}{\diagup \diagdown} CH - (CH_2)_7 CO_2 CH_3 \qquad (I)$$

Component 5 was identified as methyl eicosanoate (C20).

The major peak (component 6) gave a spectrum whose interpretation was much more complex. Part of the spectrum of component 6 is shown in Figure 8.10. The three upper tracings are the sample spectrum and the two lower tracings are the spectrum of perfluorokerosene in the second source. The sample spectrum showed no obvious molecular ion, and the only peak distinguishable

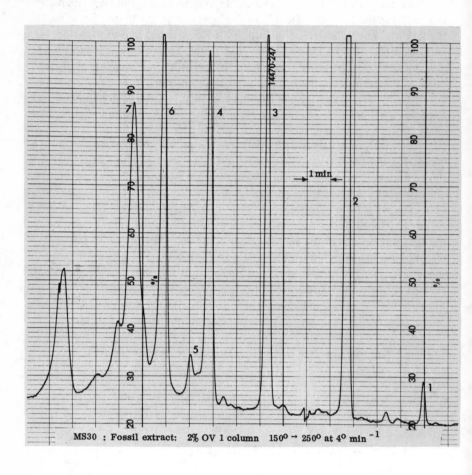

MS30 : Fossil extract: 2% OV 1 column   150° → 250° at 4° min^{-1}

FIGURE 8.9. Chromatogram of esterified fraction of ichthyosaur vertebrum fossil extract. 2% OV 1 column programmed from 150°C → 250°C at 4°C/min.

from the background at higher mass is at mass 313 (not shown in Figure 8.10). Unfortunately, the maximum intensity of the peak at mass 313 fell below the computer interface threshold, so that the datum was not obtained for accurate mass measurement. However, the spectrum did contain a number of intense peaks at lower mass which could be related by metastable transitions

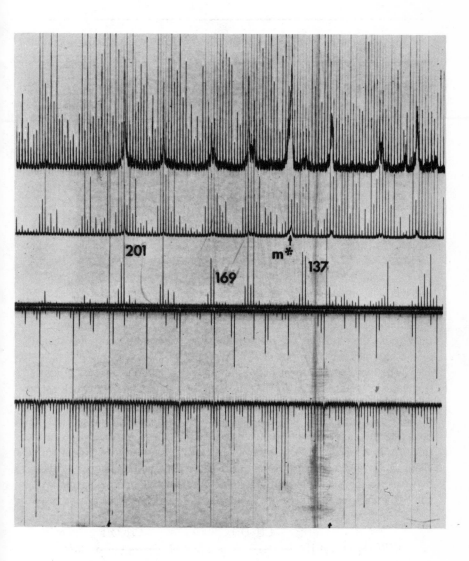

FIGURE 8.10. Double-beam mass spectrum of component 6 from Figure 8.9.

TABLE 8.6. Accurate measurement mass data for selected peaks from the mass spectrum of component 6.

Measured Mass	Calculated Mass	Composition
201.148	201.1491	$C_{11}H_{21}O_3$
187.136	187.1334	$C_{10}H_{19}O_3$
169.124	169.1228	$C_{10}H_{17}O_2$
157.160	157.1592	$C_{10}H_{21}O$
155.108	155.1072	$C_9H_{15}O_2$
137.096	137.0966	$C_9H_{13}O$
125.116	125.1330	C_9H_{17} [1]

[1] The large apparent mass measurement error for this peak indicates that the peak is probably a doublet between C_9H_{17} (125.1330) and $C_8H_{13}O$ (125.0966).

and whose elemental compositions were available from the mass measurement data. These two sets of data are given in Tables 8.6 and 8.7. If the mass measurement data had been obtained at a resolving power considerably higher than 1000, then most of the metastable peaks (intensity usually less than 1% of the base peak) would not have been detected.

The mass measurement data establishes that component 6 gives ion fragments of composition $C_{11}H_{21}O_3$, $C_{10}H_{19}O_3$, and $C_{10}H_{21}O$ which further lose methanol twice, once, and once, respectively. In the mass spectrometry of fatty acid derivatives, it is well known that methyl esters and methyl ethers both lose methanol readily on electron impact.[333] It is also well known that substituents such as a methyl ether or hydroxyl group on the main aliphatic chain cause very facile cleavage α to the substituted carbon and particularly between vicinal substituted carbons. With these considerations in mind, probable structures for the fragments at masses 201, 187, and 157 can be proposed.

$$CH(CH_2)_7CO_2CH_3 \xrightarrow{-CH_3OH} m/e\ 169 \xrightarrow{-CH_3OH} m/e\ 137$$

$$| \atop OCH_3\ m/e\ 201$$

TABLE 8.7. Metastable transitions in the mass spectrum of component 6.

Metastable Mass	Transition
142.2	$201 \rightarrow 169$
128.6	$187 \rightarrow 155$
111.1	$169 \rightarrow 137$
99.6	$157 \rightarrow 125$

$$\underset{\overset{|}{OH} \ m/e \ 187}{CH(CH_2)_7 CO_2 CH_3} \xrightarrow{-CH_3OH} m/e \ 155$$

$$\underset{\overset{|}{OCH_3} \ m/e \ 157}{CH_3(CH_2)_7 CH} \xrightarrow{-CH_3OH} m/e \ 125$$

The origin of these fragments from a mixture of the two compounds represented by formula (II) is supported by the peak at mass 313, which would then represent an $M-31$ ion, a significant fragment in the high-mass region of the spectra of such compounds.

$$CH_3(CH_2)_7 CH(OR_1)CH(OR_2)(CH_2)_7 CO_2 CH_3 \qquad (II)$$
$$(R_1 = H, R_2 = CH_3 \ or \ R_1 = CH_3, R_2 = H)$$

The two compounds represented by (II) would not be separated under these chromatographic conditions. An estimate of the retention time for (II) based on data for the corresponding dihydroxy and dimethoxy esters gives a figure which is close to that observed for component 6.

Interpretation of the spectrum of component 7 posed similar problems. No obvious molecular ion was seen, but there were again a number of intense

TABLE 8.8. Accurate measurement mass data for selected peaks from the mass spectrum of component 7.

Measured Mass	Calculated Mass	Composition
225.150	225.1491	$C_{13}H_{21}O_3$
193.125	193.1228	$C_{12}H_{17}O_2$
187.136	187.1334	$C_{10}H_{19}O_3$
171.138	171.1385	$C_{10}H_{19}O_2$
155.107	155.1072	$C_9H_{15}O_2$
139.114	139.1123	$C_9H_{15}O$
121.104	121.1017	C_9H_{13}

peaks at lower mass related by metastable transitions. The accurate mass data and metastable data from this scan are given in Tables 8.8 and 8.9.

Some of the fragment ions involved in the metastable transitions listed in Table 8.9 are not listed in Table 8.8 (e.g., mass 203) and were virtually nonexistent in the mass spectrum. The major peaks appearing in the spectrum were those at masses 225, 187, and 171. On the basis of the data in the tables, structures for these fragments were proposed as before, viz:

$$m/e\ 257 \xrightarrow{\ -CH_3OH\ } CH{=}CH.CH{=}\underset{\underset{\displaystyle OH\ m/e\ 225}{|}}{C}(CH_2)_7CO_2CH_3 \xrightarrow{\ -CH_3OH\ }$$

$$m/e\ 193 \xrightarrow{\ -H_2O\ } m/e\ 175$$

$$\underset{\underset{\displaystyle OH\ m/e\ 187}{|}}{C}H(CH_2)_7CO_2CH_3 \xrightarrow{\ -CH_3OH\ } m/e\ 155$$

TABLE 8.9. Metastable transitions in the mass spectrum of component 7.

Metastable Mass	Transition
197.1	257 → 225
165.6	225 → 193
158.8	193 → 175
144.2	203 → 171
128.6	187 → 155
113.1	171 → 139
105.4	139 → 121

$$\text{m/e 203} \xrightarrow{-CH_3OH} \begin{array}{c} OH \\ | \\ CH_3(CH_2)_4CH \cdot CH \cdot CH=CH \\ | \\ OCH_3 \ \ \text{m/e 171} \end{array} \xrightarrow{-CH_3OH}$$

$$\text{m/e 139} \xrightarrow{-H_2O} \text{m/e 121}$$

These fragments can be fit together to suggest formula III as the original structure, compound 7.

$$CH_3(CH_2)_4CH(OCH_3)CH(OH)CH_2CH(OCH_3)CH(OH)(CH_2)_7CO_2CH_3 \qquad \text{(III)}$$

Interchanging some of the hydroxyl and methoxyl groups in formula (III) would provide a structure that is equally acceptable on the basis of the available data. Similarly, we cannot be sure that the compound as analyzed was not formed by loss of a molecule of water from (III). Nevertheless, it is reasonable to deduce that the compound was derived from linoleic acid and contains methoxyl and hydroxyl groups substituted on the sites of the original unsaturated carbons. In the same way, although the exact structure for component 6 is not defined, the recorded spectrum represents a compound derived from oleic acid with methoxyl and hydroxyl groups substituted on the site of the original unsaturated carbons. Thus, components 6 and 7 would seem to be artifacts of either the analytical process or natural oxidation of oleic and linoleic acids after the death of the live animal.

In summary, it can be said that the double-beam method gives in a single scan the benefit of both types of conventional mass spectrometer operation, i.e. high sensitivity and metastable data of the low-resolution mode and accurate mass measurements equivalent to those normally obtained in the high-resolution mode. In this example, complementary use of the metastable and accurate mass data facilitated deductions about chemical constitution.

Editor's Note. The value of the double-beam mode of operation is well illustrated in this example, particularly when the accurate mass data is used in combination with metastable ion data. As is emphasized, the method is very useful when high sensitivity is important and, other factors being equal, this mode of operation requires an order of magnitude less sample. The important limitation is noted in Table 8.6, namely, that the measured peak must be a singlet to assure sufficient accuracy for determination of molecular formula.

C. BIOCHEMISTRY

1. Sequence Determination of Proteins and Peptides by GCMS– Computer Analysis of Complex Hydrolysis Mixtures.

H. J. Förster, J. A. Kelley, H. Nau, and K. Biemann
Department of Chemistry
Massachusetts Institute of Technology
Cambridge, Massachusetts 02139

Over the past dozen years, considerable effort has been devoted to the development of new and efficient instrumental techniques for the amino acid sequencing of peptides and proteins.[334] Mass spectrometry gave great promise because of its high sensitivity and structural specificity for linear molecules. These mass spectral approaches centered around the interpretability of mass spectra of peptide derivatives but neglected the real problem in protein sequencing, viz. working with very complex mixtures of degradation peptides on a submicromolar level. Recently, the authors have developed a technique which addresses itself to this problem.

[1]Presented at 3rd American Peptide Symposium, Boston, June 1972
and published in "Chemistry and Biology," J. Meienhofer, ed.,
Ann Arbor Science Publications, 1972, p. 679.

Gas chromatography was chosen as the most efficient and sensitive separation technique in conjunction with mass spectrometry. For various other investigations the authors had already developed a sophisticated gas chromatograph-mass spectrometer-computer (GCMS-computer) system. A crucial aspect of the current approach is the generation of a mixture of small peptides that represents a reasonably complete record of all the peptide bonds present in the protein or primary degradation peptide. The identification technique must be able to handle all peptides regardless of the nature of the amino acids present with one single reaction sequence. Thus, the main emphasis is placed on a complete identification of small peptides rather than on as large a degradation peptide as possible, a process which is often limited by the presence of certain amino acids. At the outset, the authors aimed at unambiguous identification of all possible di- and tripeptides in complex mixtures, but it turned out that the system can handle most tetra- and some pentapeptides as well. This substantially improves the confidence in the reassembly of the original structure.

Reduction of peptide derivatives produces polyamino alcohols which are well suited to both gas chromatographic separation and mass spectrometric sequencing.[335, 336] The earlier difficulties with polyfunctional amino acids have now been overcome by O-trimethyl-silylation of the polyamino alcohols and the efficient manipulation of peptide mixtures on a very small scale. A major effort was devoted to the generation of these peptide mixtures but, because of the diversity of the characteristics of proteins, one could not expect to find one mode of degradation applicable to all. Three different approaches were developed: (1) partial acid hydrolysis, (2) enzymatic cleavage utilizing a single enzyme or a set of enzymes, and (3) di-peptidylaminopeptidase I (cathepsin C) digestion before and after one Edman degradation step. The differences in specificity of these three govern the choice based on the amino acid composition and genesis of the original protein or peptide. The other major aspect of the problem is the development of interpretative techniques that could deal with the vast amount of data generated by analysis of an entire peptide mixture. These computer-assisted techniques are outlined later.

Extensive enzymatic and partial acid hydrolysis studies have been conducted on ribonuclease S-peptide to determine the feasibility of generating mixtures of overlapping oligopeptides which will allow reconstruction of the original peptide sequence. Mixtures have been generated by partical acid hydrolysis with 6 N HCl for 15 min to 16 hr and by employing enzymes such as chymotrypsin, papain, pepsin, pronase, subtilisin, trypsin-pepsin and trypsin-chymotrypsin. In a typical experiment 3.0 mg (1.37 μ mole) of S-peptide was hydrolyzed. The hydrolyzate mixture was esterified, acetylated and reduced with $LiAlD_4$ to yield the corresponding polyamino alcohols. ($LiAlD_4$ is preferred over $LiAlH_4$ because it minimizes the occurrence of different sequence ions of the same mass.) Silylation with pyridine/trimethylsilyldiethylamine (2:1) yields a corresponding mixture of O-silyl ethers that possesses excellent gas chromatographic separability and produces easily interpretable mass spectra.

The peptide derivative mixtures were analyzed by a GCMS-computer system that consisted of a Perkin Elmer 990 gas chromatograph and Hitachi RMU 6L mass spectrometer operated on-line with an IBM 1800 computer. The chromatographic effluent is continuously scanned every 4 seconds through the mass ranges 28–455 or 28–743. At the termination of the GCMS experiment, the computer generates a total ion chromatogram (Figure 8.11a), the mass spectrum for every scan, and selected mass chromatograms (such as Figures 8.11b and 8.11c). (See page 304.) These are stored on magnetic tape for future data manipulation, and are available to the user as microfilm copies filmed from an oscilloscope display.[297] Thus, the data are available for manual interpretation of individual mass spectra or analysis of selected chromatographic peaks by mass chromatograms. In addition, there is available partially automatic interpretation by computer-assisted data manipulation[337] and automatic identification and sequence assembly programs.[338] Alternatively, a technique of controlled sample fractional vaporization from a probe (page 12) directly into the ion source of the mass spectrometer can be used with the same program for identification of large peptides which are not volatile enough for gas chromatography.

The silylated polyamino alcohols possess a repetitive ethylene-diamine backbone with carbon-carbon bonds which cleave upon electron impact to yield ions stabilized by neighboring nitrogen atoms and to produce mass spectra composed almost entirely of sequence-indicating ions (Figure 8.12). Retention of charge on carbon atoms toward the N-terminal end gives rise to ions representing the sequence from the N-terminal side of the molecule (A_1, A_2, A_3, . . .), while retention of charge at the other carbon atoms generates ions corresponding to the sequence from the C-terminal end of the molecule (Z_1, Z_2, Z_3, . . .). The silylated derivative produces a fairly abundant $M-15$ ion (loss of methyl from the molecule) which indirectly indicates the molecular weight (page 236).

Inspection of the mass chromatograms corresponding to the mass values of possible sequence-determining ions enables one to identify rapidly the peptide derivatives. The importance of this technique for the identification of components of very complex mixtures cannot be overemphasized since it allows location of amino acid and peptide derivatives containing the same

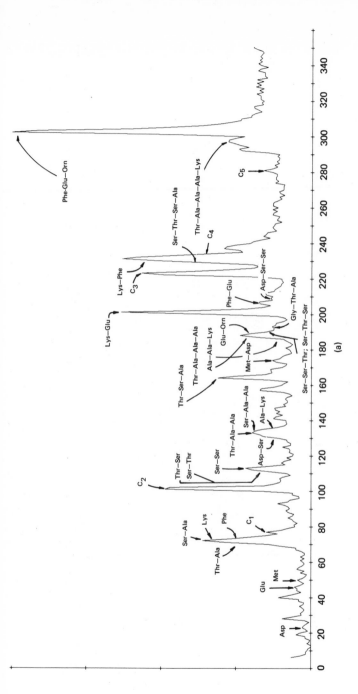

FIGURE 8.11. (a) Total ionization plot of silylated amino alcohols obtained by digestion of ribonuclease s-peptide by trypsin and pepsin followed by hydrazinolysis and subsequent derivatization. C_1, 2-6-di-t-butyl-4-methyl phenol; C_2, unidentified; C_3, C_4, TMS-sucrose; C_5, results from elimination of THSOH from the derivative of Phe-Glu-Orn.

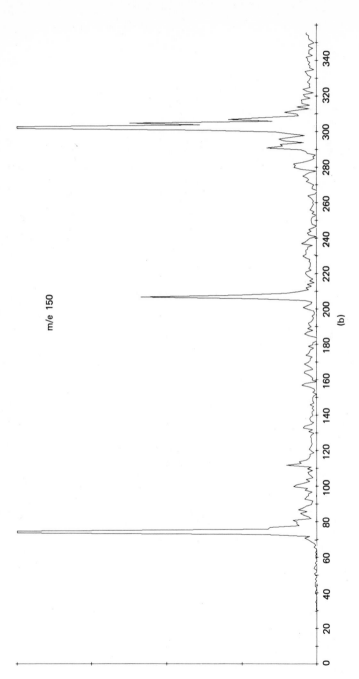

m/e 150

(b)

FIGURE 8.11. (b) Mass chromatogram of mass 150 of the GCMS experiment in Figure 8.11a.

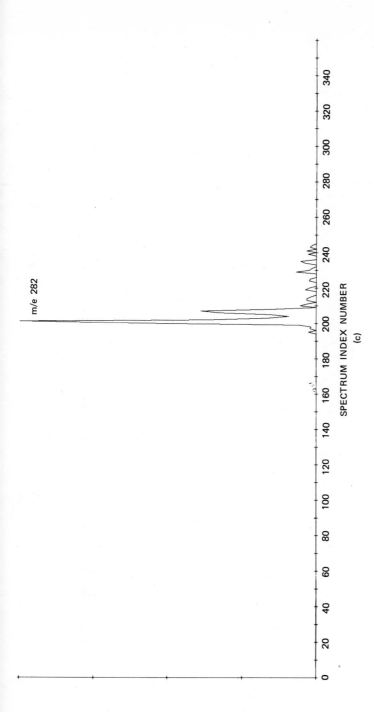

FIGURE 8.11. (c) Mass chromatogram of mass 282 of the GCMS experiment in Figure 8.11a.

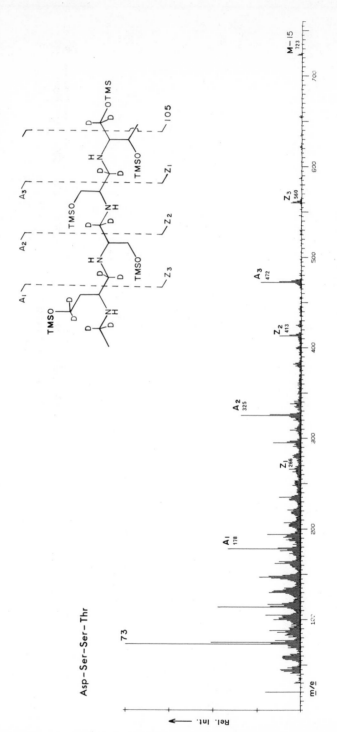

FIGURE 8.12. Mass spectrum of the silylated polyamino alcohol corresponding to Asp-Ser-Ser-Thr.

N-terminal amino acid. For example, a plot of mass 150 (A_1 of Phe), as shown in Figure 8.11b, locates Phe, Phe-Glu and Phe-Glu-Orn. Similarly, peptide derivatives with a certain C-terminal amino acid are located by inspection of the corresponding mass chromatogram (e.g., mass 282 for the Z_1 of Glu, Figure 8.11c). The coincidence of maxima in mass chromatograms of sequence ions makes it possible to locate and identify any peptide derivative in the gas chromatogram. For example, Phe-Glu, a relatively minor component of the mixture resulting from trypsin-pepsin degradation of S-peptide, can be readily located in the total ionization plot by mass chromatograms of mass 150 and 282 (see Figures 8.11 at spectrum index number 208). Mass chromatograms also allow facile location of peptide derivatives in the presence of the unavoidable nonpeptide artifacts ($C_1 - C_4$ in Figure 8.11a) and give artificial chromatographic resolution of incompletely separated components.

As a further parameter, the gas chromatographic characteristics, expressed by retention indices[156-158] are used to great advantage. They can be calculated from values that are determined experimentally for each amino acid side chain. Since the reconstruction of the peptide structure from the sequence-determining ions is a matter of simple arithmetic and the data already resides in the computer, it is possible to have automatic interpretation and eventual assembly of identified overlapping peptides into the complete protein structure. Work has progressed well in these areas and several interpretative programs already exist for routine use.[368]

Table 8.10 shows the peptide derivatives that were identified by GCMS and fractional vaporization experiments of the derivatized peptide hydrolyzate produced by trypsin and pepsin digestion of S-peptide (hydrazinolysis is employed to transform arginine to ornithine).[339] All original peptide bonds are represented in the fragments except those involving one glutamine. Consideration of the results of one such experiment permits a proper choice of the cleavage method which would complete the overlap, if necessary. Indeed, after hydrazine treatment, partial acid hydrolysis of S-peptide produces, among others, Orn-Glu and Glu-His-Met.

A promising third alternative method uses cathepsin C which sequentially cleaves dipeptides from the N-terminal end of a peptide, except those with N-terminal arginine or lysine, and until it encounters proline.[340, 341] This enzyme appears to be ideal for tryptic peptides where two cathepsin digests, one on the intact peptide and one on the Edman-degraded peptide, produce two sets of overlapping dipeptides. Identification of all the dipeptides then allows complete or partial reconstruction of the original sequence. The strategy, which was recently explored by others,[342-344] was tested on glucagon and several tryptic peptides from rabbit skeletal muscle actin. GCMS-computer analysis of the

TABLE 8.10. Amino acid sequence determined by GCMS-computer system on hydrolyzate of S-peptide.

Lys - Glu - Thr - Ala - Ala - Ala - Lys - Phe - Glu - Arg - Glu - His - Met - Asp - Ser - Ser - Thr - Ser - Ala - Ala

Lys - Glu
Glu - Thr - Ala
Thr - Ala
Thr - Ala - Ala
Thr - Ala - Ala - Lys
Thr - Ala - Ala - Lys
Ala - Ala - Lys
Ala - Lys
Lys - Phe
Phe - Glu
Phe - Glu - Orn
Glu - Orn
His - Met - Asp[1]
Met - Asp
Asp - Ser
Asp - Ser - Ser
Ser - Ser
Ser - Ser - Thr
Ser - Thr
Ser - Thr - Ser
Ser - Thr - Ser - Ala
Thr - Ser
Thr - Ser - Ala
Ser - Ala
Ser - Ala - Ala

[1] By fractional vaporization into ion source.

silylated oligopeptide amino alcohol mixtures is an ideal way to characterize these dipeptides unambiguously, especially since a large body of data has been accumulated for a wide range of dipeptide derivatives examined singly and in model mixtures.

The techniques outlined allow sequence determination using complex mixtures of degradation peptides. The large amount of data can be interpreted with the assistance of a computer which also aids in the assembly of the complete protein sequence. At the present time, individual interpretation is involved to a considerable extent, but the system possesses all the features necessary for complete automation. It is expected that data and experience obtained during application of this technique to more and more naturally occurring polypeptides will make the approach even more reliable, efficient, and sensitive.

The tryptic peptide samples were donated by M. Elzinga, Boston Biomedical Research Institute, Boston, Mass., and received as lyophilized samples accompanied by amino acid analysis results.[345]

Editor's Note. This beautiful piece of research is a result of three considerations: application of good chemistry, proper realization of computer ability, and full understanding of the significance of mass spectral fragmentation. Several papers have described the interpretation of the mass spectra of small peptides in terms of their amino acid sequence but there are very few instances that have used mass spectra to determine the sequence of unknowns. By using the GCMS technique, this paper has demonstrated the capability of obtaining sequence structure from very complex degradation mixtures thus indicating practical application in the determination of protein structures.

---

2. Analysis of the Branched Fatty Acids of the Human Skin Surface by GCMS

N. Nicolaides
Department of Medicine
University of Southern California Medical School
Los Angeles, California

The products formed by the skin provide a window through which to look at internal metabolism as do products of blood or urine. Deviations from normal metabolism—in a word, disease—must also be reflected in some manner by these products: some component goes up or down, is absent, or is replaced by some other. Thus, skin products have a potential for diagnosis, but first we must identify them and learn how they are biosynthesized.

The surface lipid of human skin is an especially easy sample to obtain. One need only wipe the surface clean with cotton soaked with solvent, or soak a limb or the scalp in solvent to obtain an adequate sample for analysis.

Skin products, of course, will be contaminated by bacterial products and whatever else is on the skin surface. As is well known, people have been encouraged to put all sorts of things on their skin by the cosemtic, soap, and pharmaceutical industries. Fortunately, at least as far as this analysis is concerned, most of these things are quite different from the natural oils. Indeed, the uniqueness of human skin lipids compared to lipids of internal tissue is a striking characteristic. Analysis of one group, namely the branched chain fatty acids, is the subject of this paper. An earlier report of this study has already been made.[346]

Not only did we want to know the structures of these acids, but we also wanted to know from what part of the skin they originated, for there are two major sources of skin surface lipid: the sebaceous gland ($\sim$ 85–95%), and the epidermis ($\sim$ 15–5%). By choosing the wax esters of vernix caseosa as the sample, both questions are answered. Vernix caseosa is the greasy matter that covers the newborn. It is a mixture of dead epidermal cells and sebaceous gland excretions (sebum). It has the virtue of being free from bacteria and other contaminants. Furthermore, the wax esters (fatty acids esterified to fatty alcohols) are sebum components, since epidermis free of sebaceous glands, such as from palm or sole skin, does not form wax esters.

A mixture of wax esters plus sterol esters was obtained from vernix caseosa lipid by silicic acid column chromatography. The wax esters were separated from the sterol esters by chromatography on magnesium oxide.[346, 347] The wax esters were saponified and methylated, and separated into saturated, monoenic, and dienoic fractions on an $AgNO_3$–SiO_2 column. The saturated fraction, which had most of the branched material, was analyzed by GCMS.

Figure 8.13 shows the gas chromatogram of the methyl esters of the saturated fatty acids from the wax esters of vernix caseosa, and the points on the peaks where the mass spectrometric scans were taken (i.e., S–1, S–2, etc.). It also gives the equivalent chain lengths (ECL)[348] of the tops of the peaks, many of which are obvious mixtures. Figures 8.14a, b, and c are typical mass spectra taken for scans 9, 18, and 28. Interpretation of spectra was based largely on the work of Ryhage and Stenhagen,[349] and of Odham, who analyzed the preen gland lipids (sebaceous-type glands) from many birds (see reference 350 and the references listed there). Because the position of methyl branching

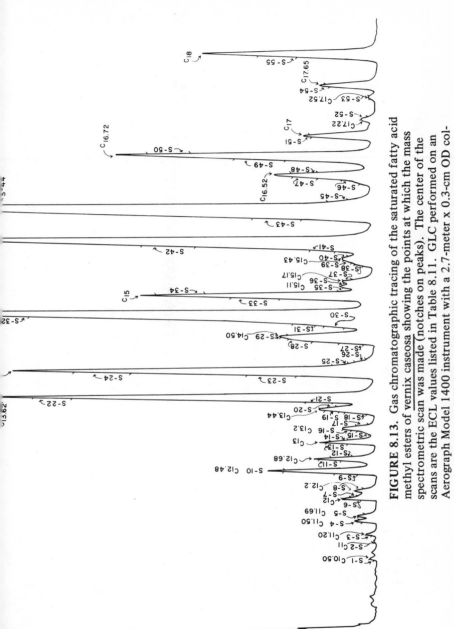

FIGURE 8.13. Gas chromatographic tracing of the saturated fatty acid methyl esters of vernix caseosa showing the points at which the mass spectrometric scan was made (notches on peaks). The center of the scans are the ECL values listed in Table 8.11. GLC performed on an Aerograph Model 1400 instrument with a 2.7-meter x 0.3-cm OD column packed with 3% OV—101 on 100—120 mesh Gas Chrom Q. Programmed from 150—270°C at 2°C/min with He flow of 40 cm^3 atm/min. Although esters to C_{28} emerged, only the peaks to C_{18} are shown.

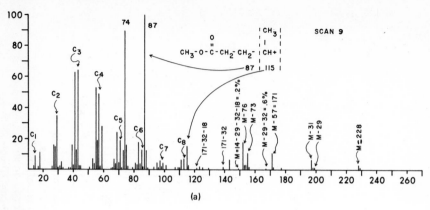

(a)

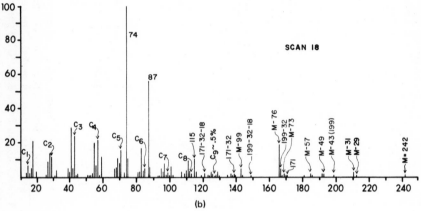

(b)

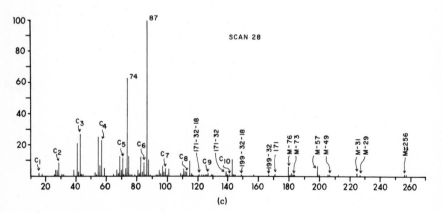

(c)

FIGURE 8.14. (a) Mass spectrum of scan 9. (b) Mass spectrum of scan 18. (c) Mass spectrum of scan 28.

influences the nature of the rearrangements and fragmentation of the ionized esters, it was possible to find evidence for several different branched esters in each mass scan.

Table 8.11 lists the structures of the monomethyl branched fatty acids (as methyl esters), the main identifying masses on which these structures are based, and the "ECL" of the center of the scan for scans 9, 18, and 28. The data for these scans is typical, and most of the scans showed the existence of three or four components in each peak. Except for some of the iso derivatives, the methyl is usually on an even carbon. The chromatographic retention data (Table 8.12) also sheds some light in this direction. Note that the 4 methyl isomer has a higher ECL (18.51) than any of the other isomers where the position of the methyl branch is closer to the carboxyl group than the 14 methyl isomer.

Besides branched material at ECL's of 0.40–0.51, the region of monomethyl branches, there was also some material at ECL's of 0.10–0.20. The lower parent ion intensity and other spectral features of the latter material, plus the fact that compounds with these molecular weights had such low retention times in GLC, indicated dimethyl branched compounds. In general, the mass spectra of these substances were similar to those of the monomethyl compounds with methyl branches appearing only on the even C-atoms. Since they were mixtures, it was not possible to tell which of the methyl branches were present in any molecule.

Human sebum, which makes up a major part of vernix caseosa, differs significantly in lipid composition from the preen gland lipid of most birds.[350, 351] However, similar mechanisms for the biosynthesis of the branched chain acids may be involved, because the methyl branches of the fatty acids found in this study are, as in preen gland lipids, mainly on the even-numbered carbon atoms. In branched preen gland fatty acids, however, the methyl branch is primarily at the 2 position,[350] whereas in vernix caseosa it is mainly at the 4 position, indicating an important difference in biosynthesis. Any methyl group in the 2 position in vernix caseosa fatty chains is of very low abundance if it is present at all.

The occurrence of branched methyl groups almost entirely on the even C-atoms suggests a means for their biosynthesis. If methyl malonyl-CoA substitutes occasionally for malonyl-CoA, then a methyl branch would appear whenever this substitution occurred:

TABLE 8.11. Typical monomethyl branched esters identified from <u>vernix caseosa</u> by GCMS.

	Structure	ECL[1]	Mol. Wt.	Main Identifying Masses (% base peak)
Scan 9				
	$4-MeC_{12}$	12.46	228	87, 115(16), M−57(11), M−73(11)
	$6-MeC_{12}$			M−76(18)
	$8-MeC_{12}$			171(11), 139(1.5), 121(0.5)
Scan 18				
	$4-MeC_{13}$	13.41	242	87(57), 115(11), M−57(9), M−49(23), M−73(2.1)
	$6-MeC_{13}$			M−76(19)
	$8-MeC_{13}$			171(0.7), 139(2), 121(0.5)
	$10-MeC_{13}$			199(0.9), 167(4.7), 149(0.7)
Scan 28				
	$4-MeC_{14}$	14.47	256	87, 115(10), M−57(6.7), M−49(0.7), M−73(0.6)
	$6-MeC_{14}$			M−76(6.3)
	$8-MeC_{14}$			171(1.1), 139(2.7), 121(1.2)
	$10-MeC_{14}$			199(0.9), 167(0.6), 149(0.6)

[1]Equivalent chain length.

TABLE 8.12. Equivalent chain length values of various mono methyl branched methyl octadecanoates on OV−101[1]

Position of methyl branch on C_{18} chain	ECL
2	18.33
3	18.39
4	18.51
6	18.42
8	18.40
9	18.41
12	18.44
14	18.52
15	18.58
16 (anteiso)	18.73
17 (iso)	18.63

[1] GLC conditions: 1.8 meter x 0.6 cm OD column packed with 3% OV−101 on Gas Chrom Q 100−120 mesh. He flow, 40 cm^3 atm/min, isothermal at 200°C.

from malonyl−CoA

from methyl malonyl−CoA

Such syntheses of fatty chains are known to occur.[352, 353] Thus, from the nature of our results we have a bonus, namely, insight into the mechanism of biosynthesis.

Traces of material with the same ECL values appear in the hydrogenated monoenoic fatty acids of the wax esters, and in the saturated fatty acids (up to C_{27}) of the sterol esters of <u>vernix caseosa</u>. They also occur in the fatty acids of the glycerides of <u>vernix caseosa</u> and in the total acids of adult human skin surface lipid. Their occurrence in <u>vernix caseosa</u> establishes them as bona fide human products.

Acknowledgements. I wish to thank Mr. Victor Adams of DuPont Instruments, Monrovia, California, for making available the GCMS system, and Dr. James Cason for many position isomers of the C_{18} monomethyl branched acids. Mr. Phillip Kanner provided technical assistance, and the National Institute of Arthritis, Metabolism and Digestive Diseases, USPHS Grant No. Am–10100, provided financial support.

Editor's Note. There are several important points to emphasize in this work.

1. The author points out that the GCMS data was obtained in one afternoon at the DuPont Instruments Applications Laboratory. However, the success of such cooperative endeavors, as illustrated in the above, is not automatic, but results from careful planning and preliminary testing of the whole analytical method.

2. The preliminary separation into a saturated ester fraction was an important step to keep the GCMS data from being excessively cluttered.

3. The identification of the many branched esters was possible only due to considerable prior work on the fragmentation of branched esters.[349] Because the data was obtained from multicomponent peaks, the identifications based solely on mass spectral data are weak, but the use of chromatographic data (ECL) considerably strengthens this position.

The author has also realized that packed columns were not efficient enough to resolve the chromatographic peaks, but the time of analysis was short enough to run the sample in an industrial application laboratory in one afternoon. Subsequently, suitable capillary columns have been obtained and many of the indicated esters have been separated by a 0.075 cm ID, 150 meter column. Further work is in progress using the better columns.

D. CLINICAL AND FORENSIC CHEMISTRY

1. Identification of Volatile Compounds in Human Urine

K. E. Matsumoto, A. B. Robinson, and Linus Pauling
Stanford University, Stanford, California

and

R. A. Flath, T. R. Mon, and Roy Teranishi
Western Marketing and Nutrition Division
U. S. Department of Agriculture
Albany, California

The human organism involves a complex system of interconnected biochemical reactions. The state of this system determines the health of the organism, and conversely, the health of the organism is reflected in the state of this system. A quantitative examination of the compounds involved in these biochemical processes should give an insight into the state of the system. The composition of the urine gives an indication as to the amounts of the compounds present in the system, and because of the availability of a urine sample, urine is a convenient material the quantitative examination of which should tell something about the health of the organism.[354-356]

As has been shown in the field of biochemical analysis,[357-361] gas chromatography is a powerful tool for studying complex mixtures of organic compounds. This technique has been adapted for use with the volatile compounds in urine.[362, 363] The use of this relatively fast and inexpensive method for examining an individual's metabolic characteristics might permit the diagnosis of many diseases in a more effective and efficient manner than is now possible. It might greatly expedite the trial-and-error process generally required to make decisions regarding the medical procedures desired for maintaining health and treating disease.[356]

To identify some of the compounds present in urine vapor, we have chosen the technique of GCMS. Other investigators have used GCMS for studying urine components.[361, 364-366] Zlatkis[366] has looked at the volatile compounds in urine with use of an extraction-distillation procedure for preparing his sample. We have used a modified head space technique for collecting our sample. The urine of ten male subjects was collected and combined. Helium was bubbled through the buffered, stirred sample at 85°C, and the effluent was passed through a Chromosorb−101 precolumn to remove the organic materials from the water-saturated gas.[367] The organic compounds were back flushed off the precolumn with heating into a cold trap before injection into the gas chromatograph. The chromatography column was a 300-m x 0.075-cm stainless steel open tubular column coated with methyl silicone oil SF–96(50) containing 5% Igepal CO–880. The oven temperature was programmed from 25–172°C. The mass spectra were obtained on a quadrupole mass spectrometer scanning either 0–120 mass units in one second or 0–240 mass units in 1.5 seconds and recorded on a Datagraph light deflecting galvanometer recorder. Preliminary identifications were made using Cornu's compilation of mass spectra[270] and reference spectra available in the USDA laboratory. Final identities, as given in Table 8.13, were confirmed, in most cases, by comparison of both relative retention times and mass spectra with those of authentic compounds obtained on the same system. Twenty-two of the compounds have not been previously identified in urine.

TABLE 8.13. Volatile compounds identified in normal human male urine.

Acetaldehyde	4-Methyl-2-pentanone[1]
Trimethylamine[2]	Pent-3-en-2-one[1]
2-Propanone	2,3-Dithiabutane[1]
2-Methylpropanal	3-Methyl-2-pentanone[1]
Butanal	2-Pentenal[2]
2-Butanone	Toluene
2,3-Butanedione	3-Hexanone[1]
2-Methylfuran	2-Hexanone[1]
Hexane	2-Propylfuran
Chloroform	4-Methylpent-3-en-2-one[1]
Propylene sulfide	Pyrrole[1]
Benzene	4-Heptanone[1]
3-Methylbutanal	3-Heptanone[1]
3-Methyl-2-butanone[1]	2-Heptanone[1]
2-Methylbutanal	2-Butylfuran
Thiophene	4-Octanone
2-Pentanone[1]	3-Octanone[1]
3-Pentanone	Benzaldehyde[1]
2,3-Pentanedione	2-Pentylfuran
2-Ethylfuran	2-Nonanone
2,5-Dimethylfuran	Carvone[1]

[1] Also identified by Zlatkis[366]
[2] Identified by mass spectral comparisons only.

Editor's Note. Analysis of body fluids for volatile organic compounds opens up interesting possibilities in clinical chemistry. The GCMS technique now available can be applied routinely. One annoying consideration in the analysis of many biological samples is the removal of excess water. The method described here[367] is extremely effective and should be considered whenever large quantities of water must be eliminated.

2. Identification of Drugs in Body Fluids, Particularly in Emergency Cases of Acute Poisoning

C. E. Costello, T. Sakai, and K. Biemann
Department of Chemistry
Massachusetts Institute of Technology
Cambridge, Massachusetts 02139

For about a year, the NIH Mass Spectrometry Facility of the Chemistry Department at MIT has provided an emergency service for hospitals in the Boston area and the Boston Poison Information Center for the rapid identification of drugs in the body fluids of comatose patients when the hospital laboratory or local commercial firms are unable to make an unambiguous identification. There are about 1–3 such cases per day at some fifteen local hospitals, with the majority being admitted to five of these. Results are reported by telephone within 1–2 hours after receipt of the sample at the laboratory.

The specimens include gastric contents, blood, or urine. After filtration or centrifugation (and neutralization if necessary), the body fluid is extracted twice with a fivefold volume of methylene chloride, which retrieves all drugs thus far encountered, except quaternary salts and morphine. (Morphine can be extracted into 9:1 chloroform/isopropyl alcohol after the addition of solid sodium bicarbonate. Quaternary salts are not detectable by this approach, without prior degradation.) The extract is evaporated to a small volume before analysis.

The instrumentation consists of a Perkin Elmer 990 Gas Chromatograph interfaced to a Hitachi RMU–6L Mass Spectrometer using a Watson-Biemann effusive separator. The mass spectrometer operates on-line with an IBM 1800 data acquisition and control system.[241, 337] The computer has 32,768 words of core memory, six IBM 2315 magnetic disks, and two IBM 2401 magnetic tape drives. All this equipment is available as part of the NIH Mass Spectrometry Facility at MIT; its use is pre-empted when an emergency sample arrives.

A single run consists of up to 400 mass spectra taken continuously at 4-sec intervals. Both gas chromatographic effluents (150 cm, 3% OV−17 column, programmed from 80−270°C at 12°C/min) and direct insertion samples (see page 12) may be introduced into the mass spectrometer and their spectra recorded. Immediately after a run, time-to-mass conversions are made by the computer with calibration against a perfluoro-alkane reference. The spectra are displayed as bar plots during processing on a Tektronix 611 oscilloscope and are microfilmed with a Bolex Cine Camera.[297] Mass chromatograms[368] are also generated for each mass value and are similarly filmed. Processing with a Kodak Prostar developer in the laboratory makes the films available immediately for examination of the data in a Kodak Microstar reader.

Because rapid identification is of utmost importance in this particular application, selected spectra or the entire run are compared to a reference collection of relevant mass spectra, currently consisting of spectra of about 225 drugs, metabolites, and other materials frequently encountered in body fluids. Output of the library search may be either a listing of the ten closest matching spectra together with their similarity indexes (see page 313) or a listing of the closest fit for each spectrum printed along the contours of the total ion plot (Figure 8.15), together with a summary of the major components identified during the run, which is printed at the end.[274] Any unassigned spectra can also be compared to the general library of about 8000 spectra. For locating minor but significant components which elute at the same time as other components present in much larger concentrations, mass chromatograms are often useful (see page 304).

If new drugs are encountered and identified on the basis of their spectra, authentic samples are obtained and the spectra added to the library. For example, the drug phencyclidine, an anesthetic used in veterinary medicine, was first encountered as a substance confiscated from illegal sources, but it has since figured in several poisonings. In each instance it was correctly identified by a library search of the data from a urine extract. The 1-phenyl cyclohexene is a common contaminant in illegally manufactured samples of this drug, and its spectrum has also been added to the collection for easy recognition. Similar additions are made wherever appropriate.

The utility of the rapid, repetitive scanning feature of the system is illustrated in Figure 8.16. In this case, the extract of urine from an infant

24 hours old presented an extremely complex mixture of drugs, metabolites, and artifacts. In the scans shown, which are separated by only 4 sec, two entirely different spectra were obtained. Scan 161 (taken on the side of the chromatographic peak) is that of a contaminant, tri-2-butoxyethyl phosphate, originating from the stopper of the "B. D. Vacutainer" used as a container for the sample. The next scan (number 162) clearly corresponds to that of promazine, a drug which had been administered to the mother.

Frequently, especially when the body fluid is urine, metabolites rather than the original drug are present in the extract. As an example, in a case involving the drug Mellaril and two of its metabolites (Figure 8.17), the search successfully deduced the type of drug (phenothiazine) represented by metabolite A, from scan number 202. The drug itself, structure B, was correctly identified from scan number 275. The spectrum of the second metabolite C, in scan number 280, retained enough of the characteristics of the initial drug to produce a "best fit" for the Mellaril, although the similarity index was lower than that found for scan 275 because the mass spectrum of metabolite C retains only part of the characteristics of the original drug. This structure-related character of electron impact spectra is often preferred over the one-peak approach of chemical ionization if previously unknown or unexpected substances are to be identified on the basis of their metabolites. (See Dzidick et al. page 401.)

The authors' experience has indicated that the wide variety of toxic drugs and the complexity of the mixtures often present in body fluids justify the approach which is being taken for cases where the patient's condition is very serious. Often the available information from other analytical methods is inadequate. Rapid identification of the toxic material leads to prompt and proper treatment of the individual. A negative finding is also useful because it directs the doctors' attention to other probable causes of the coma.

The data indicate that the most susceptible age groups are 2 year olds whose natural curiosity leads them to sample indiscriminately, and 13–15 year olds whose relatively narrow experience with illegal sources of drug supply often causes them to ingest misrepresented materials in harmful doses.

The success of the program depends to a large extent on the communication channels with the medical community. In the Boston area, this is assisted by a listing compiled by the Poison Bureau, which summarizes all the facilities available for analytical services and their capabilities. Use of this information sheet assures that the mass spectrometry laboratory receives only those samples with which it should be concerned and that these samples are rapidly dispatched. Continuing contact between the laboratory and the patient's doctor results in further studies of the cases of special interest. The response to the service has

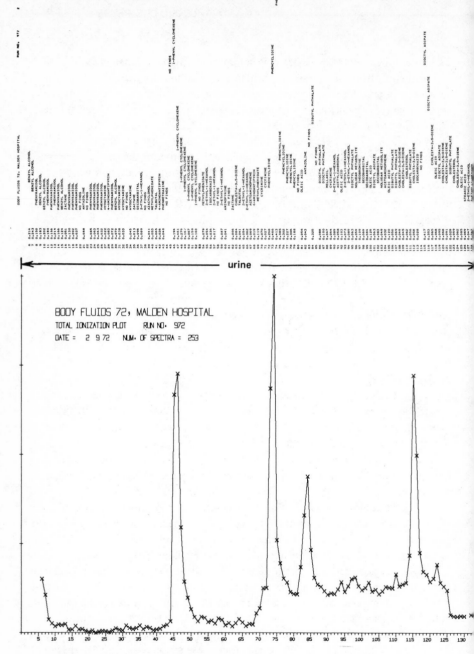

FIGURE 8.15. (a) Contour printout of library search results of a GCMS run on urine and blood extract from a teenage patient who had ingested phencyclidine.[369]

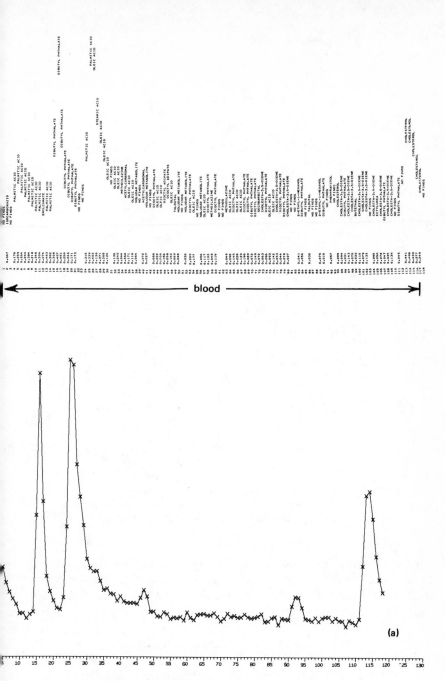

FIGURE 8.15. (a) cont.

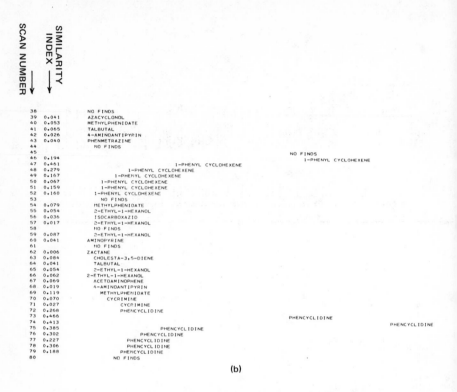

SCAN NUMBER ↓

SIMILARITY INDEX ↓

38		NO FINDS	
39	0.041	AZACYCLONOL	
40	0.053	METHYLPHENIDATE	
41	0.065	TALBUTAL	
42	0.026	4-AMINOANTIPYRIN	
43	0.040	PHENMETRAZINE	
44		NO FINDS	
45			NO FINDS
46	0.194		1-PHENYL CYCLOHEXENE
47	0.461	1-PHENYL CYCLOHEXENE	
48	0.279	1-PHENYL CYCLOHEXENE	
49	0.167	1-PHENYL CYCLOHEXENE	
50	0.067	1-PHENYL CYCLOHEXENE	
51	0.159	1-PHENYL CYCLOHEXENE	
52	0.160	1-PHENYL CYCLOHEXENE	
53		NO FINDS	
54	0.079	METHYLPHENIDATE	
55	0.054	2-ETHYL-1-HEXANOL	
56	0.036	ISOCARBOXAZID	
57	0.017	2-ETHYL-1-HEXANOL	
58		NO FINDS	
59	0.087	2-ETHYL-1-HEXANOL	
60	0.041	AMINOPYRINE	
61		NO FINDS	
62	0.006	ZACTANE	
63	0.084	CHOLESTA-3,5-DIENE	
64	0.041	TALBUTAL	
65	0.054	2-ETHYL-1-HEXANOL	
66	0.062	2-ETHYL-1-HEXANOL	
67	0.069	ACETOAMINOPHENE	
68	0.019	4-AMINOANTIPYRIN	
69	0.119	METHYLPHENIDATE	
70	0.070	CYCRIMINE	
71	0.027	CYCRIMINE	
72	0.268	PHENCYCLIDINE	
73	0.466		PHENCYCLIDINE
74	0.413		PHENCYCLIDINE
75	0.385	PHENCYCLIDINE	
76	0.302	PHENCYCLIDINE	
77	0.227	PHENCYCLIDINE	
78	0.306	PHENCYCLIDINE	
79	0.188	PHENCYCLIDINE	
80		NO FINDS	

(b)

FIGURE 8.15. (b) Enlarged section of above showing identification of 1-phenyl cyclohexene (scan no. 46) and phencyclidine (scan no. 74). Note low similarity index for base line "identifications."

demonstrated that it is fulfilling a need which exists in the community. Similar needs probably exist in other urban areas, and one could expect that wider applications will develop as physicians and mass spectrometrists come into closer contact through this type of activity.

Editor's Note. This paper illustrates a very sophisticated form of GCMS in which the chromatogram itself is printed out with the probable identity of the compound at each stage of the chromatographic process. The

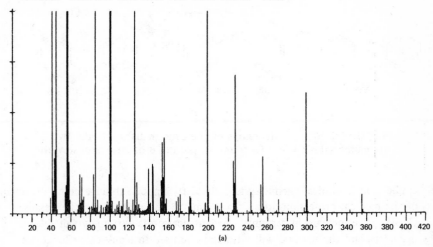

URINE 132, CHMC

TOTAL ION· =0·14114E 07   NUM· LINES = 227   NORM· FACTOR = 0·0306

(a)

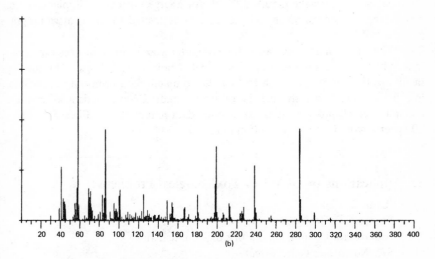

URINE 132, CHMC

TOTAL ION = 0.84205E 06 NUM. LINES = 207   NORM. FACTOR = 0.0305

(b)

FIGURE 8.16. Two mass spectra taken from the GCMS run on a urine extract of 24-hour old infant showing drug withdrawal symptoms.[369] (a) Mass spectrum of scan 161, B. D. Vacutainer contaminant. (b) Mass spectrum of scan 162, promazine.

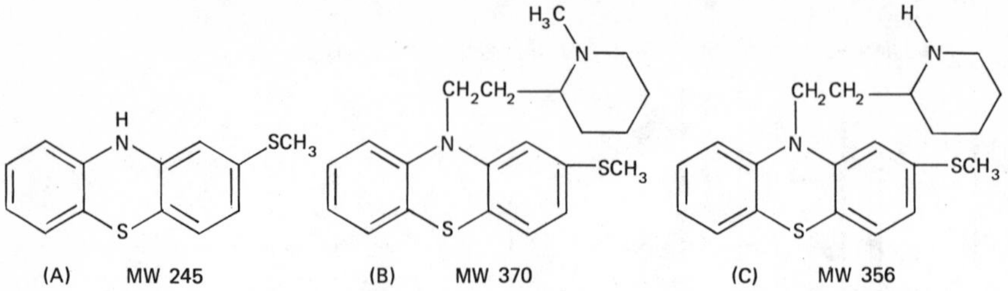

FIGURE 8.17.   Structures of the drug mellaril (B) and two of its metabolites (A and C) found in the urine of an overdose victim.

application in clinical chemistry is particularly important since many members of the medical profession do not have time to be familiar with the details of GCMS. The increase in certainty of identification is extremely valuable for diagnosis, and the method will ultimately be used in many other locations besides Boston.

One should note that in the present form of the method, the equipment is quite expensive and an experienced operator must be available at all times. Such demands will make it very difficult for a large hospital to implement the technique, and general usage will probably be restricted to cooperative programs as described.

The file of mass spectra for the computer search program needs to contain only a few hundred entries of anticipated poisons and drugs. This greatly facilitates the computer search so that the complete data print-out is available in a few minutes. A restricted file of a few hundred compounds also facilitates a manual search process. Even so, manual data reduction and file search of 20 spectra would average 60–90 minutes.

3. Applications of GCMS to Toxicological Problems

Bryan S. Finkle
Forensic Toxicologist
Santa Clara County Laboratory of Criminalistics
875 North San Pedro Street
San Jose, California 95110

The necessity for rapid identification of submicrogram amounts of unknown organic materials extracted from human physiological samples is one of the most difficult daily problems encountered by toxicologists. The circumstances in whic the problem occurs may range from hospital emergency rooms needing accurate analyses to aid diagnosis of comatose patients, to urine analysis for narcotics or drugs-of-abuse on patients associated with methadone clinics and probation departments, to forensic science analyses to detect drugs in drivers suspected of being "under the influence," and to the complex analytical problems involved in post-mortem toxicology where cause of death is the critical issue.

GCMS is often considered the most applicable technique to solve these problems because it is direct, fast, and sensitive, and provides a result which is seldom in dispute.

In the cases described below, a Finnigan Model 3000–03 Gas Chromatograph Peak Identifier GCMS system was used complete with an oscilloscope, oscillographic recorder, and strip chart recorder for total ion monitoring. The quadrupole mass spectrometer is capable of scanning with unit resolution to mass 500, which is adequate to embrace most materials of concern to a toxicologist. The mass spectrometer and gas chromatograph are interfaced with an all-glass jet separator.

The gas chromatography method is based upon that of Finkle et al.[370] and employs two columns each packed with 25% SE 30 on 80/100 mesh Chromsorb G. The columns are glass, 60-cm x 3-mm ID and 90-cm x 3-mm ID, respectively, and operate isothermally in accordance with the parameters detailed in reference 370. Helium is used as the carrier gas, and the chromatographic separation and retention times are observed by means of a total ion monitor and strip chart recorder.

Unknown spectra from case samples are identified by comparative search (manually or by computer) of a library consisting of standard reference spectra. Such a library, spectrum coding system, and method of search is published[273] and was used to identify the drugs in the following case illustrations.

1. A forty-one year old male was found dead, lying on the floor of the bathroom at home. A crudely made syringe and needle and spoon and a rubber balloon containing white powder were near the body. The victim had a history of drug abuse and narcotics use. He had fresh needle marks on his left arm. Apart from the needle marks and pulmonary edema, there were no significant anatomical findings at autopsy. The white powder in the balloon was identified as heroin. The syringe contained traces of morphine (probably from heroin, hydrolysed by heating on the spoon to affect solution), and procaine, a common heroin "cutting agent."

It should be appreciated that inasmuch as a known narcotics user or heroin addict might well have at least trace amounts of morphine in his body, the detection of morphine following an injected dose of heroin, in the bile or urine, is not adequate evidence of an acute, fatal episode. Detection in the blood and/or the site of injection is necessary to demonstrate a dose just prior to death. In this particular case, gas chromatography analysis indicated the possible presence of codeine or morphine in the blood, bile, urine, and veins at the injection site. The GCMS system was used to identify definitively the very small amounts of suspected drugs. Blood concentrations of morphine are in the range 0.1–0.8 μgm/ml in fatal situations. GCMS analysis indicated that the blood and urine contained codeine, and the bile and veins contained morphine. (See Figures 8.18 and 8.19.)

2. A male, age 25, was stopped by the highway patrol for driving erratically at excessive speed on a busy freeway. He was a bartender and smelled strongly of alcohol. He admitted drinking and taking drugs to "pep me up" but felt "ok." The suspect exhibited symptoms of intoxication, slurred speech, and a staggered walk. A blood sample contained 0.12% w/v ethanol and 10 μgm/ml barbiturate. Gas chromatographic analysis indicated the barbiturate to be either amobarbital or talbutal. (These two drugs cannot be satisfactorily separated on commonly used, conventional columns.)

The extract was dissolved in methanol and analyzed by GCMS. The spectrum clearly indicated the drug to be amobarbital. It is very important, however, that all of the analytical data be evaluated in finalizing the identification—that is, chemical extraction and chromatographic relative retention time, as well as the mass spectrum. Each part of the analysis must be in consonance for the identification to be complete. (See Figure 8.20.)

3. A male, age 34, was found dead approximately four hours after he was seen alive and well by a friend. The victim was a known narcotics user and was receiving medical help for this problem.

A fatty, cirrhotic liver was the only significant finding at autopsy. Secobarbital was found in the blood and liver, but in concentrations less than 2 μgm/ml. The urine contained methadone and its cyclic metabolite. Gas chromatographic analysis indicated that the blood might contain submicrogram amounts of methadone, but the sensitivity and specificity of GCMS were needed to confirm this. Figure 8.21 is a spectrum of methadone obtained from the blood extract, and the metabolite 1,5-dimethyl-3,3-diphenyl-2-ethylidene pyrrolidine from the urine extract. Recognition of the cyclic metabolite is particularly important in surveillance work in support of methadone clinics, where it is important to know that a patient is taking the drug as directed by the physician. The presence of only the metabolite and absence of the parent drug is an important indication that the patient had not ingested methadone for at least 24 hours.

4. Identification of metabolites and naturally occurring artifacts is essential to the toxicologist if urine and tissue analyses are to be interpreted correctly. Propoxyphene (Darvon) is a drug commonly prescribed for pain.

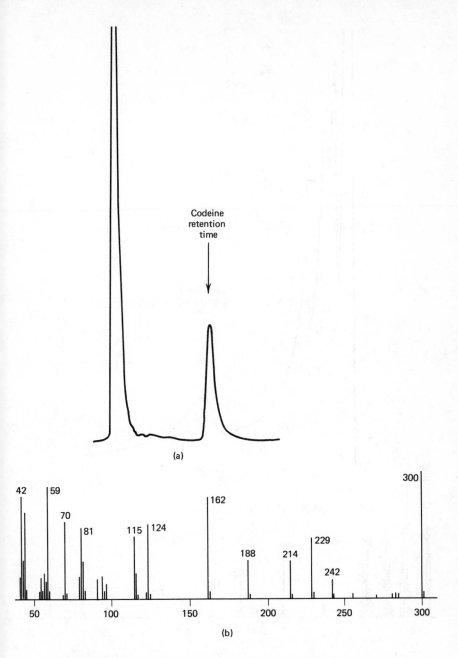

FIGURE 8.18. (a) Chromatogram of urine extract from deceased drug overdose victim. (b) Mass spectrum of codeine obtained from GCMS run on urine extract.

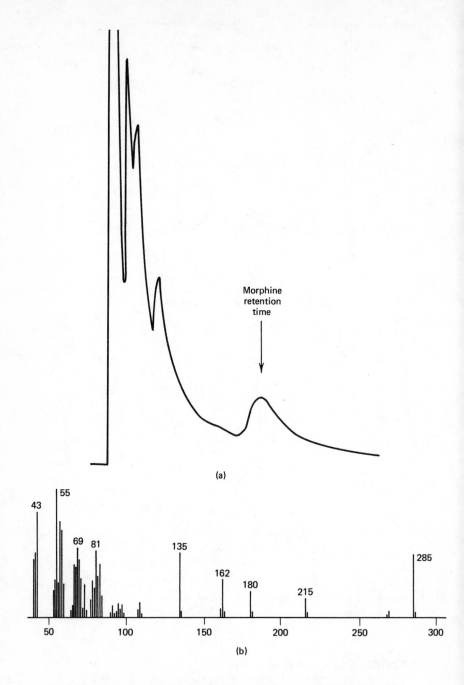

FIGURE 8.19. (a) Chromatogram of bile extract from deceased drug overdose victim. (b) Mass spectrum of morphine obtained from GCMS run on bile extract.

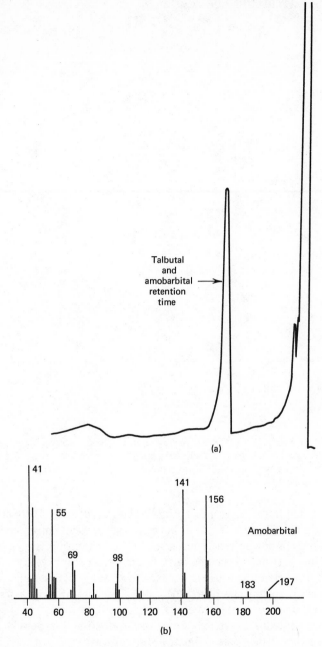

FIGURE 8.20. (a) Chromatogram from blood sample of drug suspect. (b) Mass spectrum of amobarbital from GCMS run on blood sample.

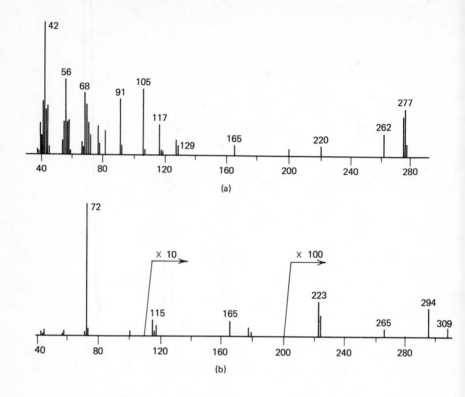

FIGURE 8.21. (a) Methadone spectrum from blood extract of deceased drug user. (b) Methadone metabolite spectrum (1,5-dimethyl-3,3-diphenyl-2-ethylidene pyrrolidine) from urine extract of deceased drug user.

It is excreted in the urine, principally as two metabolites. The metabolites have been identified by GCMS as norpropoxyphene and norpropoxyphene amide, and are now routinely recognized by gas chromatography. Patients undergoing postsurgery therapy (e.g., taking 65 mg propoxyphene four times a day) will excrete the drug almost exclusively in metabolic form. GCMS is an important analytical tool to establish the presence of propoxyphene metabolites extracted from the urine of such patients.

    5. A juvenile automobile driver was involved in a single-vehicle accident. His demeanor at the scene and past history of drug abuse led the investigating officer to suspect that alcohol and drugs may have contributed to the occurrence of the accident.

General toxicological analyses of blood and urine samples from the suspect were negative for alcohol and drugs. However, a chromatographic peak having a relative retention time matching phenaglycodol, a sedative-tranquilizer drug, was noted in the neutral-organic extract from the blood. Further work and color reactions seemed to confirm the suspicion of phenaglycodol. A GCMS analysis was carried out. The result unequivocally negated the possibility of the material being phenaglycodol and indicated it to be a fatty acid ester. This illustration is very important as an example of GCMS in one of its more powerful practical forms, namely, providing indisputable integrity to an analytical result, whether it be positive or negative.

GCMS is now a proven tool in analytical toxicology and is daily contributing to a more effective understanding of toxicity and the physiological disposition of drugs and their metabolites.

Editor's Note. The application of GCMS to forensic chemistry is long overdue, and too often important analyses have been based on more ambiguous methods. The neglect has unfortunately resulted from a lack of financial support, but as the examples show, the results fully justify the expense. The paper also illustrates that large, expensive equipment is not essential. Furthermore, as indicated by the previous paper (Costello et al.), the important aspects of the file search are to obtain or eliminate known drugs and poisons. Hence, a convenient search of a few hundred compounds is possible without computer aid provided that an hour or so is not a serious delay.

E. ECOLOGY STUDIES

1. Analytical Chemistry at Sea

E. J. Bonelli
Finnigan Corporation
Sunnyvale, California 94086

On a recent cruise, the University of Washington's oceanographic research ship, Thomas G. Thompson, sailed with a Finnigan Model 3000 gas chromatograph/mass spectrometer and data system installed below deck. The instrument was involved in a research project studying the inhibition of phytoplankton growth by chlorinated hydrocarbons and airborne transport of these species within the vicinity of industrial and municipal outfalls in the Southern California Bight and San Francisco Bay area. The studies were directed by Dr. S. P. Pavlou, Research Associate of Oceanography at the University of Washington.

One important question to be answered was the feasibility of successfully operating such a complex and sensitive instrument under the adverse conditions encountered during oceanographic experimentations.

Dr. Pavlou and his research team used the GCMS system to analyze synthetic organic residues in marine phytoplankton and airborne particulate samples collected during the cruise. Using a water sampling system designed by the University of Washington personnel (and nicknamed "Phoebe"), water samples were collected at various depths. Phoebe is a large tubular container which is lowered from the stern of the ship and preset to automatically admit four gallon of sea water at the selected depth. Phytoplankton samples are collected by passing the sea water through a glass fiber filter. Subsequent filter grinding followed by standard extraction procedures is used for recovery of the synthetic organic compounds of interest.

A nephelometer mounted on the vessel's bridge monitored airborne particle concentrations. Air was sampled for simultaneous collection of gaseous and particulate fractions via a single-stage variable-cut impactor. Mr. Dennis Schuetzle of the chemistry department was responsible for the air-sampling program. Particles were trapped on a specially-treated glass fiber filter and extracted with hexane in a Soxhlet apparatus. The extract was ultimately concentrated to 0.3 ml, and 5 μl aliquots were injected for GCMS analysis. Methanol extracts of the aerosol samples were similarly analyzed.

Water samples were analyzed for polychlorinated biphenyls (PCB's) and chlorinated hydrocarbon pesticides by both electron capture gas chromatography and GCMS. Because of overlapping peaks, a simple chromatographic analysis based only on retention time data is wholly inadequate for PCB and pesticide identification. However, mass spectrometric analysis of the eluted chromatographic fractions easily distinguishes the pesticides from the PCB's even though the compounds were not separated by the GC column.

Figure 8.22a is a total ion current chromatogram plotted by the data system of a typical water sample. Figure 8.22b (spectrum 46) was identified as DDE, a metabolite of DDT. Table 8.14 summarizes the results found at various sampling stations. In general, the concentrations of PCB's and DDE in the sea water were found to be less than 10 ppb.

The composition of two different types of particulate material was studied—an urban aerosol sample (Figure 8.23a) was collected near Pasadena, California (approximately 25 miles inland), and a marine aerosol sample (Figure 8.23b) was collected fifty miles off the coast of Big Sur, California (approximately 250 miles north of Los Angeles).

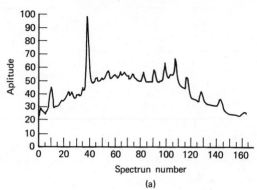

1–2 water sample from 6 filters 5510
Suml. tem.pr. 150–210

Aplitude

Spectrun number

(a)

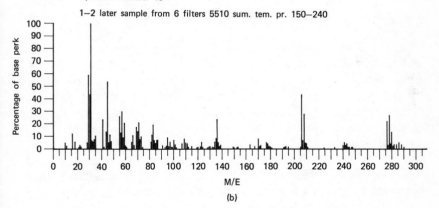

Specimin number 46

1–2 later sample from 6 filters 5510 sum. tem. pr. 150–240

Percentage of base perk

M/E

(b)

FIGURE 8.22. (a) Total ion chromatogram of a typical seawater sample. (b) Mass spectrum of DDE identified in seawater sample.

TABLE 8.14. Summary of phytoplankton analyses.[1]

Station	Depth, m	Compound
47	50	DDE (major); 7 chlorine PCB
55	10	DDE, 4–6 chlorine PCB
70	10	DDE, 4–6 chlorine PCB
72 (zoo-plankton)	10	DDE

[1]Cruise: OUTFALL–1.

The hexane extract of Pasadena air consisted mainly of substituted aromatics, alkanes, and alkenes. Hydroquinone and several phthalate esters were also found; however, the esters were also found in blank samples. Hexane soluble materials accounted for 8.3% (by weight) of the aerosols collected. The marine aerosol sample consisted mostly of alkanes, alkenes, fatty acid esters, and fatty acids. Methanol soluble materials accounted for 0.7% (by weight) of this sample.

The GCMS system operated continuously during the two-week cruise, with no failures in spite of the rather casual and temporary installation of the equipment. During the "graveyard shift," the scientists particularly appreciated the fact that, after the samples were run through the GCMS, the data system operated unattended to derive and plot the output spectra. Data were compared against calibration spectra of the compounds of interest collected before and during the cruise.

Dr. Pavlou described the instrument's value in oceanographic research vessels as follows: "Most of the problems we encounter today in chemical oceanography are, to a large extent, analytical in nature. We need sensitive instrumentation and real-time analytical data to guide our sampling and experimental program during the cruise. The GCMS-computer system was a very useful instrument to have on board because it helped us to identify the compounds we sought quickly without risking degradation of samples from long-term storage prior to analysis. It seems that the GCMS system will be a valuable tool in studying the distribution and chemical pathways of natural and man-made organic compounds in the oceans."

Air sample Pasadena—53 temp. pr 80—230 at 8D—min

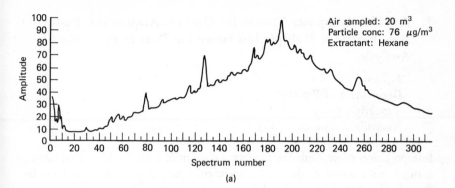

(a)

Air sample—50 miles at sea—temp. pr. 80—230 at 8D—min.

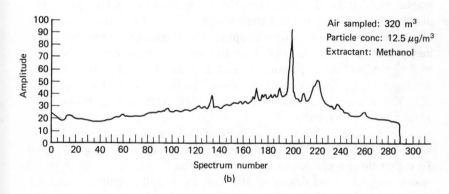

(b)

FIGURE 8.23. Total ion chromatograms of the extract of particulate aerosol samples. (a) Collected near Pasadena, California. (b) Collected 50 miles off the coast of Big Sur, California.

F. SPECIAL TECHNIQUES

1. A GCMS-Computer System for On-Line Acquisition, Reduction, and Display of Multiple Ion Detection Data in Prostaglandin Analysis.

J. Throck Watson
Department of Pharmacology
Vanderbilt University
Nashville, Tennessee 37232

Interpretation of multiple ion detection data from complex biological samples is often complicated or ambiguous when conventional oscillographic recording is used. (See page 252.) The data record is a series of transient sweeps caused by interruption of the recorder as the accelerating voltage is switched between the selected operating values. The useful information, namely the profiles indicated by the endpoints of the transient sweeps, is occasionally obliterated by the streaking effect. Recognition of these profiles is especially difficult when trace quantities are being sought in the presence of more abundant components. Figure 8.24a is a photographic reproduction of an actual oscillographic multiple ion detection record in which three ion profiles should be discernible. The trace resulted from analysis of 1/10 the extract from 5 ml of plasma after column chromatography. The extract was converted to the methyl ester and trimethyl silyl derivative and co-injected with the derivatives of 9 ngm of authentic deuterium-labeled d_4-prostaglandin B_2 and 0.9 ngm of prostaglandin B_2 (PGB_2). In this analysis, the most abundant ions of mass 321, 323, 325 for PGB_2–ME–TMS, PGB_1–ME–TMS, and d_4–PGB_2–ME–TMS[372] were monitored.

Even if the various profiles can be easily resolved, it is always difficult to identify unequivocally the individual profiles on the oscillographic record. To cope with this problem, a computer interface[373] has been built to permit on-line acquisition and display of MID data by a small laboratory computer, in this case a PDP–12 with 8K core. The computer interface uses a multiplexor to acquire and store the data from each ion profile. Thus, the operator may easily request and identify each channel as demonstrated by the photographs of the oscilloscopic screen shown in Figure 8.24b, c, d. Figure 8.24b shows the ion profile composite acquired from the same sample that generated the oscillographic record in Figure 8.24a. Although the three mass channels 325, 323, 321 are indicated on the composite display, the question still remains,

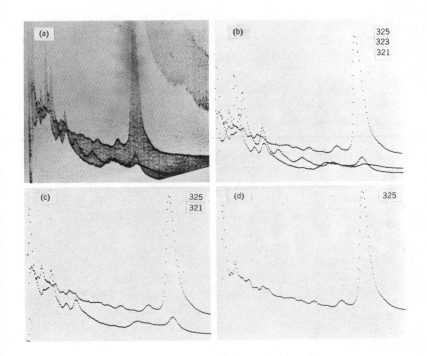

FIGURE 8.24. Comparison of oscillographic recording (a) and computer display (b) of MID data of the prostaglandin sample. Sequence going from (b) to (c) to (d) shows consecutive photographs of the computer oscilloscope as the operator removes the profile of m/e 323 and then that of m/e 321 from the composite display.

which profile is which? The operator merely manipulates switches on the computer console to remove a given profile and its corresponding mass value indicator from the composite display. In the sequence Figure 8.24b, c, d, the operator has stripped away the display of the mass 323 profile in panel c, and finally that of mass 321 in panel d so that only the profile of mass 325 remains.

To further facilitate the interpretation of the MID data, the investigator may display a time cursor on the screen to establish the retention time of various peaks. In the photograph of the computer-oscilloscopic screen shown

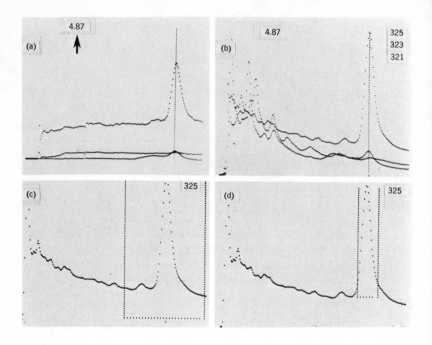

FIGURE 8.25. Upper panels illustrate use of time cursor adjustable from computer console to determine retention time (arrow) of prostaglandin standards alone (a) and in presence of biological matrix (b). Lower panels show use of the calculation option with area-limiting cursors; (c) before and (d) after operator has adjusted them for automatic calculation of the major peak under the m/e 325 profile.

in Figure 8.25a, the operator has adjusted the time cursor, as controlled from the computer console, to the center of the peaks resulting from injection of 9 ngm d_4–PGB$_2$–ME–TMS (mass 325) and 0.9 ngm PGB$_2$–ME–TMS (mass 321) to establish a retention time of 4.87 minutes (see arrow). The retention time of these standards as coinjected with an aliquot (1/10) of extract from 5 ml of plasma (Figure 8.25b) agrees with that of injection of the standards alone onto a 2 meter x 2.5 mm, 1.5% OV–17 on Chromosorb–G–HP 80/100 mesh at 265°C.

In addition to aiding identification of various ion profiles, the computer facilitates calculation of peak areas. This software option displays a selected mass profile on the oscilloscope along with three adjustable cursors which resemble three sides of a rectangle (Figure 8.25c). The operator adjusts the position of each of the three cursors to the peak baseline, the leading edge, and the trailing edge of the peak and activates the "C" key on the Teletype. The value of the peak area within the designated limits is printed. Small peaks may be magnified on the computer screen to facilitate judgment on the positioning of the area-limiting cursors.

The results of an actual biological sample are analyzed in Figure 8.26. The extract from 5 ml of human renal venous blood (platelets not removed) was purified over Sephadex LH–20 followed by TLC. The overall recovery was 25% as determined by losses of a tracer of ^3H–PGE_2 added to the sample residue prior to treatment with methanolic-KOH to convert the PGE_2 to PGB_2.[372] Treatment with CH_2N_2 and BSTFA formed the PGB_2–ME–TMS derivatives which are suitable for quantification by MID. The composite display (Figure 8.26a) shows a discernible peak on each profile at the expected retention times (indicated by arrow). The profile of mass 325 represents d_4–PGB_2–ME–TMS (Figure 8.26c) and indicates that the internal standard is the smallest of the three peaks of interest. Comparison of Figure 8.26b and d indicates that the biggest peak is the mass 321 ion profile which represents PGB_2–ME–TMS. The profile of mass 323 represents PGB_1–ME–TMS which has one less double bond than PGB_2–ME–TMS. Calculation and comparison of the respective peak areas indicates that the original sample of human renal venous blood contained approximately 41 ngm PGE_2 per ml, which is in excess of those values approximately 41 ngm PGE_2 per ml, which is in excess of those values reported for peripheral venous blood.[374, 375] The high value might be explained by the fact that the venous blood taken directly from the kidney had not yet been in contact with degradative enzymes which are localized in the lung. Furthermore, the presence of platelets in the sample could be a rich source of PGE_2. While the results of this analysis are not meant to be representative of normal renal venous blood, the sample serves as a good example to demonstrate the importance of MID in biological analyses.

Editor's Note. Applications of multiple ion detection will become increasingly common as GCMS is applied to the more sensitive biological compounds. The ability to observe one or more specific ions facilitates analysis of materials in complex mixtures that are not always separated by the chromatographic process. Furthermore, it permits addition of an excess of a similar material, often in a deuterated form, which acts as a carrier and

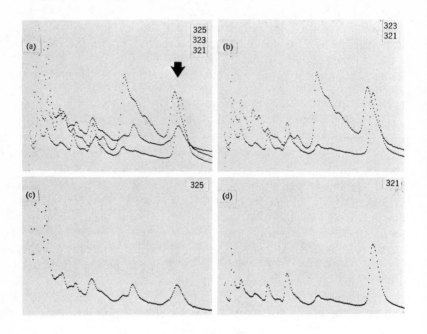

FIGURE 8.26. Photographs of computer screen displaying MID data after analysis of plasma extract for prostaglandins PGB_1−ME−TMS and PGB_2−ME−TMS. Panel (a) is a composite display indicating a major peak under each profile at the expected retention time (arrow). In (b), m/e 325 profile has been stripped away; (c) shows the m/e 325 profile only, while that of m/e 321 only is displayed in (d).

internal standard, and saturates active sites that would capture the trace of unknown. The process of alternately selecting each designated mass offers an additional one to two orders of magnitude increase in sensitivity. This is the most significant difference between the multiple ion detection method and the computer-generated mass chromatogram obtained from conventional GCMS runs. Use of the computer as illustrated will increase the general utility of the method and simplify data interpretation and measurement.

2. Detection of Drugs and Drug Metabolites Using Chemical Ionization GCMS

I. Dzidic, E. C. Horning, and M. G. Horning
Institute for Lipid Research
Baylor College of Medicine
Houston, Texas 77025

GCMS-computer techniques have been used extensively in this laboratory for the study of metabolic profiles in blood and urine.[376] Methods based on gas chromatography are used for separation purposes, mass spectrometry is used for identification and establishment of the molecular structure, and computer techniques are employed for conversion of instrumental data into qualitative and quantitative chemical information. The mass spectrometer can be operated either in the electron impact or chemical ionization mode. It has been shown that the main advantage of chemical ionization compared with electron impact is the ability of the former to enhance the production of ions in the molecular ion mass region, mainly as MH^+ ions. (See page 24.) Since the chemical ionization spectra generally contain very few ions, the spectra are easily interpreted, and computer recognition of individual components present in biological samples is greatly facilitated. Furthermore, the sensitivity appears to be higher relative to electron impact partly because the total ion current is distributed among only a few ions.

Blood and urine profile procedures developed in this laboratory can be used in studies of drugs and drug metabolites. Drug administration usually results in the appearance of new peaks in the gas chromatogram of urine or blood profile. However, if the drug(s) and/or its metabolites are present in relatively low concentration, visual inspection of the chromatographic profile may not lead to detection of foreign compounds. To detect and estimate drugs and drug metabolites present in low concentrations in blood and/or urine, it is necessary to combine chromatographic profile methods with the detection capabilities of the mass spectrometer. Here we describe the potential use of chemical ionization mass spectrometry[377] in a GCMS-computer system.

Figure 8.27 shows a profile of urinary metabolites present in a rat after administration of phenobarbital. The data were obtained with a GCMS-computer system based on a Finnigan quadrupole mass spectrometer equipped with a chemical ionization source. The chart is a reconstructed gas chromatogram of the total ion current resulting from approximately 360 consecutive scans of a temperature-programmed gas chromatographic separation. All major peaks were readily identified from their electron impact mass spectra. Figure 8.28 illustrates the methane chemical ionization mass spectrum of the ME–TMSi

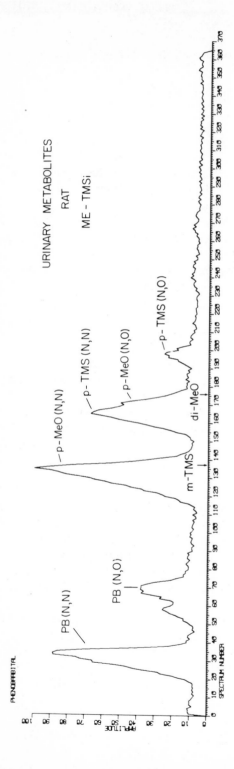

FIGURE 8.27. Metabolic profile of neutral drug metabolites in urine from a rat dosed with phenobarbital. The total ion chromatogram was obtained using a Finnigan quadrupole mass spectrometer in a chemical ionization mode. Methane used as carrier gas. Compounds are: PB(N,N), phenobarbital as the N,N-dimethyl derivative; PB(N,O), phenobarbital as the N,O-dimethyl derivative; p-MeO(N,N), phenobarbital as the N,O-dimethyl derivative; p-MeO(N,N), p-hydroxyphenobarbital as the N,N-Me derivative; p-MeO(N,O), p-hydroxyphenobarbital as the N,O-Me derivative; p-TMS(N,N), p-hydroxyphenobarbital as the N,N-MeTMS derivative; p-TMS(N,O), p-hydroxyphenobarbital as the N,O-MeTMS derivative; m-TMS, m-hydroxyphenobarbital as the N,N-MeTMS derivative; di-MeO, a dihydroxyphenobarbital as the N,N-Me derivative containing 4 methyl groups.

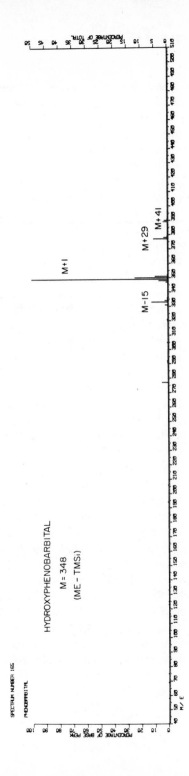

FIGURE 8.28. Chemical ionization mass spectrum of the N,N Me-TMSi derivative of *p*-hydroxyphenobarbital (M = 348), taken as spectrum no. 165 from the separation in Figure 8.27.

derivative of p-hydroxyphenobarbital identified in Figure 8.27. The spectrum in Figure 8.28 contains essentially only molecular ion addition products such as MH^+, $MC_2H_5^+$, and $MC_3H_5^+$.[378] The detection of metabolites present in low concentration, which may not have been observed previously, and which are not evident in the profile separation, is possible by employing a computer search of each spectrum for MH^+, the most common ion occurring in the chemical ionization spectrum. (See page 24.) Figure 8.29 shows a record (mass chromatogram) resulting from the search of all spectra for ions at mass 349, corresponding to $MH^+ = 349$ for the ME-TMSi derivative of a mono-hydroxyphenobarbital. Three peaks are evident. The N,N-dimethyl-p-hydroxy-(N,N-p) and N,O-dimethyl-p-hydroxy- (N,O-p) isomers were evident in the profile separation (Figure 8.27). The small peak, which was not visible in that plot, is believed to be the N,N-dimethyl derivative of m-hydroxypheno-barbital. When a search was made for the fully methylated derivative of dihydroxyphenobarbital using mass 321, the result in Figure 8.30 was obtained. The position of these trace components is noted in Figure 8.27 with arrows. The examples illustrate an important method for detecting small amounts of different compounds in biological samples by using a chemical ionization-computer system.

There are a number of compounds, however, which do not give MH^+ but yield fragment ions in a manner similar to electron impact ionization, when methane, the most common reagent gas, is used. In that case, one has the choice of using a different reagent gas which will provide milder ionizing conditions for the compound to be detected, and thus produce MH^+ ions without fragmentation.[379] Figure 8.31a, b gives the chemical ionization spectrum of atropine when methane and ammonia are used as reagent gases. Protonation of the ester group in the atropine molecule by CH_5^+ ions is followed by fragmentation producing the m/e 124 ions in the absence of a molecular MH^+ ion. However, NH_4^+ from ammonia cannot protonate a carbonyl or ether oxygen, since the gas phase basicity of these groups is lower than that of ammonia. The NH_4^+ thus leads only to protonation of the tertiary amine nitrogen. The MH^+ ion formed is stable and does not undergo fragmentation. Since many drugs and drug metabolites contain functional groups such as $-NH_2$, $-NHCH_3$, and $-N(CH_3)_2$, these groups can be protonated in the presence of NH_4^+ ions and form stable MH^+ ions. Currently, we are using ammonia routinely as a chemical ionization reagent gas for detection of traces of basic compounds in samples of biological origin.

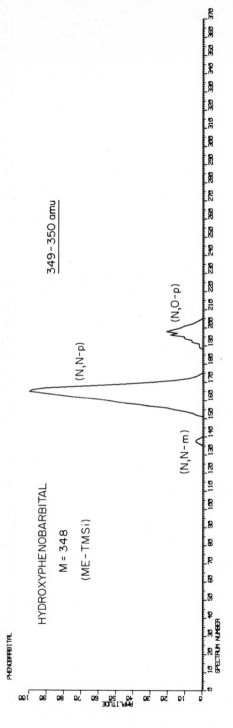

FIGURE 8.29. Mass chromatogram, based upon the separation in Figure 8.27, and showing the occurrence of compounds yielding ions at 349–350 amu. In addition to the expected N,N and N,O ME-TMSi derivatives of *p*-hydroxyphenobarbital, a small amount of another derivative was found. This has been tentatively identified as the isomeric N,N Me-TMSi derivative of *m*-hydroxyphenobarbital.

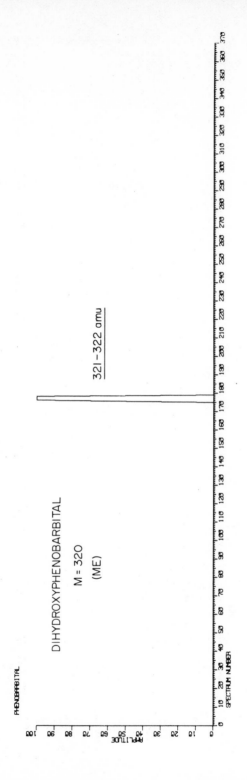

FIGURE 8.30. Mass chromatogram, based on the separation in Figure 8.27 and showing the occurrence of a compound yielding ions at 321–322 amu. This has been tentatively identified as the N,N ME derivative of 3,4-dihydroxyphenobarbital.

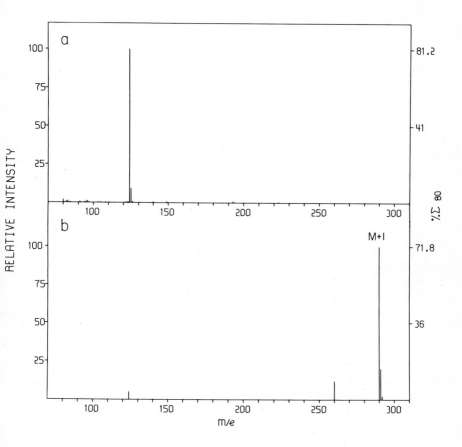

FIGURE 8.31. Chemical ionization mass spectrum of atropine using (a) methane and (b) ammonia as reagent gases.

***

Editor's Note. This application shows how useful the preferred formation of a few ions can be in searching for traces of expected metabolites not separated from major components. The same search can be performed on electron impact spectra but the multiplicity of fragment ions may interfere with the recognition of the selected ion especially when searching for a minor metabolite.

The paper also points out the very significant differences that can occur in chemical ionization spectra when different gases are used. As shown, the technique of using two or more reactive gases can yield valuable structural information. However, a knowledge of fragmentation pathways is essential for interpreting chemical ionization as well as electron impact spectra.

---

3. Photographic vs. Electrical Recording in High-Resolution GCMS

K. Habfast, K. H. Maurer, and G. Hoffman
Varian MAT GMBH
Bremen, Germany

Although electrical recording of mass spectra is far more convenient than photographic recording, it has been claimed that the integrating properties of the photographic plate permit analysis of much smaller samples than is possible by electric methods.[131, 132] This paper gives the results of an experimental comparison of both recording techniques in a special application. The experiments were executed under standardized conditions, and all parts of the experimental procedure (sample, sample introduction, ion source, ion optics, etc.) were exactly the same except for the recording method. The aim of these experiments was to determine the detection limit and the achievable precision in mass determination for both techniques in order to decide in which applications the more elaborate and more costly photographic recording technique is to be preferred. Our earlier paper[132] indicated that photographic recording should be more successful in cases where a low amount of sample is available for only a short time. Since these are typical conditions for GCMS, all measurements were made with sample-introduction via a gas chromatograph.

The data was obtained using a double-focusing Mattauch-Herzog instrument, the Varian MAT 731, equipped for both electric and photographic recording (Figure 8.32). (See page 42.)   The mass spectrometer was connected on-line to a SpectroSystem 100, consisting of an 8K 16-bit Varian 620 computer, an interface, magnetic tape, display, plotter, and Teletype. The same data system was also connected on-line to a precision comparator,[380] type Leitz. Sample introduction was via a Varian 1400 chromatograph, using a glass capillary column[381] coupled directly to the ion source without molecular separator.[382] Flow rate through the column was 1.5 cm^3 atm/min. The sample of methyl stearate resulted in a peak of about 10 sec half-width. The operating conditions were set so that the sensitivity of the mass spectrometer for the molecular peak of methyl stearate was:

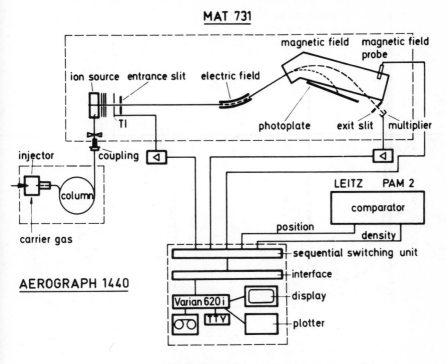

FIGURE 8.32. Schematic of measuring system.

3.5 x 10^{-12} Coul/μgm at R = 15,000 (10% valley) and

7.0 x 10^{-11} Coul/μgm at R =   1,000 (10% valley).

Typical chromatograms recorded on the total ion current detector are shown in Figure 8.33 for 100 ngm and 2 ngm injected onto the chromatographic column. Figure 8.34 shows a part of a low-resolution mass spectrum from the 2 ngm sample. The signal-to-noise ratio of this spectrum is better than 50:1 for the molecular peak.

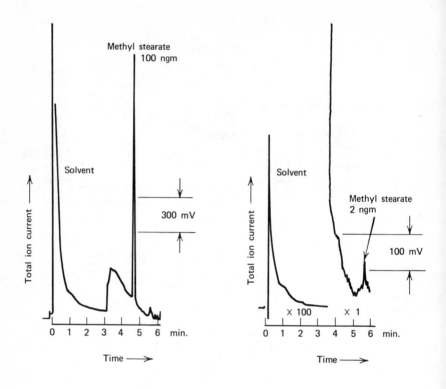

FIGURE 8.33. Total ion monitor chromatogram of (a) 100 ngm of methyl stearate and (b) 2 ngm of methyl stearate.

Figure 8.35 presents a series of high-resolution spectra on a photographic plate at a resolution of 40,000. The molecular peak can be clearly detected with 5 ngm of methyl stearate injected into the column, and the M + 1 peak (one ^{13}C) can be detected with 25 ngm injected. With 200 ngm injected, the M + 2 peak (two ^{13}C) is clearly above the detection limit. The dynamic range in this case is 1:125. For larger sample amounts, one can record two or three mass spectra from a 10-sec peak (using the remote controlled photoplate carrier system) to prevent overexposed lines and thus extend the dynamic range of the mass spectrum.

These measurements permit a comparison of the sensitivity of the two recording modes. It is seen that the detection limit for fast scan (3 sec/decade) electrical recording at resolution 1,000 is within the same order of magnitude (measured by amount of sample consumed) as the detection limit for photographic recording at resolution 40,000. Thus, for all other factors equal, the

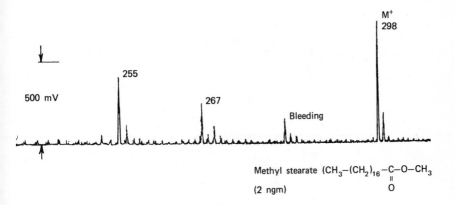

FIGURE 8.34. Part of a low-resolution mass spectrum of 2 ngm of methyl stearate.

data demonstrates the higher sensitivity of the photographic method.

Although sensitivity is always important, the main purpose of high-resolution mass spectrometry is to get the elemental composition of the molecular ion and selected fragment peaks. Thus, the achievable accuracy of mass measurement is of utmost importance. For conventional electrical recording of high-resolution mass spectral data, it is common practice to make repetitive scans to improve the precision and accuracy. This, however, is not possible under GCMS conditions, and all calculations must be made from a single scan.

Several single-scan, electrical recording runs were made under GCMS conditions for comparison with photoplate data. Scans were made at resolution 5,000 in 8 sec/decade and at resolution 10,000 in 16 sec/decade. Two brands of photoplate were used. The results are shown in Table 8.15. The standard deviation is given in millimass units. The parentheses denote the number of peaks above the detection limit which was about 5×10^{-15} amp or 25 ions/peak for a scan of 8 sec/decade. The data permit the following comparisons.

(1) Due to a lower background,[383, 384] the Ionomet plates have a wider dynamic range (i.e., more peaks above detection limit) than the Ilford–Q2 plates.

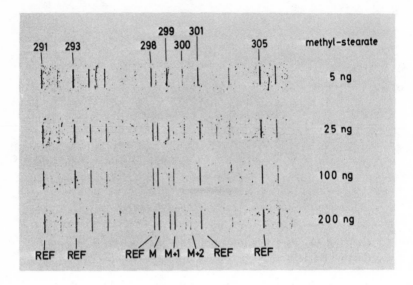

FIGURE 8.35. Photographic recording of high-resolution spectrum of methyl stearate.

(2) Accuracy of mass measurement for Ionomet plates is about 1.5 millimass units and is independent of sample amount. The standard deviation for scanning is in the range 2–3 millimass units.

(3) The number of peaks above the detection limit depends on the amount of sample for electrical recording but, within the range investigated, the number varies only slightly with amount of sample for photoplate recording.

It can be concluded that for recording times of less than 10 seconds (which is satisfactory for photoplate recording but is too fast a time window for good electrical recording) the use of a photoplate permits more accurate mass determination at up to four times higher resolution with 10–20 times less sample. For sample amounts of less than 20 nanograms and at resolution of 40,000, mass determination of 1–2 millimass units can be achieved.

TABLE 8.15. Comparison of the standard deviation of mass measurement in millimass units of methyl stearate spectrum using various recording conditions.[1]

Amount of Sample, ngm	Photographic Recording R = 15,000–40,000 Ilford–Q2	Ionomet	Electrical Recording R = 5,000, R = 10,000 8 sec/dec	16 sec/dec
5	4.8 (4)	2.0 (8)	—	—
25	1.5 (9)	1.2 (16)	—	—
50	1.0 (10)	1.4 (19)	3.0 (2)	—
100	2.3 (16)	1.5 (21)	3.0 (8)	1.6 (1)
200	2.7 (17)	1.9 (29)	3.3 (14)	3.0 (2)
400			2.5 (16)	3.1 (4)
750			2.0 (18)	2.6 (5)

[1] The number of peaks above detection limit is given in parentheses.

4. Simultaneous Use of Nitrogen, Sulfur, and Flame Ionization Detectors as an Ancillary Aid in GCMS

J. P. Walradt
International Flavors and Fragrances
Union Beach, New Jersey, 07735

Successful GCMS depends to a considerable extent on the ability of the operator to judge when components of interest are emerging from the chromatograph. Even with the use of high-resolution chromatographic columns, it is often difficult to determine the optimum point for a scan on small, incompletely resolved peaks. Since many naturally occurring compounds of interest contain nitrogen and/or sulfur, it is a logical step to combine flame ionization, sulfur, and nitrogen detectors in a preliminary screening gas chromatographic run. The ensuing paragraphs describe the modification of a commercial instrument and its use as a qualitative screening tool in flavor analysis.

A Tracor MT220 chromatograph was used, equipped with dual flame ionization and Melpar flame photometric detectors.[385] The Melpar flame photometric detector (FPD) incorporates a flame ionization electrode assembly which is utilized as the conventional FID signal source. One of the standard flame ionization detectors is converted to an alkali flame detector (AFD) by the addition of a rubidium sulfate-potassium bromide (1:1 mixture) pellet to the flame tip.[165] When hydrogen and air flows are optimized, the AFD provides a source of ionization current which is more than a hundredfold enhanced in response for nitrogen-containing compounds.

A column effluent splitter is installed in a heated connection block at the exit of the column so that 50% of the flow is directed to the nitrogen detector and 50% to the flame photometric detector. Low dead volume injector inserts and detector connections enable the system to handle 0.05 cm and 0.075 cm ID open tubular columns with negligible band broadening. The addition of a third electrometer and use of two dual pen recorders completed the instrument modifications. The recorders are connected so that one pen records the FID signal, a second pen records the FPD (sulfur) signal, and the third pen records the AFD (nitrogen) signal.

Figure 8.36 shows the chromatograms obtained from a synthetic mixture and illustrates the ease of detection of small amounts of compounds which contain sulfur or nitrogen. We routinely survey natural extracts with this three-detector system before GCMS analysis and have found that the additional information not only assists the operator in obtaining spectra at optimum points of the eluted peaks but also gives added confidence in interpretation of the low-resolution mass spectra. Previously reported flavor research work has made extensive use of this system.[167, 386, 387]

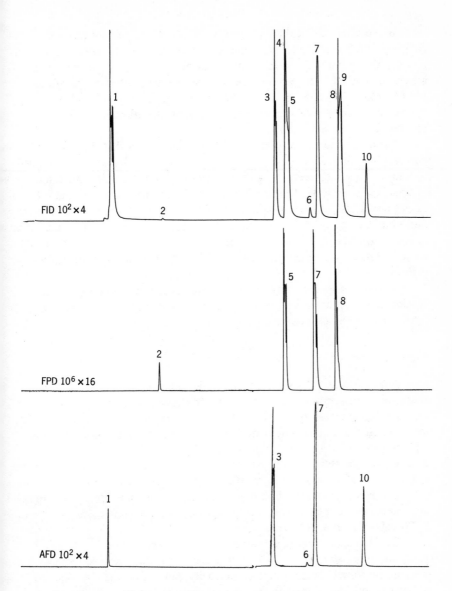

FIGURE 8.36. Chromatogram of synthetic mixture using 3-detector gas chromatographic system. (1) Ethanol. (2) Methylethyldisulfide. (3) 1-n-pentylpyrrole. (4) Furfural. (5) 2-n-pentylthiophene. (6) 2-methyl-5-vinyl pyridine. (7) 4-methyl-5-vinyl thiazole. (8) Tetrahydrothiophene-3-one. (9) 5-methyl furfural. (10) n-pentylpyrazine.

Because the AFD exhibits a greatly enhanced response for nitrogen compounds (100–500 fold depending on the compound and detector conditions), the response for large quantities of non-nitrogen components is relatively diminished. These compounds are easily distinguished by comparing the FID and AFD recorder traces as in the case of the large ethanol peak in Figure 8.36. The sulfur response of the FPD, on the other hand, rarely gives any signal for non-sulfur compounds.

Operation over a two-year period has demonstrated that shaping and boring a hole through a $RbSO_4$–KBr head to fit on the flame tip as described by Craven[165] is not necessary. In the Tracor detector it is sufficient simply to break off a small piece (approximately 2–3 mm diameter) from the 0.15 cm thick pressed disc and lean it against the flame tip so that the small hydrogen flame burns one side of the salt chip. Periodically, the bead must be repositioned or replaced to maintain sensitive, noise-free operation. Major GC instrument manufacturers now market commercial AFDs which are comparable in their performance.

Acknowledgements. My thanks to Thomas E. Kinlin who assisted in development of the three-detector system and to Miss Cynthia J. Mussinan for preparation of the chromatogram.

Editor's Note. The previous paper by Habfast et al. and the paper on double-beam mass spectrometry by Chapman showed how high-resolution mass spectrometry can be applied in GCMS. There are very few mass spectrometrists who have not wanted inaccessible high-resolution data at one time or another. It must be emphasized, however, that the mass spectrometer used for such measurements is expensive, generally costing more than $120,000, and inclusion of a computer ($60–80,000) is seldom considered optional. On the other hand, as is shown in this paper, much of the same kind of information can be obtained using relatively inexpensive specific detectors costing less than $5,000 above basic GCMS costs.

Data obtained from these detectors is fairly specific but fails to give the structural information that can be obtained by accurate mass measurement on fragment ions. The comparative sensitivity depends on the mass spectrometric method, the compound fragmentation, and the element being detected. As an approximate rule, the specific detector will be as sensitive as the high-resolution mass spectrometer. Often the specific detector will be one or two orders of magnitude more sensitive.

---

5. Field Ionization Mass Spectrometry of GLC Effluents

Joseph N. Damico
Division of Chemistry and Physics
Office of Science, Bureau of Foods
Food and Drug Administration
U. S. Department of Health, Education, and Welfare
Washington, D. C. 20204

The technique of field ionization has been discussed in Chapter 2 as an effective way of enhancing the relative abundance of molecular ions. Beckey[62] demonstrated the applicability of this phenomenon in organic analysis, and additional examples are provided in other references.[63, 67, 388-392] Dramatic examples of molecular ion enhancement observed by the author were obtained for the amino acids L-Leucine, DL-Norleucine, DL-Isoleucine, DL-Alanine, and L-Cysteine using the direct inlet system.[393] These amino acids yield the M + 1 peak as the most abundant ion in their field ionization spectra. (This is commonly observed in the FI spectra of highly polar compounds.) Conversely, electron impact ionization of the aforementioned amino acids does not give a molecular ion or an M + 1 peak, and even for the corresponding ethyl esters the molecular ion abundance is relatively small.[394]

The advantages of field ionization are apparent, and one would expect a large interest in field ionization GCMS. However, to the author's knowledge at this writing, his is the only published work[63] utilizing field ionization for GCMS, although Professor Beckey's group has carried out field ionization GCMS studies of multicomponent mixtures containing several hundred unresolved components.[395] Perhaps this lack of activity can be attributed to the problems associated with field ionization that were discussed in Chapter 2 making it unattractive to most routine analytical purposes.

Conventional GCMS is widely used for the analysis of chromatographic effluents as evidenced in this book. Nevertheless, one of the most important pieces of information that can be obtained from mass spectral data is the molecular weight of a compound, but from electron impact the relative intensity of the molecular ion is often very small for many compounds. Since components of interest in natural products are often present in the parts per million range, the total ion yield is small and the molecular ion intensity is too low to distinguish it from column bleed and/or background. The primary object of this work is to demonstrate that field ionization has utility in combined GCMS despite the experimental disadvantages of the technique.

In general, the sensitivity (total ion yield) of field ionization is markedly less than that of electron impact, but inspection of Figure 8.37 shows that the

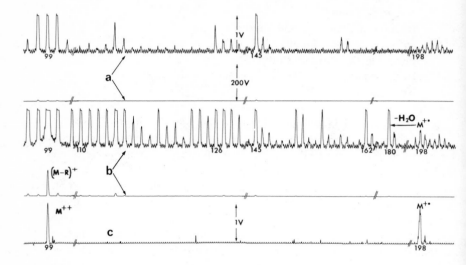

FIGURE 8.37. Selected portions of the electron impact and field ionization spectra of Δ-dodecalactone. (a) EI background. (b) EI spectrum. (c) FI spectrum.

field ionization spectra can complement the conventional spectra. Both spectra were obtained from a 2 μgm injection to the chromatograph, and the signal-to-noise ratio for the molecular ion response is seen to be greater in the field ionization mode than in the electron impact mode. Similar results are obtained for $C_{10}\Delta$-lactone and $C_{14}\Delta$-lactone. Another important feature of the field ionization spectrum is that no background peaks are present from column bleed. Also, the absence of the $M-H_2O$ peaks of $C_{10}\Delta$-, $C_{12}\Delta$-, and $C_{14}\Delta$-lactones indicates that the intense $M-H_2O$ peaks observed in the electron impact spectra for lactones[396] of this type are not due to thermal processes.

In all the GCMS studies described here, a membrane separator was used to interface the column to the mass spectrometer. The total ion monitor, widely used as the chromatographic detector for conventional GCMS, is inadequate for field ionization GCMS because the total ion current is very small. For example, the maximum total ion current observed on the total ion monitor from the field ionization of 2 μgm of $C_{10}\Delta$-lactone was about

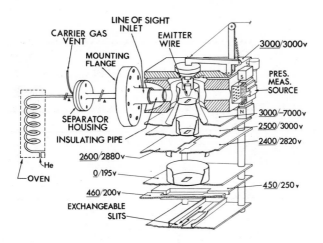

FIGURE 8.38. Detailed drawing of EFO-4B Field Ionization Source. Underlined ion optic potentials are for electron impact mode.

10^{-14} amp, whereas the maximum response for the same amount of sample using a pressure measuring source (PMS) as the GLC detector was about 10^{-10} amp. (The PMS source is an electron impact total ion collection source with the electron energy and emission current variable between 0–70 eV and 0–100 μamp. See page 260.) A detailed cut-away drawing of the source components is shown in Figure 8.38. Wires of 2.5 μm diameter were used as anodes. They were activated as described previously[392] except that maximum sensitivity was approached in only 4–5 hours.

An interesting application of field ionization GCMS occurred in analysis of extracts from decomposed eggs. Incubator egg rejects are forbidden in interstate commerce. Gas chromatographic analysis of decomposed egg extracts found a component that was always present in incubator egg rejects and absent from fresh eggs and accordingly, this component was designated as an index to test for incubator egg rejects. Even though there is no discernible odor or color, any egg liquid, pasteurized or unpasteurized, can be tested for egg rejects by the presence of this characteristic component. Conventional GCMS of this unique component (acid esterified for analysis) gave a mass

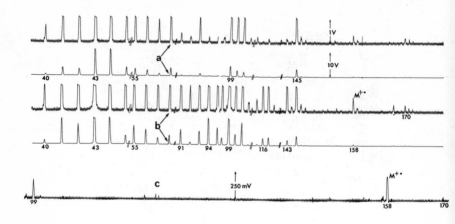

FIGURE 8.39. Electron impact and field ionization spectra of n-propyl levulenate from egg extract. (a) FI background. (b) EI spectrum. (c) FI spectrum.

spectrum with the highest mass peak at 131. If mass 131 was the molecular ion, this would mean that the unknown must contain an odd number of nitrogen atoms. However, field ionization GCMS showed that the molecular weight was most probably mass 146. The esterified component was subsequently identified as the n-propyl ester of 3-hydroxybutyric acid (mol. wt. 146, base peak 131).

Another compound identified in the GLC effluent of the esterified egg extract was n-propyl levulenate.[63] The electron impact and field ionization spectra are compared in Figure 8.39. The electron impact spectrum shows only a small peak at mass 158, which was corroborated as the molecular ion by field ionization. Without the additional data, it would not have been possible to assign a molecular weight with complete confidence. Often it is possible to identify a compound without the molecular weight by matching characteristic peaks with reference spectra. However, in this case the relatively intense peaks at masses 91 and 94 in the electron impact spectrum were present due to impurities, and accordingly, any attempt to match the spectrum is made more difficult and more uncertain.

Another interesting application of field ionization GCMS was with a flavor sample associated with milk. A flavor and standards committee (ASDA

Committee on Flavor Nomenclature and Standards) is charged with preparing synthetic chemical standards to simulate off-flavors which develop in milk and its products. Two standard samples were received by the Agricultural Research Service to be added to milk for taste panel tests. The purpose was to duplicate a cardboard off-flavor (descriptive term for oxidized off-flavor in milk) that develops in raw milk during cold storage prior to shipment from the farm to the dairy plant.

Chromatographic studies of the first sample showed that two compounds were present in almost equal amounts.[397] One component was 1-octen-2-one (vinyl ketone), and the other is a contaminant. Conventional GCMS data[397] of the contaminant showed a quintet of peaks in the 160 mass range (161–165). These peaks were of very low intensity, the most intense peak (mass 162) of the group being about 5% of the base peak. Field ionization GCMS of the contaminant showed two peaks at masses 162 and 164 with intensities corresponding to the isotopic distribution for one chlorine atom. Based on a molecular weight of 162 and one chlorine atom, the mass spectrum was now easily interpreted to be consistent with that of 2-chloro-3-octanone.

The interest in the contaminant comes from the fact that the same odor is detected in evaporated milk.[397] However, to date, sufficient amount of sample for GCMS has not been obtained to corroborate the observation. It should be mentioned that the flavor threshold for the vinyl ketone is about 1 ppb.

The above cases illustrate useful applications of field ionization GCMS, but unfortunately, the technique does not give additional information for all compounds. For example, in a chemical study of the structure of the sex attractant of the Banded Cucumber Beetle,[398], a conventional GCMS run failed to give an unequivocal molecular weight. A field ionization study seemed to be indicated. However, even using 2–3 μgm of isolated sample, a field ionization GCMS run failed to give any ions, much less the parent ion. This quantity is in the range required for field ionization of most carbonyl compounds,[63] but as has been pointed out (page 32), the energy available in a field ionization source is relatively low, and some compounds fail to respond. The shortcoming must be kept in mind by the research worker when utilizing field ionization GCMS for analysis of trace compounds.

Editor's Note. The author has illustrated several applications of field ionization GCMS and at the same time has fairly appraised the shortcomings. Although absolute sensitivity is low, the intensity of the molecular ion may be higher relative to noise or background signals and hence can be more clearly distinguished.

The advantage of field ionization is the clear observation of a molecular ion signal, but this cannot always be counted on. Adjustment of the high voltage is not generally practical, and even this will not often give the desired molecular ion. In contrast, in chemical ionization (see paper of Dzidic et al. page 401), selection of a different reaction gas may result in a different type of energy exchange and thus result in observation of the quasi-molecular ion.

References

1. J. C. Holmes and F. A. Morrell, Appl. Spec., 11, 86 (1957).

2. Gregor A. Junk, Int. J. Mass Spectrom. Ion Phys., 8, 1 (1971).

3. A. L. Burlingame and G. A. Johanson, Anal. Chem. Annual Reviews, 44, 337 R (1972).

4. K. Biemann, Mass Spectrometry; Organic Chemistry Applications, McGraw-Hill, New York, 1962.

5. J. H. Beynon, Mass Spectrometry and Its Applications to Organic Chemistry, Elsevier, Amsterdam, 1960.

6. J. H. Beynon, R. A. Saunders, and A. E. Williams, The Mass Spectra of Organic Molecules, Elsevier, New York, 1968.

7. F. W. McLafferty, Ed., Mass Spectrometry of Organic Ions, Academic Press, New York, 1963.

8. F. W. McLafferty, Interpretation of Mass Spectra, W. H. Benjamin, Inc., New York, 1966.

9. H. Budzikiewicz, C. Djerassi, and D. H. Williams, Interpretation of Mass Spectra of Organic Compounds, Holden-Day, San Francisco, 1964. Mass Spectra of Organic Compounds, Holden-Day, San Francisco, 1967.

10. G. R. Waller, Ed., Biochemical Applications of Mass Spectrometry, Wiley-Interscience, New York, 1972.

11. M. C. Hamming and N. G. Foster, Interpretation of Mass Spectra of Organic Compounds, Academic Press, New York, 1972.

12. D. B. Harrington, Fifth Annual Conference on Mass Spectrometry and Allied Topics, New York, 1957.

13. W. Donner, T. Johns, and W. S. Gallaway, Fifth Annual Conference on Mass Spectrometry and Allied Topics, New York, 1957.

14. R. S. Gohlke, Anal. Chem., 31, 535 (1959).

15. K. L. Stevens, J. Bomben, A. Lee, and W. H. McFadden, Agri. and Food Chem., 14, 249 (1966)

16. A. T. James and A. J. P. Martin, Biochem. J., 50, 679 (1952).

17. L. S. Ettre and W. H. McFadden, Eds., Ancillary Techniques of Gas Chromatography, Wiley-Interscience, New York, 1969.

18. D. A. Leathard and B. C. Shurlock, Identification Techniques in Gas Chromatography, Wiley-Interscience, London, 1970.

19. C. T. Malone and W. H. McFadden, in reference no. 17, p. 341.

20. R. Teranishi, R. E. Lundin, W. H. McFadden, and J. R. Scherer, in The Practice of Gas Chromatography, L. S. Ettre and A. Zlatkis, Eds., Wiley-Interscience, New York, 1967.

21. J. W. Amy, E. M. Chait, W. E. Baitinger, and F. W. McLafferty, Anal. Chem., 37, 1265 (1965).

22. J. N. Damico, N. P. Wong, and J. A. Sphon, Anal. Chem., 39, 1045 (1967).

23. W. E. Harris, W. H. McFadden, and R. G. McIntosh, J. Phys. Chem. 63, 1784 (1959).

24. R. W. Kiser, Introduction to Mass Spectrometry and Its Applications, Prentice-Hall, Englewood Cliffs, New Jersey, 1965.

25. J. Roboz, Introduction to Mass Spectrometry, Wiley-Interscience, New York, 1968.

26. C. A. McDowell, Ed., Mass Spectrometry, McGraw-Hill, New York, 1963.

27. H. C. Hill, Introduction to Mass Spectrometry, Heyden, London, 1966.

28. R. I. Reed, Ion Production by Electron Impact, Academic Press, New York, 1962.

29. A. L. Burlingame, Ed., Topics in Organic Chemistry, Wiley-Interscience, New York, 1970.

30. G. W. A. Milne, Ed., Mass Spectrometry, Wiley-Interscience, New York, 1971.

31. D. H. Williams, Ed., Mass Spectrometry, Volume 1, The Chemical Society, Burlington House, London, 1971.

32. H. A. Bondarovich and S. K. Freeman, in Interpretive Spectroscopy, S. K. Freeman, Ed., Reinhold, New York, 1965.

33. Organic Mass Spectrometry, Heyden and Sons, Spectrum House, Alderton Crescent, London NW4, 3XX.

34. Journal of Mass Spectrometry and Ion Physics, Elsevier Publishing Company, Amsterdam.

35. W. H. McFadden in Advances in Chromatography, J. C. Giddings and R. A. Keller, Eds., Marcel Dekker, Inc., New York, 1967.

36. W. H. McFadden, Separation Science, $\underline{1}$, 723 (1966).

37. J. T. Watson, in reference no. 17, p. 145.

38. J. T. Watson, in reference no. 10, p. 23.

39. R. Ryhage and S. Wikström, in reference no. 30, p. 91.

40. C. J. W. Brooks, in reference no. 31, p. 288.

41. C. J. W. Brooks and B. S. Middleditch, Clinica Chim. Acta, $\underline{34}$, 145 (1971).

42. D. Henneberg and G. Schomberg, in Advances in Mass Spectrometry, A. Quayle, Ed., Vol. V, Institute of Petroleum, London, 1971, p. 605.

43. Jan Sjovall, in Bile Acid Metabolism, L. Schiff, J. B. Carey, Jr., and J. M. Dietschy, Eds., Charles C. Thomas Publisher, Springfield, Ill., 1969.

44. E. C. Horning, in Gas Phase Chromatography of Steroids, K. B. Eik-Nes and E. C. Horning, Springer Verlag, Berlin, 1968.

45. C. Merritt, Jr., Appl. Spectrosc. Rev., 3, 263 (1970).

46. R. A. Flath, in Guide to Modern Methods of Instrumental Analysis, T. H. Gouw, Ed., Wiley-Interscience, New York, 1972.

47. J. A. Vollmin, W. Simon, and R. Kaiser, Z. Anal. Chem., 229, 1 (1967).

48. GC/MS Abstracts, Science and Technology Agency, 3 Dyers Building, London EC 1.

49. P. M. Krueger and J. A. McCloskey, Anal. Chem., 41. 1930 (1969).

50. W. K. Rohwedder, E. Selke, and E. D. Bitner, Appl. Spectrosc., 18, 134 (1964).

51. H. M. McNair, C. R. Dobbs, L. H. Areng, D. C. Damoth, and A. J. Luckte, 19th Annual Conference on Mass Spectrometry and Allied Topics, Atlanta, Georgia, June, 1971, p. 212.

52. H. M. Fales, G. W. A. Milne, and M. L. Vestal, J. Am. Chem. Soc., 91, 3682 (1969).

53. F. H. Field, Accounts Chem. Res., 1, 42 (1968).

54. D. M. Schoengold and B. Wenson, Anal. Chem., 42, 1811 (1970).

55. H. M. Fales, G. W. A. Milne, and T. Axenrod, Anal. Chem., 42, 1432 (1970).

56. H. M. Fales, G. W. A. Milne, and M. L. Vestal, J. Am. Chem. Soc., 91, 3682 (1969).

57. A. A. Kiryushkin, H. M. Fales, T. Axenrod, E. J. Gilbert, and G. W. A. Milne, Org. Mass Spec., 5, 19 (1971).

58. R. L. Flotz, Nineteenth Annual Conference on Mass Spectrometry and Allied Topics, Atlanta, Georgia, June, 1971, p. 142.

59. G. P. Arsenault, J. J. Dolhun, and K. Biemann, Anal. Chem., 43, 1720 (1971).

60. G. P. Arsenault and J. J. Dolhun, 19th Annual Conference on Mass Spectrometry and Allied Topics, Atlanta, Georgia, June, 1971.

61. R. Gomer and M. G. Inghram, J. Chem. Phys., 22, 1279 (1954).

62. H. D. Beckey, Field Ionization Mass Spectrometry, Pergamon Press, Oxford, 1971.

63. J. N. Damico and R. P. Barron, Anal. Chem., 43, 17 (1971).

64. H. R. Schulten and H. D. Beckey, 20th Annual Conference on Mass Spectrometry and Allied Topics, Dallas, Texas, June, 1972, paper G3.

65. E. M. Chait, W. O. Perry, G. E. Van Lear, and F. W. McLafferty, Int. J. Mass Spectrom. Ion Phys., 2, 141 (1969).

66. Peter Brown, Arizona State University, Tempe, private communication, 1973.

67. J. P. Pfeifer, A. M. Falick, and A. L. Burlingame, 19th Annual Conference on Mass Spectrometry and Allied Topics, Atlanta, Georgia, June, 1971, p. 52.

68. J. Mattauch and R. Herzog, Z. Physik, 89, 786 (1934).

69. J. Mattauch, Phys. Rev., 50, 617, 1089 (1936).

70. E. G. Johnson and A. O. Nier, Phys. Rev. 91, 10 (1953).

71. A. O. Nier and T. R. Roberts, Phys. Rev., 81, 507 (1951).

72. J. T. Watson and K. Biemann, Anal. Chem., 36, 1135 (1964).

73. J. Roboz, D. Kruman, and F. Hutterer, 20th Annual Conference on Mass Spectrometry and Allied Topics, Dallas, Texas, June, 1971, p. 76.

74. A. J. Campbell and J. S. Halliday, 13th Annual Conference on Mass Spectrometry and Allied Topics, St. Louis, Missouri, June, 1965, p. 200.

75. M. Barber, J. R. Chapman, B. N. Green, T. O. Merren, and A. Riddock, 18th Annual Conference on Mass Spectrometry and Allied Topics, San Francisco, June, 1970, p. B299.

75a. J. R. Chapman, T. O Merren, and H. J. Limming, 19th Annual Conference on Mass Spectrometry and Allied Topics, Atlanta, Georgia, June, 1971.

76. A. L. Burlingame, D. H Smith, T. O. Merren, and R. W. Olsen, in Computers in Analytical Chemistry, C. H. Orr and J. A. Norris, Eds., Plenum Press, New York, 1970.

77. A. E. Cameron and D. E. Eggers, Rev. Sci. Instr., 19, 605 (1948).

78. W. H. Bennet, J. Appl. Phys., 21, 143 (1950).

79. H. Sommer, H. A. Thomas, and J. A. Hipple, Phys. Rev., 76, 1877 (1949).

80. S. A. Goudsmit, Phys. Rev., 74, 622 (1948).

81. L. G. Smith, Rev. Sci. Instr., 22, 115 (1951).

82. W. C. Wiley and I. H. McLaren, Rev. Sci. Instr., 26, 1150 (1955).

83. M. H. Studier, 11th Annual Conference on Mass Spectrometry and Allied Subjects, San Francisco, 1963, p. 142.

84. W. Paul and H. Steinwedel, Z. Naturforsch., 8A, 448 (1953).

85. E. J. Bonelli, M. S. Story, and J. B. Knight, Dynamic Mass Spectrometry, 2, 177 (1971).

86. P. H. Dawson and N. R. Whetton, in Advances in Electronics and Electron Physics, Vol. XXVII, Academic Press, New York, 1969, p. 59.

87. J. B. Farmer, in reference no. 26, p. 7.

88. W. Paul, H. P. Reinhard, and U. von Zahn, Z. Phys., 152, 143 (1958).

89. J. M. McGuire, A. L. Alford, and M. H. Carter, 20th Annual Conference on Mass Spectrometry and Allied Topics, Dallas, Texas, 1972, p. 366.

90. U. von Zahn, Rev. Sci. Instr., 34, 1 (1963).

91. W. R. Turner, R. D. Board, W. P. Kruger, and E. F. Barnett, paper presented at Pittsburgh Conference, March 1971.

92. J. S. Allen, Phys. Rev., 55, 966 (1939), Rev. Sci. Inst., 18. 739 (1947).

93. F. A. White and T. L. Collins, Appl. Spec., 8, 169 (1954)

94. C. La Lau, in reference no. 29, p. 93.

95. W. H. Rann, J. Sci. Instr., 16, 241 (1939).

96. C. Merritt, Jr., P. Issenberg, M. L. Bazinet, B. N. Green, T. O. Merren, and J. G. Murray, Anal. Chem., 37, 1037 (1965).

97. B. N. Green, T. O. Merren, and J. G. Murray, 13th Annual Conference on Mass Spectrometry and Allied Topics, St. Louis, Missouri, May, 1965, p. 204.

98. W. J. McMurray, B. N. Green, and S. R. Lipsky, Anal. Chem., 38, 1194 (1966).

99. H. G. Boettger, 15th Annual Conference on Mass Spectrometry and Allied Topics, Denver, Colorado, May, 1967, p. 90.

100. W. H. McFadden and E. A. Day, Anal. Chem., 36, 2362 (1964).

101. F. W. McLafferty, Science, 151, 641 (1966).

102. K. Biemann, P. Bommer, and D. M. Desiderio, Tetrahedron Letters, 26, 1725 (1964).

103. K. Habfast, K. H. Maurer, and G. Hoffman, 20th Annual Conference on Mass Spectrometry and Allied Topics, Dallas, Texas, June, 1972, p. 414.

103a. K. Habfast and K. H. Maurer, 17th Annual Conference on Mass Spectrometry and Allied Topics, Dallas, Texas, May, 1966, p. 217.

104. F. A. Leemans and J. A. McCloskey, J. Am. Oil Chem. Soc., 44, 11 (1967).

105. D. M. Desiderio, in reference no. 30, p. 11.

106. W. H. McFadden, Lecture presented at Washington University Extension Course on Combined GCMS, St. Louis, Missouri, June, 1971.

107. R. P. W. Scott, I. A. Fowlis, D. Welti, and T. Wilkens, in Gas Chromatography 1966, A. B. Littlewood, Ed., Inst. of Petroleum, London, 1967, p. 318.

108. A. A. Ebert, Jr., Anal. Chem., 33, 1865 (1961).

109. T. H. Schultz, R. Teranishi, W. H. McFadden, P. W. Kilpatrick, and J. Corse, J. Food Sci., 29, 790 (1964).

110. R. S. Gohlke, Anal. Chem., 34. 1333 (1962).

111. J. H. Purnell, Gas Chromatography, Wiley-Interscience, New York, 1962, p. 105.

112. S. Dal Nogare and R. S. Juvet, Gas Chromatography; Theory and Practice, Wiley-Interscience, New York, 1962.

113. A. B. Littlewood, Gas Chromatography, 2nd Edition, Academic Press, New York, 1970.

114. J. H. Purnell, Ed., Progress in Gas Chromatography, Wiley-Interscience, New York, 1968.

115. Rudolph Kaiser, Capillary Chromatography, Butterworths, Washington, 1963.

116. L. S. Ettre, Open Tubular Columns, Plenum Press, New York, 1965.

117. W. E. Harris and H. W. Habgood, Programmed Temperature Gas Chromatography, Wiley-Interscience, New York, 1966.

118. I. I. Domsky and J. A. Perry, Eds., Recent Advances in Gas Chromatography, Marcel Dekker, New York, 1971.

119. R. A. Jones, An Introduction to Gas-Liquid Chromatography, Academic Press, New York, 1970.

120. D. E. Durbin, 7th International Symposium on Advances in Chromatography, Las Vegas, December, 1971.

121. O. L. Hollis, Anal. Chem., 33, 352 (1961).

122. T. R. Mon, Res. Develop., 22, 14 (1971).

123. R. Teranishi, I. Hornstein, P. Issenberg, and E. L. Wick, Flavor Research, Marcel Dekker, New York, 1971.

124. R. G. Buttery, R. E. Lundin, and Louisa Lang, Agri. and Food Chem., 15, 58 (1967).

125. I. Halász and C. Horváth, Anal. Chem., 36, 2226 (1964).

126. I. Halász and E. Heine, in Advances in Chromatography, J. C. Giddings and R. A. Keller, Eds., Vol. 4, Marcel Dekker, New York, 1967.

127. L. S. Ettre, J. E. Purcell, and K. Billeb, in Separation Techniques in Chemistry and Biochemistry, R. A. Keller, Ed., Marcel Dekker, New York, 1967.

128. L. Rohrschneider, Z. Anal. Chem., 211, 18 (1965).

129. L. Rohrschneider, J. Chromatogr., 22, 6 (1966).

130. G. J. Peirotti, C. H. Deal, E. L. Derr, and P. E. Porter, J. Am. Chem. Soc., 78, 2989 (1965).

131. L. Rohrschneider, Z. Anal. Chem., 170, 256 (1959).

132. D. E. Martire and L. Z. Pollara, in Advances in Chromatography, Vol. 1, J. C. Giddings and R. H. Keller, Eds., Marcel Dekker, New York, 1966, p. 355.

133. R. Teranishi, R. G. Buttery, W. H. McFadden, T. R. Mon, and J. Wasserman, Anal. Chem., 36, 1509 (1964).

134. M. Novotny and A. Zlatkis, J. Chromatogr., 56, 353 (1971).

135. C. Chen and D. Gacke, Anal. Chem., 36, 72 (1964).

136. G. Hesse, Zeit. Anal. Chem., 211, 5 (1965).

137. R. Teranishi, private communication, 1965.

138. C. Horváth, in The Practice of Gas Chromatography, L. S. Ettre and A. Zlatkis, Eds., Wiley-Interscience, New York, 1967, p. 209.

139. A. B. Littlewood, in Gas Chromatography 1958, D. H. Desty, Ed., Academic Press, New York, 1958, p. 23.

140. R. L. Levy, H. D. Gesser, T. S. Herman, and F. W. Hougen, Anal. Chem., 41, 1480 (1969).

141. W. H. Elliott, in reference no. 10, p. 291.

142. M. Novotny and A. Zlatkis, Chromatogr. Reviews, 14, 1 (1971).

143. W. H. McFadden, R. Teranishi, J. Corse, D. R. Black, and T. R. Mon, J. Chromatogr., 18, 10 (1965).

144. J. Andersson and E. von Sydow, Acta Chem. Scand., 18, 1105 (1969).

145. R. G. Arnold, L. M. Libbey, and R. C. Lindsay, J. Agr. Food Chem., 17, 390 (1969).

146. R. Ryhage, J. Dairy Res., 34, 115 (1967).

147. G. Eglinton, in Advances in Organic Geochemistry, A. Schenk and B. Havenaar, Eds., Pergamon Press, Oxford, 1969, p. 1.

148. R. A. Hites and K. Biemann, Anal. Chem., 42, 855 (1970).

149. M. G. Horning, E. C. Chambaz, C. J. Brooks, A. M. Moss, E. A. Boucher, E. C. Horning, and R. M. Hill, Anal. Biochem., 31, 512 (1969).

150. R. Ryhage and E. von Sydow, Acta Chem. Scand., 17, 2025 (1963).

151. W. Averill, in Gas Chromatography, N. Brenner, J. E. Callen, and M. D. Weiss, Eds., Academic Press, New York, 1962, p. 1.

152. D. M. Ottenstein, J. Gas Chromatogr., 1, 11 (1963).

153. T. R. Mon, R. R. Forrey, and R. Teranishi, J. Gas Chromatogr., 5, 497 (1967).

154. J. L. Franklin, J. G. Dillard, H. M. Rosenstock, J. T. Herron, K. Draxl, and F. H. Field, Ionization Potentials and Heats of Formation of Gaseous Positive Ions, NSRDS–NBS26, U. S. Government Printing Office, Washington, D. C., 20402, June, 1969.

155. L. S. Ettre, in The Practice of Gas Chromatography, L. S. Ettre and A. Zlatkis, Eds., Wiley-Interscience, New York, 1967, p. 373.

156. W. O. McReynolds, Gas Chromatography Retention Data, Preston Technical Abstracts Co., Evanston, Illinois, 1966.

157. E. Kovats, in Advances in Chromatography, Vol. 1, J. C. Giddings and R. A. Keller, Eds., Marcel Dekker, New York, 1966, p. 229. E. Kovats, Helv. Chem. Acta, 41, 1915 (1958).

158. H. Van den Dool and P. D. Kratz, J. Chromatogr., 11, 463 (1963).

159. T. H. Schultz, T. R. Mon, and R. R. Forrey, J. Food Sci., 35, 165 (1970).

160. R. A. Flath, D. R. Black. D. G. Guadagni, W. H. McFadden, and T. H. Schultz, J. Agri. Food Chem., 15, 29 (1967).

161. J. D. Winefordner and T. H. Glenn, in Advances in Gas Chromatography, Vol. 5, J. C. Giddings and R. A. Keller, Eds., Marcel Dekker, New York, 1968, p. 263.

162. D. M. Coulson, L. A. Cavanagh, E. deVries, and B. Walther, J. Agri. Food Chem., 8, 399 (1960).

434   REFERENCES

163. C. A. Bache and D. J. Lisk, in Lectures in Gas Chromatography 1966, L. R. Mattick and H. A. Szymanski, Eds., Plenum Press, New York, 1967, p. 17.

164. L. Guiffrida and J. Bostwick, J. Assoc. Offic. Anal. Chemists, 49, 8 (1966).

165. D. A. Craven, Anal. Chem., 42, 1679 (1970).

166. R. V. Hoffman, in Lectures in Gas Chromatography 1966, L. R. Mattick and H. A. Szymanski, Eds., Plenum Press, New York, 1967, p. 137.

167. T. E. Kinlin, R. Muralidhara, A. O. Pittet, A. Sanderson, and J. P. Walradt, J. Agri. Food Chem., 20, 1021 (1972).

168. D. E. Oaks, H. Hartmann, and K. P. Dimick, Anal. Chem., 36, 1563 (1964).

169. Aerograph Research Notes "Previews and Reviews," Walnut Creek, California, April, 1965.

170. See standard physical chemistry texts. For example, S. Glasstone, Textbook of Physical Chemistry, D. Van Nostrand, 2nd edition, 1946, p. 274.

171. A. E. Barrington, High Vacuum Engineering, Prentice-Hall, New York, 1962

172. S. Dushman, Vacuum Techniques, 2nd edition, Chapman Hall, London, 1962.

173. G. Lewin, Fundamentals of Vacuum Science, McGraw-Hill, New York, 1965.

174. L. B. Loeb, Kinetic Theory of Gases, 3rd edition, Dover Publications, Inc., New York, 1961.

175. L. D. Hall, Rev. Sci. Inst., 29, 367 (1958).

176. W. Becker, Vacuum Technik, 7, 149 (1958).

177. K. A. Spangenberg, Vacuum Tubes, McGraw-Hill Co., New York, 1948.

178. D. Alpert and R. S. Buritz, J. Appl. Phys., 25, 202 (1954).

179. A. Guthrie and R. K. Wakerling, Vacuum Equipment and Techniques, McGraw-Hill Co., New York, 1949.

180. L. P. Lindeman and J. L. Annis, Anal. Chem., 32, 1742 (1960).

181. J. A. Dorsey, R. H. Hunt, and M. J. O. O'Neal, Anal. Chem., 35, 511, (1963).

182. C. Brunnée, L. Jenkel, and K. Kronenberger, Zeit. Anal. Chem., 189, 50 (1962).

183. W. H. McFadden, R. Teranishi, D. R. Black, and J. C. Day, J. Food Sci., 28, 316 (1963).

184. W. H. McFadden and R. Teranishi, Nature, 200, 329 (1963).

185. A. Copet and J. Evans, Org. Mass Spectrom., 3, 1457 (1970).

186. R. F. Cree, Pittsburgh Conference on Analytical Chemistry and Applied Spectroscopy, March, 1967. Abstract of papers, P. 96, No. 188.

187. M. Blumer, Anal. Chem., 40, 1590 (1968).

188. C. Brunnée, H. J. Bultemann, and G. Kappus, 17th Annual Conference on Mass Spectrometry and Allied Topics, Dallas, 1969, paper No. 46.

189. S. R. Lipsky, C. G. Horváth, and W. J. McMurray, Anal. Chem., 38, 1585 (1966).

190. P. G. Simmonds, G. R. Schoemake, and J. E. Lovelock, Anal. Chem., 42, 881 (1970).

191. D. P. Lucero and F. C. Haley, J. Gas Chromatogr., 6, 477 (1968).

192. P. M. Llewellyn and D. P. Littlejohn, Pittsburgh Conference on Analytical Chemistry and Applied Spectroscopy, February, 1966.

193. D. R. Black, R. A. Flath, and R. Teranishi, J. Chromatogr. Sci., 7, 284 (1969).

194. R. Ryhage, Anal. Chem., 36, 759 (1964).

195. E. J. Bonelli, M. S. Story, and J. B. Knight, Dynamic Mass Spectrometry, 2, 177 (1971).

196. M. A. Grayson and C. J. Wolf, Anal. Chem., 39, 1438 (1967).

197. R. Ryhage, Arkiv Kemi, 26, 305 (1967).

198. M. C. ten Noever de Braw and C. Brunnée, Z. Anal. Chem., 229, 321 (1967).

199. S. P. Markey, Anal. Chem., 42, 306 (1970).

200. M. A. Grayson and R. L. Levy, J. Chromatogr. Sci., 9, 687 (1971). M. A. Grayson and J. J. Bellina, Jr., Anal. Chem., 45, 487 (1973).

201. Private communication, R. H. Allen Co., Boulder, Colorado, 1970.

202. C. Brunnée, L. Delgmann, K. Habfast, and S. Meier, 18th Annual Conference on Mass Spectrometry and Allied Topics, San Francisco, June, 1970, paper No. L 11.

203. E. W. Becker, "The Separation Jet," in Separation of Isotopes, H. London, Ed., George Newnes, London, 1961, p. 360.

204. V. H. Reis and J. B. Fenn, J. Chem. Phys., 39, 3240 (1963).

205. S. A. Stern, P. C. Waterman, and T. F. Sinclair, J. Chem. Phys., 33, 805 (1960).

206. R. R. Chow, University of California IER Tech. Rept. HE–150–175.

207. R. Ryhage, S. Wikström, and G. R. Waller, Anal. Chem., 37, 435 (1965).

208. M. Novotny, Chromatographia, 2, 350 (1969).

209. S. R. Lipsky, W. J. McMurray, and C. G. Horváth, in Gas Chromatography 1966, A. B. Littlewood, Ed., The Institute of Petroleum, London, 1966, p. 229.

210. M. A. Grayson and C. J. Wolf, Anal. Chem., 42, 426 (1970).

211. J. E. Hawes, R. Mallaby, and V. P. Williams, J. Chromatogr. Sci., 7, 690 (1969).

212. R. M. Teeter and E. J. Gallegos, private communication, 1966.

213. J. E. Lovelock, K. W. Charlton, and P. G. Simmonds, Anal. Chem., 41, 1048 (1969).

214. Viking Project Documents No. M73–101–5 and No. M73–112–0, Langley Research Center, NASA, 1969.

215. M. A. Grayson, McDonnell Douglas Corp., St. Louis, Mo., private communication, 1972.

216. W. K. Seifert and W. G. Howells, Anal. Chem., 41, 554 (1969).

217. R. G. Buttery, D. R. Black, and Mary P. Kealy, J. Chromatogr., 18, 399 (1965).

218. W. D. MacLeod, Jr., and B. Nagy, Anal. Chem., 40, 841 (1968).

219. B. Samuelsson, M. Hamberg, and C. C. Sweeley, Anal. Biochem., 38, 301 (1970).

220. T. F. Gaffney, C. G. Hammar, B. Holmstedt, and R. E. McMahon, Anal. Chem., 43, 307 (1971).

221. Birgitta Sjoquist and E. Anggard, Anal. Chem., 44, 2297 (1972).

222. D. Roach and C. W. Gehrke, J. Chromatogr., 44, 269 (1969).

223. E. Gelpi, W. A. Koenig, J. Gilbert, and J. Oro, J. Chromatogr. Sci., 7, 604 (1969).

224. C. W. Gehrke, H. Nakamoto, and R. W. Zumwalt, J. Chromatogr., 45, 24 (1969).

225. A. G. Sharkey, R. A. Friedel, and S. H. Langer, Anal. Chem., 29, 770 (1957).

226. H.M. Fales and T. Luukkainen, Anal. Chem., 37, 955 (1965).

227. J. A. McCloskey, "Mass Spectrometry of Lipids and Steroids" in Lipids, Vol. XIV, J. M. Lowenstein, Ed., Academic Press, New York, 1969.

228. W. L. Gardiner and E. C. Horning, Biochem. Biophys. Acta, 115, 524 (1966).

229. J.Karkainen, Carbohydrate Res., 11, 227 (1969).

230. C. J. W. Brooks and J. Watson in Gas Chromatography 1968, C. L. A. Harbourn, Ed., Institute of Petroleum, London, 1969, p. 129.

231. R. W. Kelly, Steroids, 13, 507 (1969).

232. R. Roper and T. S. Ma, Microchem. J., 1, 245 (1957).

233. J. C. Cavagnol and W. R. Betker, in The Practice of Gas Chromatography, L. S. Ettre and A. Zlatkis, Eds., Wiley-Interscience, New York, 1967, p. 71

234. R. A. Morrissette and W. E. Link, J. Gas Chromatogr., 3, 67 (1965).

235. W. W. Wells, C. C. Sweeley, and R. Bentley, in Biomedical Applications of Gas Chromatography, H. A. Szysmanski, Ed., Plenum Press, New York, 1964, p. 169.

236. B. H. Kennett, Anal. Chem., 39, 1506 (1967).

237. A. E. Banner, J. Sci. Inst., 43, 138 (1966).

238. A. E. Banner, Thirteenth Annual Conference on Mass Spectrometry and Allied Topics, St. Louis, Missouri, 1965, paper no. 38.

239. B. N. Green, T. O. Merren, and J. G. Murray, Thirteenth Annual Conference on Mass Spectrometry and Allied Topics, St. Louis,Missouri, 1965, paper no. 40.

240. F. W. Karasek, W. H. McFadden, and W. E. Reynolds, <u>GC/MS-Computer Techniques</u>, ACS Short Course Publication, December 1970, Supplement, January 1971.

241. R. A. Hites and K. Biemann, Anal. Chem., <u>39</u>, 965 (1967); <u>40</u>, 1217 (1968).

242. D. H. Smith, R. W. Olsen, F. C. Walls, and A. L. Burlingame, Anal. Chem., <u>43</u>, 1796 (1971).

243. P. Issenberg, M. L. Bazinet, and C. Merritt, Jr., Anal. Chem., <u>37</u>, 1074 (1965).

244. W. E. Reynolds, J. C. Bridges, R. B. Tucker, and T. B. Coburn, Sixteenth Annual Conference on Mass Spectrometry and Allied Topics, Pittsburgh, Pennsylvania, 1968, paper no. 33.

245. P. D. Olsen and E. V. W. Zschau, Sixteenth Annual Conference on Mass Spectrometry and Allied Topics, Pittsburgh, Pennsylvania, 1968, p. 20.

246. W. E. Reynolds, T. B. Coburn, J. Bridges, and R. Tucker, IRL Report No. 1062, Department of Genetics, Stanford University School of Medicine, NASA Accession No. N68–11869, November, 1967.

247. D. Henneberg and G. Schomburg, <u>Gas Chromatography 1962</u>, M. van Swaay, Ed., Butterworths, London, p. 191.

248. C. C. Sweeley, W. H. Elliott, Ian Fries, and R. Ryhage, Anal. Chem., <u>38</u>, 1549 (1966).

249. C. G. Hammar, B. Holmstedt, and R. Ryhage, Anal. Biochem., <u>25</u>, 532 (1968).

250. O. Borga, L. Palmer, A. Linnarsson, and B. Holmstedt, Anal. Lett., <u>4</u>, 837 (1971).

251. L. Bertilsson, A. J. Atkinson, Jr., J. R. Althaus, A. Härfast, J. E. Lindgren, and B. Holmstedt, Anal. Chem., <u>44</u>, 1434 (1972).

252. A. K. Cho, B. Lindeke, Barbara J. Hodshon, and D. J. Jenden, Anal. Chem., <u>45</u>, 570 (1973).

253. J. M. Strong and A. J. Atkinson, Jr., Anal. Chem., 14, 2287 (1972).

254. J. T. Watson, D. Pelster, B. J. Sweetman, and J. C. Frolich, 20th Annual Conference on Mass Spectrometry and Allied Topics, Dallas, Texas, June, 1972, p. 82, also Anal. Chem., 1973, in press.

254a. J. R. Chapman, K. R. Compson, D. Done, T. O. Merren, and P. W. Tenent, 20th Annual Conference on Mass Spectrometry and Allied Topics, Dallas, Texas, June, 1972, p. 166.

255. B. S. Middleditch and D. M. Desiderio, Anal. Chem., 45, 806 (1973).

256. J. Carter Cook, University of Illinois, Urbana, private communication, 1972.

257. C. Brunnée, L. Delgmann, K. Habfast, and S. Meier, 18th Annual Conference on Mass Spectrometry and Allied Topics, San Francisco, California, June, 1970, paper no. L 11.

258. S. A. Ryce and W. A. Bryce, Can. J. Chem., 35, 1293 (1957).

259. R. E. Elskin and W. H. McFadden, USDA, Albany, California, unpublished data, 1965.

260. R. S. Gohlke, Anal. Chem., 34, 1332 (1962).

261. D. R. Black, R. A. Flath, and R. Teranishi, in Advances in Chromatography A. Zlatkis, Ed., Preston Tech. Abstracts Co., Evanston, Illinois, 1969, p. 203.

262. M. S. Story, Finnigan Corporation, Sunnyvale, California, private communication, 1971.

263. H. H. Bovee and L. E. Monteith, in Advances in Mass Spectrometry, Vol. 5, A. Quale, Ed., Elsevier Publishing Co., New York, 1971, p. 295.

264. D. H. Smith, R. W. Olsen, F. C. Walls, and A. L. Burlingame, Anal. Chem., 43, 1796 (1971).

265. American Petroleum Institute Catalog of Mass Spectral Data, Project 44, Texas A & M University, College Station, Texas, 1948 to date.

266. Manufacturing Chemists Association Catalog of Mass Spectral Data, Texas A & M University, College Station, Texas, 1959 to date.

267. ASTM Committee E–14 Uncertified Mass Spectral Data.

268. Uncertified Dow Chemical Mass Spectral Data, Dow Chemical Company, Midlands, Michigan.

269. E. Stenhagen, S. Abrahamsson, F. W. McLafferty, Eds., Atlas of Mass Spectral Data, Wiley-Interscience, New York, 1969. Archives of Mass Spectral Data, Wiley-Interscience, New York, 1970.

270. A. Cornu and R. Massot, Compilation of Mass Spectral Data, Heyden Publishing Company, London, 1966. First Supplement, 1967. Second Supplement, 1972.

271. Eight Peak Index of Mass Spectra, Mass Spectrometry Data Center, AWRE, Aldermaston, Reading, RG7–4PR, U.K. In U. S. A., British Informational Services, 845 Third Ave., New York, 10022.

272. G. W. A Milne, H. M. Fales, and T. Axenrod, Anal. Chem., 43, 1815 (1971).

273. B. S. Finkle, D. M. Taylor, and E. J. Bonelli, J. Chromatogr. Sci., 10, 312 (1972).

274. H. S. Hertz, R. A. Hites, and K. Biemann, Anal. Chem., 43, 681 (1971). H. S. Hertz, D. A. Evans, and K. Biemann, Org. Mass Spectrom., 4, 452 (1970).

275. H. Buzikiewicz, C. Djerassi, and D. H. Williams, Structure Elucidation of Natural Products, Volumes 1 and 2. Holden Day, San Francisco, 1964.

276. D. Desiderio and K. Biemann, Twelfth Annual Conference on Mass Spectrometry and Allied Topics, Montreal, Quebec, 1964, p. 433.

277. R. A. Hites and K. Biemann, Adv. Mass Spectrom., 4, 37 (1968).

278. W. J. McMurray, S. R. Lipsky, and B. N. Green, Adv. Mass Spectrom., 4, 77 (1968).

279. C. C. Sweeley, B. D. Ray, W. I. Wood, J. F. Holland, and M. I. Krichevsky, Anal. Chem., 42, 1505 (1970).

280. W. E. Reynolds, V. A. Bacon, J. C. Bridges, T. C. Coburn, B. Halpern, J. Lederberg, E. C. Levinthal, E. Steed, and R. B. Tucker, Anal. Chem., 42, 1122 (1970).

281. S. Abrahamsson, Science Tools, 14, 29 (1967).

282. B. Hedfjall, P. A. Jansson, Y. Marde, R. Ryhage, and S. Wikström, J. Sci. Instrum., 2, 1031 (1969).

283. P. A. Jansson, S. Melkersson, R. Ryhage, and S. Wikström, Arkiv Kemi, 31, 565 (1969).

284. C. H. Orr and J. Norris, Eds., Computers in Analytical Chemistry, Plenum Press, New York, 1969.

285. K. Habfast, Adv. Mass Spectrom., 4, 3 (1968).

286. J. R. Chapman, Chem. In Britain, 5, 563 (1969).

287. R. Venkataraghavan, R. J. Klimowski, and F. W. McLafferty, Accounts Chem. Res., 3, 158 (1970).

288. D. H. Smith, R. W. Olsen, and A. L. Burlingame, Sixteenth Annual Conference on Mass Spectrometry and Allied Topics, Pittsburgh, Pennsylvania, May, 1968, p. 101.

289. A. J. Campbell and J. S. Halliday, 13th Annual Conference on Mass Spectrometry and Allied Topics, St. Louis, 1965, p. 200.

290. R. S. Klimowski, R. Venkataraghavan, F. W. McLafferty, and E. B. Delany, Org. Mass Spectrom., 4, 17 (1970).

291. B. R. Simoneit and D. A. Flory, "Apollo 11, 12, and 13 Organic Contamination Monitoring History," NASA, Lunar and Earth Sciences Division Note MSCO4350 (May 1971).

292. AEI Scientific Apparatus, Inc. Application Bulletin, June, 1971.

293. W. E. Haddon, USDA, Albany, California, private communication, 1972.

294. E. J. Bonelli, American Laboratory, February, 1971.

295. L. R. Crawford and J. D. Morrison, Anal. Chem., 40, 1464 (1968).

296. B. A. Knock, I. C. Smith, D. E. Wright, R. G. Ridley, and W. Kelly, Anal. Chem., 42, 1516 (1970).

297. J. E. Biller, H. S. Hertz, and K. Biemann, Nineteenth Annual Conference on Mass Spectrometry and Allied Topics, Atlanta, Georgia, June, 1971, p. 85.

298. S. L. Grotch, Anal. Chem., 42, 1214 (1970).

299. S. L. Grotch, Twentieth Annual Conference on Mass Spectrometry and Allied Topics, Dallas, Texas, June, 1972.

300. L. E. Wangen, W. S. Woodward, and T. L. Isenhour, Anal. Chem., 43, 1606 (1971).

301. S. R. Heller, Anal. Chem., 44, 1951 (1972).

302. Barbro Peterson and R. Ryhage, Anal. Chem., 39, 790 (1967).

303. V. V. Raznikov and V. L. Talroze, Dokl. Akad. Nauk, SSSR, 170, 379 (1966).

304. L. R. Crawford and J. D. Morrison, Anal. Chem., 40, 1469 (1968).

305. L. R. Crawford and J. D. Morrison, Anal. Chem., 41, 994 (1969).

306. P. C. Jurs, B. R. Kowalski, T. L. Isenhour, and C. N. Reilley, Anal. Chem., 41, 690 (1969).

307. B. R. Kowalski, P. C. Jurs, T. L. Isenhour, and C. N. Reilley, Anal. Chem., 41, 695 (1969).

308. P. C. Jurs, Anal. Chem., 43, 22 (1971).

309. D. H. Smith, Anal. Chem., 44, 536 (1972).

310. D. H. Smith, B. G. Buchanan, R. S. Engelmore, A. M. Duffield, A. Yeo, E. A. Feigenbaum, J. Lederberg, and C. Djerassi, J. Am. Chem. Soc., 94, 5962 (1972). (Paper No. VIII. Reference to other papers given in above.)

311. R. G. Buttery, R. M. Seifert, R. E. Lundin, D. G. Guadagni, and Louisa C. Ling, Chem. and Ind. (London), 490 (1969).

312. K. E. Murray, J. Shipton, and F. B. Whitfield, Chem. and Ind., 897 (1970)

313. P. Friedel, V. Krampl, T. Radford, J. A. Renner, F. W. Shephard, and M. A. Gianturco, J. Agri. Food Chem., 19, 530 (1971).

314. A. F. Bramwell, J. W. K. Burrell, and G. Riezebos, Tetrahedron Letters, No. 37, 3215 (1969).

315. P. A. Schenck and I. Havenaar, Eds., Advances in Organic Geochemistry, Pergamon Press, Oxford, 1969.

316. G. Eglinton and M. T. J. Murphy, Eds., Organic Geochemistry, Springer-Verlag., New York-Heidelberg-Berlin, 1969.

317. G. Eglinton, P. M. Scott, T. Belsky, A. L. Burlingame, and M. Calvin, Science, 145, 263 (1964).

318. J. G. Bendoraitis, B. L. Brown, and L. S. Hepner; Anal. Chem., 34, 49 (1962).

319. R. A. Dean and E. V. Whitehead, Tetrahedron Letters, 21, 768 (1961).

320. G. Mattern, P. Albrecht, and G. Ourissin, Chem. Comm., 1570 (1970).

321. W. Henderson, W. E. Reed, G. Steel, and M. Calvin, Nature, 231, 308 (1971).

322. W. K. Seifert, E. J. Gallegos, and R. M. Teeter, Angewandte Chemie Intr. Edition 10, 747 (1971).

323. A. L. Burlingame, P. Haug, T. Belsky, and M. Calvin, Proc. Nat. Acad. Sci. U. S., 54, 1406 (1965).

324. W. Henderson, V. Wollrab, and G. Eglinton, Chem. Comm., 710 (1968).

325. P. C. Anderson, P. M. Gardner, E. V. Whitehead, D. E. Anders, and W. E. Robinson, Cosmochim. Acta, 33, 1304 (1969).

326. E. J. Gallegos, Anal. Chem., 43, 1151 (1971).

327. D. E. Anders and W. E. Robinson, Geochim. Cosmochim. Acta, 35, 661 (1971).

328. I. R. Hills and E. V. Whitehead, Nature, 209, 977 (1966).

329. A. Treibs, Ann. Chem., 509, 103 (1934); Angew. Chem., 49, 682 (1936).

330. J. J. Cummins and W. E. Robinson, J. Chem. Engr. Data, 9, 304 (1964).

331. M. T. J. McCormick and G. Eglinton, Science, 157 (1967).

332. I. R. Hills, E. V. Whitehead, D. E. Anders, J. J. Cummins, and W. E. Robinson, Chem. Commun., 752 (1966).

333. R. Ryhage and E. Stenhagen, Arkiv Kemi, 15, 545 (1960).

334. K. Biemann in Reference 10, page 405.

335. K. Biemann, F. Gapp, and J. Seibl, J. Amer. Chem. Soc., 81, 2274 (1959).

336. K. Biemann, Chimia, 14, 393 (1960).

337. J. E. Biller, Ph.D. Thesis, Massachusetts Institute of Technology, Cambridge, Mass. (1972).

338. J. E. Biller, H. Nau, T. Smith, and K. Biemann, submitted for publication.

339. M. M. Shemyakin, Yu. A. Ovchinnikov, E. I. Vinogradova, M. Yu. Feigina, A. A. Kiryushkin, N. A. Aldanova, Yu. B. Alakhov, V. M. Lipkin, and B. V. Rosinov, Experientia, 23, 428 (1967).

340. J. K. McDonald, P. X. Callahan, S. Ellis, and R. E. Smith, in Tissue Proteinases, A. J. Barrett and J. T. Dingle, Eds., North Holland Publishing Co., Amsterdamn, 1971, p. 69.

341. J. K. McDonald, P. X. Calalhan, B. B. Zeitman, and S. Ellis, J. Biol. Chem., 244, 6199 (1969).

342. R.-A. A. Valyulis and V. M. Stepanov, Biokhimiya, 36, 866 (1971).

343. H. Lindley, Biochem. J., 126, 683 (1972).

344. Yu. A. Ovchinnikov and A. A. Kiryushkin, FEBS Letters, 21, 300 (1972)

345. D. G. Smyth, W. H. Stein, and S. Moore, J. Biol. Chem., 238, 227 (1963)

346. N. Nicolaides, Lipids, 6, 901 (1971).

347. N. Nicolaides, J. Chrom. Sci., 8, 717 (1970).

348. T. D. Miwa, K. L. Mikolajczak, F. R. Earle, and I. A. Wolff, Anal. Chem., 32, 1739 (1960).

349. R. Ryhage and E. Stenhagen, Arkiv Kemi, 15, 291 (1960).

350. H. Karlsson and G. Odham, Arkiv Kemi, 31, 143 (1969).

351. N. Nicolaides, H. C. Fu, and G. R. Rice, J. Invest. Derm., 51, 83 (1968).

352. R. N. Nobel, R. L. Stjernholm, D. Mercier, and E. Lederer, Nature, 199, 600 (1963).

353. D. L. Friedman and J. R. Stern, Biochem. Biophys. Res. Com., 4, 226 (1961).

354. Linus Pauling, Science, 160, 265 (1968).

355. Linus Pauling, et al., in Chapter 2 of Orthomolecular Psychiatry, D. Hawkins and L. Pauling, Eds., W. H. Freeman and Company, San Francisco, California, 1973.

356. Arthur B. Robinson and Linus Pauling, in Chapter 3 of Orthomolecular Psychiatry, D. Hawkins and L. Pauling, Eds., W. H. Freeman and Company San Francisco, California, 1973.

357. R. G. Buttery and Roy Teranishi, J. Agr. Food Chem., 11, 504 (1963).

358. R. G. Buttery and Roy Teranishi, Anal. Chem., 33, 1439 (1961).

359. R. A. Flath, R. R. Forrey, and Roy Teranishi, J. Food Sci., 34, 382 (1969).

360. A. J. Burbott and W. D. Loomis, Plant Physiol., 42, 20 (1967).

361. For example: E. C. Horning and M. G. Horning, in "Methods in Medical Research," Volume 12, R. E. Olson, Ed., Year Book Medical Publishers, Inc., Chicago, Ill., 1970, Section V.

362. Linus Pauling, Arthur B. Robinson, Roy Teranishi, and Paul Cary, Proc. Nat. Acad. Sci. U. S. A., 68, 2374 (1971).

363. Roy Teranishi, T. R. Mon, Arthur B. Robinson, Paul Cary, and Linus Pauling, Anal. Chem., 44, 18 (1972).

364. E. C. Horning and M. G. Horning, J. Chromatogr. Sci., 9, 129 (1971).

365. E. Jellum, O. Stokke, and L. Eldjarn, Scan. J. Clin. Lab. Invest., 27, 273 (1971).

366. A. Zlatkis and H. M. Liebich, Clin. Chem., 17, 592 (1971). A. Zlatkis, W. Bertsch, H. A. Lichtenstein, A. Tishbel, F. Shunbo, H. M. Leibich, A. M. Coscia, and N. Fleischer, Anal. Chem., 45, 763 (1973).

367. T. H. Schultz, R. A. Flath, and T. R. Mon, J. Agr. Food Chem., 19, 1060 (1971).

368. R. A. Hites and K. Biemann, Anal. Chem., 42, 858 (1970).

369. C. E. Costello, T. Sakai, and K. Biemann, Twentieth Annual Meeting on Mass Spectrometry and Allied Topics, Dallas, Texas, June, 1972.

370. B. S. Finkle, E. J. Cherry, D. M. Taylor, J. Chromatogr. Sci., 9, 393 (1971).

371. E. J. Bonelli, Anal. Chem., 44, 603 (1972).

372. B. J. Sweetman, J. C. Frolich, and J. T. Watson, Prostaglandins, 3, 75 (1973).

373. D. R. Pelster and J. T. Watson, submitted for publication.

374. W. G. Unger, I. F. Stamford, and A. Bennet, Nature, 233, 336 (1971).

375. B. Samuelson, E. Granstrom, K. Green, and M. Hamberg, Ann. New York Acad. Sci., 180, 138 (1971).

376. E. C. Horning and M. G. Horning, Methods in Medical Research, 12, 369 (1970).

377. E. C. Horning and M. G. Horning, Clin. Chem., 17, 802 (1971).

378. F. H. Field, Accounts Chem. Res., 1, 42 (1968).

379. I. Dzidic, submitted to J. Am. Chem. Soc.

380. K. Heinicke, Z. Instr., 74 (1966).

381. K. Grob, Helvetia Chim. Acta, 51, 718 (1968).

382. P. Schulze and K. Haus Kaiser, Chromatographia, 4, (1971).

383. J. I. Masters, Nature, 223, 611 (1969).

384. C. Hignite and K. Biemann, Org. Mass. Spectrom., 2, 1215 (1969).

385. S. S. Brody and J. E. Chaney, J. Gas Chromatogr., 4, 42 (1966).

386. C. Mussinan, R. Wilson, and I. Katz, Paper No. 21, 164th National ACS Meeting, New York, September, 1972.

387. R. A. Wilson, C. Mussinan, I. Katz, and A. Sanderson, Paper No. 22, 164th National ACS Meeting, New York, September, 1972.

388. H. D. Beckey, Int. J. Mass Spectrom. Ion Phys., 2, 101 (1969).

389. D. F. Barofsky and E. W. Muller, ibid., p. 125.

390. E. M. Chait, W. O. Perry, G. E. Van Lear, and F. W. McLafferty, ibid., p. 141.

391. M. Barber, R. M. Elliott, and T. R. Kemp, ibid., p. 157.

392. J. N. Damico, R. P. Barron, and J. A. Sphon, ibid., p. 161.

393. J. N. Damico, Sixth Middle Atlantic Regional Meeting of the American Chemical Society, Baltimore, Md., Feb. 3–5, 1971, Paper no. B10 17.

394. K. Biemann, J. Seibl, and F. Gapp, J. Amer. Chem. Soc., 83, 3795 (1961).

395. H. D. Beckey, University of Bonn, Bonn, Germany, private communication, 1970.

396. W. H. McFadden, E. A. Day, and M. J. Diamond, Anal. Chem., 37, 89 (1965).

397. O. Parks, Agriculture Research Service, Eastern Marketing and Nutrition Research Division, Dairy Products Laboratory, Washington, D. C., 1970, private communication

398. M. Schwartz, M. Jacobson, and F. P. Cuthbert, Jr., J. Econ. Entomol., 64, 769 (1971).

Index

C

Cajon high vacuum couplings 154
Capillary columns 75, 82, 374, 376, 408
Caprate, methyl 202
Caproate, methyl 202
Caprylate, methyl 202
Carbowax 20M 91, 335
    TPA 91
Carotenes, in oil shale 342, 348
Carrier gas flow 75, 80, 101
    in GC 81, 84, 86, 163, 182, 206, 214, 223, 225, 227
    to MS 5, 27, 75, 81, 86, 122, 161, 163, 180, 206, 209, 218, 220, 223, 227
    background spectrum 114
Carvone 377
Caryophyllene 336
Catalogues, see mass spectral data files
Catalytic reduction, in H$_2$/Pd separator 205
Cathepsin C 359
Center of area, of mass peak 297
Center of gravity, of mass peak 297
Ceramic separator 173, 176
Cerebral spinal fluid, analysis of 253
Chemical ionization 19, 24, 401
    vacuum requirements 27
    GCMS mode 28
    reactant gas 24, 28, 103, 404
    analysis of gastric contents 31
Cholene 348
Cholestane 343, 348
Cholesterol 71, 233
Chlorinated hydrocarbons 391
Chloroform 377
Chromatographic columns
    bleed characteristics 90, 92, 137, 225
    capillary 75, 82, 374, 376, 408
    conditioning process 91, 101
    open tubular, see capillary
    operating characteristics 86, 90
    operating temperature 225
    packed 81
    preparation of capillary 82, 101
    sample load 81
    SCOT, support coated 83

Chromatographic detectors
    hydrogen flame 256, 262
    thermistor 257
    see also, auxiliary ion chamber
        computer regeneration of chromatograms
        multiple ion detection
        oscilloscope monitoring
        specific detectors
        total ion monitor
Chromatographic efficiency in GCMS 83, 97, 203, 208, 218
Chromatographic effluent flow, see carrier gas
Chromatographic method 222
    analysis time 87
    flow rate in GC, see carrier gas
    flow rate in MS, see carrier gas
    maximum sample load 80, 87
    optimization for GCMS 81
    tailing 100
Chromosorb 100 series 91, 375
Chromosorb G 385, 398
Chymotrypsin 359
Clinical chemistry applications 375
Codeine 386
Cold trap 124, 137
Collision diameter, for ion molecules 118
    for neutral molecules 117
Column bleed 90, 137, 225
Column chromatography 223, 368
Columns, see chromatographic columns
Comatose patients 376, 385
Comparator 68, 408
Computer applications in GCMS
    in biochemistry 359, 401
    in clinical chemistry 375
    in ecological studies 391
    in flavor research 334
    in geochemistry 351
Computer data acquisition 38, 247, 294, 334, 351, 359, 378, 391, 401
    high resolution mass spectra 45, 278, 281, 287, 351, 408
Computer data interpretation 67, 271, 282, 307, 318, 360, 378
    file search 307, 336, 378

Steranes, in oil shale   342, 348
Steroids, in oil shale   342
Stigmastane   343
Strawberry extract, analysis of   97
Stream splitter, in injector   105
  in interface   158, 160, 226
Subtilisin   359
Succinate, diethylene glycol   91
Sulfur detector   4, 414
Support coated open tubular col-
  umns   83
Surfactants, as stationary phase addi-
  tive   101
Swagelok coupling   153

T

Tailing, chromatographic   100
  from cold spots   232, 350
  in membrane separator   200, 218
  see also dead volumes
Talbutal   386
Teflon separator   166, 188
Temperature of interface components   229, 350
  of ion source   49, 231
Terpanes   342
Terpenes   76, 100, 336
Tergitol NP–35   92
α-Terpineol   99
4-Terpineol   337
Tetrahydrothiophene-3-one   415
Theoretical plates, HETP in chroma-
  tographic process   83, 97, 203,
  208, 218
Thermocouple gauge, pressure mea-
  surement   145
Thiazole, 4-methyl-5-vinyl   415
Thiophene   377
  pentyl   415
Threshold signal, for computer recog-
  nition   290, 352
Time-of-flight mass spectrometer   46,
  55, 254, 261
  resolution of   48
Toluene   377
Total ion chromatogram, see below

Total ion monitor   18, 49, 54, 59,
  201, 257, 260, 298, 337, 343, 350,
  360, 378, 385, 401, 418
Toxicology, GCMS applications   384
Tri-2-butoxyethyl phosphate   379
Trimecaine   254
Trimethylamine   377
Trimethylchlorosilane   237
Trimethylsilyldiethylamine   360
Triterpanes   344
Trypsin-chymotrypsin   359
Trypsin-pepsin   359
Turbomolecular pump   139
Tween 20   97
Two-stage separators   165, 185, 192,
  215, 220

U

Ultraviolet recorder   64, 267, 283,
  351, 376, 385
Ultraviolet spectrometry   4
Undecanoate, methyl   76, 202
Urine analysis   375, 377, 378, 385,
  401

V

Vac-ion pump   138
Vacuum
  hardware   146
    couplings   153
    gauges   140
    valves   146, 158
  measurement
    gauge calibration   144, 209,
      229
    high   140, 209
    low   145
  operating level, see pressure, opera-
    ting
  pumps
    diffusion   134, 258
    fore   131
    high-speed   163